Aspects of Brazilian Floristic Diversity

Maria Franco Trindade Medeiros
Bárbara de Sá Haiad
Editors

Aspects of Brazilian Floristic Diversity

From Botany to Traditional Communities

Springer

Editors
Maria Franco Trindade Medeiros
Departamento de Botânica
Laboratório Interativo em
Etnobotânica (LinE)
Museu Nacional, Universidade Federal do
Rio de Janeiro (MN/UFRJ)
Rio de Janeiro, RJ, Brazil

Bárbara de Sá Haiad
Departamento de Botânica
Laboratório de Anatomia Vegetal
Museu Nacional, Universidade Federal do
Rio de Janeiro (MN/UFRJ)
Rio de Janeiro, RJ, Brazil

ISBN 978-3-031-07452-3 ISBN 978-3-031-07453-0 (eBook)
https://doi.org/10.1007/978-3-031-07453-0

This Springer imprint is published by the registered company Springer Nature Switzerland AG
The registered company address is: Gewerbestrasse 11, 6330 Cham, Switzerland

About the Book

This initial presentation of the book will lead us to understand its eclectic proposal through the description of its structural line.

Aspects of Brazilian Floristic Diversity: From Botany to Traditional Communities presents several botanical aspects with a broader view through the chapters of each part. Certainly, our intention is not to exhaust all possibilities of approach within botany but to show aspects of Brazilian floristic diversity.

The sequence of parts of the book will act as a thematic guide through which we will permeate various areas of botany, allowing us the cognitive understanding that will constitute the unit of the book as a whole.

Thus, within each part, we will have the unique opportunity to come across different botanical areas and the specific approach to each one of them. The common theme of all chapters lies in the treatment of diversity and conservation, whether algae or angiosperms, as well as material and immaterial culture.

The themes are grouped into parts aiming to show the vast and deep research possibilities in the field of botany. The intention is precisely to offer a comprehensive approach to botany, emphasizing its *spectrum* of knowledge, as well as a pedagogical strand that respects each different line of botanical research, considering its own language and content development and, as it were, respecting the values and positions of the researchers/authors and their epistemological structure.

Following this thought, an opening commentary is presented in each part of the book, giving us a good perspective of what we will find in the next pages that constitute the book as a whole.

Following this thought, first of all, we will have a brief scenario at the opening of each part, in which a good perspective of what will be found in the chapters of that part will be presented. Thus, throughout the parts that will compose the book as a whole, we can be well situated in the *Aspects of Brazilian Floristic Diversity: From Botany to Traditional Communities.*

Finally, wishing a good reading, this is an invitation to a deeper experience in the field of botany.

Contents

Contributors

Donato Seiji Abe Associação Instituto Internacional de Ecologia e Gerenciamento Ambiental, São Carlos, SP, Brazil

Ângelo Giuseppe Chaves Alves Departamento de Ecologia, Universidade Federal Rural de Pernambuco, PE, Brasil

Angela Arruda e Albuquerque Departamento de Botânica, Laboratório de Biologia Reprodutiva de Angiospermas, Museu Nacional, Universidade Federal do Rio de Janeiro (MN/UFRJ), Rio de Janeiro, RJ, Brazil

Davi Almeida Barreto Departamento de Botânica, Laboratório de Ficologia, Museu Nacional, Universidade Federal do Rio de Janeiro (MN/UFRJ), Rio de Janeiro, Brazil

Cristine Rodrigues Benevides Departamento de Botânica, Laboratório de Biologia Reprodutiva de Angiospermas, Museu Nacional, Universidade Federal do Rio de Janeiro (MN/UFRJ), Rio de Janeiro, RJ, Brazil

Simone Cartaxo-Pinto Departamento de Botânica, Laboratório de Palinologia, Museu Nacional, Universidade Federal do Rio de Janeiro (MN/UFRJ), Rio de Janeiro, RJ, Brazil

Maria da Graça Sophia Departamento de Botânica, Laboratório de Ficologia, Museu Nacional, Universidade Federal do Rio de Janeiro (MN/UFRJ), Rio de Janeiro, Brazil

Emanoela Mano Muniz da Silva Departamento de Botânica, Laboratório de Biologia Reprodutiva de Angiospermas, Museu Nacional, Universidade Federal do Rio de Janeiro (MN/UFRJ), Rio de Janeiro, RJ, Brazil

Lúcia Helena Sampaio da Silva Departamento de Botânica, Laboratório de Ficologia, Museu Nacional, Universidade Federal do Rio de Janeiro (MN/UFRJ), Rio de Janeiro, Brazil

Ana Tereza Araújo Rodarte Departamento de Botânica, Laboratório de Biologia Reprodutiva de Angiospermas, Museu Nacional, Universidade Federal do Rio de Janeiro (MN/UFRJ), Rio de Janeiro, RJ, Brazil

Mariana Reis de Brito Departamento de Biologia, CCBS-Centro de Ciências Biológicas e da Saúde, Pontifícia Universidade Católica do Rio de Janeiro (PUC-Rio), Rio de Janeiro, Brazil

Elen de Lima Aguiar-Menezes Departamento de Entomologia e Fitopatologia, Instituto de Ciências Biológicas e da Saúde, Universidade Federal Rural do Rio de Janeiro (UFRRJ), Seropédica, RJ, Brazil

Programa de Pós-Graduação em Fitotecnia, Universidade Federal Rural do Rio de Janeiro (UFRRJ), Seropédica, RJ, Brazil

Heloisa Alves de Lima Departamento de Botânica, Laboratório de Biologia Reprodutiva de Angiospermas, Museu Nacional, Universidade Federal do Rio de Janeiro (MN/UFRJ), Rio de Janeiro, RJ, Brazil

Vera Lúcia de Moraes Huszar Departamento de Botânica, Laboratório de Ficologia, Museu Nacional, Universidade Federal do Rio de Janeiro (MN/UFRJ), Rio de Janeiro, Brazil

Patrick de Oliveira Departamento de Botânica, Laboratório de Biologia Reprodutiva de Angiospermas, Museu Nacional, Universidade Federal do Rio de Janeiro (MN/UFRJ), Rio de Janeiro, RJ, Brazil

Bárbara de Sá-Haiad Departamento de Botânica, Laboratório de Anatomia Vegetal, Museu Nacional, Universidade Federal do Rio de Janeiro (MN/UFRJ), Rio de Janeiro, RJ, Brazil

Lygia Dolores Ribeiro de Santiago-Fernandes Departamento de Botânica, Laboratório de Anatomia Vegetal, Museu Nacional, Universidade Federal do Rio de Janeiro (MN/UFRJ), Rio de Janeiro, RJ, Brazil

Luci de Senna-Valle Departamento de Botânica, Museu Nacional, Universidade Federal do Rio de Janeiro (MN/UFRJ), Rio de Janeiro, RJ, Brazil

Gabriel Henrique Gomes de Souza Freitas Teixeira Departamento de Botânica, Laboratório de Palinologia, Museu Nacional, Universidade Federal do Rio de Janeiro (MN/UFRJ), Rio de Janeiro, RJ, Brazil

Raquel Maria Batista Souza de Souza Departamento de Botânica, Laboratório de Palinologia, Museu Nacional, Universidade Federal do Rio de Janeiro (MN/UFRJ), Rio de Janeiro, RJ, Brazil

Thiago Sampaio de Souza Instituto de Investigação em Tecnologias Agrárias e do Ambiente. Faculdade de Ciências Agrárias e do Ambiente, Universidade dos Açores, Angra do Heroísmo, Ilha Terceira, Açores, Portugal

Luciana Carolina Oliveira Sepúlveda do Nascimento Departamento de Botânica, Laboratório de Biologia Reprodutiva de Angiospermas, Museu Nacional, Universidade Federal do Rio de Janeiro (MN/UFRJ), Rio de Janeiro, RJ, Brazil

Juliana Barreto Oliveira dos Santos Departamento de Botânica, Laboratório de Ficologia, Museu Nacional, Universidade Federal do Rio de Janeiro (MN/UFRJ), Rio de Janeiro, Brazil

Luiz Fernando Dias Duarte Departamento de Antropologia Social, Museu Nacional, Universidade Federal do Rio de Janeiro (MN/UFRJ), Rio de Janeiro, Brazil

Roberto Lourenço Esteves Departamento de Biologia Vegetal, Universidade do Estado do Rio de Janeiro (UERJ), Rio de Janeiro, RJ, Brazil

Vinicius José Fernandes Programa de Pós-Graduação em Fitotecnia, Universidade Federal Rural do Rio de Janeiro (UFRRJ), Seropédica, RJ, Brazil

Indhira Viana Freire Departamento de Botânica, Laboratório de Ficologia, Museu Nacional, Universidade Federal do Rio de Janeiro (MN/UFRJ), Rio de Janeiro, Brazil

Vania Gonçalves-Esteves Departamento de Botânica, Laboratório de Palinologia, Museu Nacional, Universidade Federal do Rio de Janeiro (MN/UFRJ), Rio de Janeiro, RJ, Brazil

Alexandre Verçosa Greco Departamento de Botânica, Laboratório de Biologia Reprodutiva de Angiospermas, Museu Nacional, Universidade Federal do Rio de Janeiro (MN/UFRJ), Rio de Janeiro, RJ, Brazil

Isabella Veríssimo Nader Haddad Departamento de Botânica, Laboratório de Anatomia Vegetal, Museu Nacional, Universidade Federal do Rio de Janeiro (MN/UFRJ), Rio de Janeiro, RJ, Brazil

Luciene Campos São Leão Departamento de Botânica, Laboratório de Biologia Reprodutiva de Angiospermas, Museu Nacional, Universidade Federal do Rio de Janeiro (MN/UFRJ), Rio de Janeiro, RJ, Brazil

Alice Teodorio Lixa Programa de Pós-Graduação em Fitotecnia, Universidade Federal Rural do Rio de Janeiro (UFRRJ), Seropédica, RJ, Brazil

Elysiane Barros Marinho Departamento de Botânica, Laboratório de Palinologia, Museu Nacional, Universidade Federal do Rio de Janeiro (MN/UFRJ), Rio de Janeiro, RJ, Brazil

Maria Franco Trindade Medeiros Departamento de Botânica, Laboratório Interativo em Etnobotânica (LinE), Museu Nacional, Universidade Federal do Rio de Janeiro (MN/UFRJ), Rio de Janeiro, RJ, Brazil

Thiago Ávila Medeiros Departamento de Botânica, Laboratório de Biologia Reprodutiva de Angiospermas, Museu Nacional, Universidade Federal do Rio de Janeiro (MN/UFRJ), Rio de Janeiro, RJ, Brazil

Claudia Barbieri Ferreira Mendonça Departamento de Botânica, Laboratório de Palinologia, Museu Nacional, Universidade Federal do Rio de Janeiro (MN/UFRJ), Rio de Janeiro, RJ, Brazil

Mariângela Menezes Departamento de Botânica, Laboratório de Ficologia, Museu Nacional, Universidade Federal do Rio de Janeiro (MN/UFRJ), Rio de Janeiro, Brazil

Ana Carolina Mezzonato-Pires Departamento de Botânica, Universidade Federal de Juiz de Fora (UFJF), Juiz de Fora, MG, Brazil

Marina Muniz Moreira Departamento de Botânica, Laboratório de Biologia Reprodutiva de Angiospermas, Museu Nacional, Universidade Federal do Rio de Janeiro (MN/UFRJ), Rio de Janeiro, RJ, Brazil

Leila Nunes Morgado Instituto de Investigação em Tecnologias Agrárias e do Ambiente. Faculdade de Ciências Agrárias e do Ambiente, Universidade dos Açores, Angra do Heroísmo, Ilha Terceira, Açores, Portugal

Letícia Barbosa Quesado Departamento de Botânica, Laboratório de Ficologia, Museu Nacional, Universidade Federal do Rio de Janeiro (MN/UFRJ), Rio de Janeiro, Brazil

Luiz José Soares Pinto Postgraduate Program in Biological Sciences (Botany), Museu Nacional, Universidade Federal do Rio de Janeiro (MN/UFRJ), Rio de Janeiro, RJ, Brazil

Maria Carolina Sotero Secretaria de Educação de Pernambuco, Escola Várzea Fria, São Lourenço da Mata, PE, Brasil

Renata Barreto Tostes Universidade do Estado de Minas Gerais (UEMG), Ubá, MG, Brazil

About the Editors

Maria Franco Trindade Medeiros graduated in biological and environmental sciences, holds master's and Ph.D. in biological sciences (botany), and is a postdoctoral fellow in botany and ecology. She was an educator at Pontifical Catholic University of Rio de Janeiro (2012) and Federal University of Campina Grande (2013–2018), and currently is educator at the Postgraduate Program in Biological Sciences (Botany), National Museum/ Federal University of Rio de Janeiro (since 2018). She has held the secretariat and presidency of the Brazilian Society of Ethnobiology and Ethnoecology and has been a representative of the Ethnobotany Group of the Botanical Society of Brazil. Her interdisciplinary perspective of investigation searches for a dialog between ethnoscience, with emphasis on ethnobotany, history, museology, education, and biocultural heritage.

Bárbara de Sá-Haiad holds a degree in biological sciences (ecology) from the Federal University of Rio de Janeiro, a master's degree in biological sciences (botany) from the Federal University of Rio de Janeiro, a Ph.D. in biological sciences (botany) from the Federal University of Rio de Janeiro, and postdoc in plant secretory structures from the State University of São Paulo Júlio de Mesquita Filho (2011–2014). She is a professor at the Postgraduate Program in Biological Sciences (Botany), National Museum/Federal University of Rio de Janeiro since 2013. She has experience in botany, with emphasis on anatomy, working mainly on development, anatomy, and floral secretory structures. She supervises undergraduate, master's, and doctoral students.

About the Contributors

Donato Seiji Abe received his B.S. in biological sciences from the University of São Paulo (USP), his M.Sc. in biological oceanography from the USP, and his Ph.D. in environmental engineering sciences from the USP. He did part of his Ph.D. at the Shinshu University, Japan. He did postdoctoral research at the Federal University of São Carlos, where he studied processes related to the nitrogen cycling in reservoirs. He is currently a researcher at the International Institute of Ecology and Environmental Management in São Carlos. His research interests include eutrophication, greenhouse gas emissions, and monitoring of water quality and aquatic macrophyte.

Ângelo Giuseppe Chaves Alves is an ethnoecologist and educator who works on the cultural aspects of sustainability. Since 1997, he has been an associate professor at the Federal Rural University of Pernambuco (UFRPE), Brazil. He is a Doctor of Ecology (2004), Master of Agronomy/Soil Science (1991), and a graduate in agronomy (1987). Has been a postdoctoral fellow in Lisbon (Universidade Técnica de Lisboa) and Florence (Università degli Studi di Firenze). Has been the vice president of the Brazilian Society of Ethnobiology and Ethnoecology. Most of his work is dedicated to ethnoscience as connected to ecology, agroecology, education, and epistemology.

Angela Arruda e Albuquerque holds a degree in biological sciences from Pedro II Faculty of Humanities (1982), has completed a specialization course in sciences and biology teaching from the Federal University of Rio de Janeiro (2001), and has a master's degree in biological sciences (botany) from the Federal University of Rio de Janeiro (2011). She has experience in botany and biology teaching. She is a professor at the Secretary of State for Education of Rio de Janeiro, since 1998.

Davi Almeida Barreto graduated in biological sciences and did specialization in project management at Universidade Veiga de Almeida. He holds a Master of Biological Sciences (Botany) from the National Museum-UFRJ, having developed the work on phytoplankton ecology in a eutrophic reservoir with recurrent canobacterial blooms and evaluated the mitigation of cyanobacterial blooms through the application of geoengineering.

Cristine Rodrigues Benevides holds a bachelor's degree in biological sciences from the State University of North Fluminense Darcy Ribeiro (2004), a master's degree in ecology and natural resources from the State University of North Fluminense Darcy Ribeiro (2006), a Ph.D. in biological sciences (botany) from the Federal University of Rio de Janeiro (2011), and postdoc in reproductive biology from the Federal University of Rio de Janeiro (2014). She has experience in ecology and botany. She is a volunteer researcher in the Botany Department, National Museum, Federal University of Rio de Janeiro, since 2012.

Simone Cartaxo-Pinto graduated in biological sciences at Castelo Branco University (2008). She has a master's degree in plant biology from the State University of Rio de Janeiro (2012). She is doing her Ph.D. in the postgraduate program in biological sciences (botany) at the Museu Nacional/UFRJ, developing studies in palynology of Vitaceae.

Maria da Graça Sophia graduated in biological sciences from the State University of Rio de Janeiro, and holds a master's degree and Ph.D. in biological sciences (botany) from the National Museum-UFRJ. She is currently a biologist in the Department of Botany/Phycology Laboratory/National Museum/UFRJ, and her research covers desmids and cyanobacteria and subaerial algae. So far, she has published several scientific papers on Brazilian desmids, orienting several undergraduate students.

Emanoela Mano Muniz da Silva holds a degree in biological sciences from Castelo Branco University (2009), a postgraduate in environmental management from Cândido Mendes University (2013), and a postgraduate in biosafety from Unyleya University (2019). She had done her internship in the Laboratory of Reproductive Biology of Angiosperms – Museu Nacional/Federal University of Rio de Janeiro (2007–2009). She has experience in botany and cell biology. She has been working as a laboratory technical analyst since 2011.

Lúcia Helena Sampaio da Silva is a Master of Botany from National Museum, Federal University of Rio de Janeiro (MN-UFRJ), and holds a Ph.D. in ecology from the Federal University of Rio de Janeiro. She has experience in aquatic ecology, working mainly on phytoplankton ecology and planktonic trophic interactions in reservoirs and lakes. She is an associate professor in the Department of Botany at the National Museum of the Federal University of Rio de Janeiro (MN-UFRJ), and a master's and doctoral advisor in the Graduate Program in Botany at the Museu Nacional (MN-UFRJ).

Ana Tereza Araújo Rodarte holds a bachelor's degree in forest engineering from the Federal University of Paraná (1972), master's degree in biological sciences (botany) from the Federal University of Bahia (2003), and Ph.D. in biological sciences (botany) from the Federal University of Rio de Janeiro (2008). She is currently a researcher in the field of botany at the National Museum, Federal University of Rio de Janeiro, with emphasis in phenology and floral biology. Since 2010, she is a volunteer teacher in the Postgraduate Program in Biological Sciences (Botany) at the National Museum, Federal University of Rio de Janeiro.

Mariana Reis de Brito has a bachelor's degree in ecology and botany from the Federal University of Rio de Janeiro, Brazil. She holds a master's degree and a doctorate in the graduate program in biological sciences (botany) from Museu Nacional, Federal University of Rio de Janeiro (UFRJ), Brazil. She has experience in the field of botany, with an emphasis on ethnobotany, historical ethnobotany, and history of botany. She is a professor in the Department of Biology, CCBS Center of Biological and Health Sciences, Pontifical Catholic University of Rio de Janeiro (Puc-Rio), Brazil.

Elen de Lima Aguiar-Menezes is an agronomist at Universidade Federal Rural do Rio de Janeiro (UFRRJ), and holds a Master of Agronomy (Entomology) from the Universidade Federal de Lavras (UFLA) and Ph.D. in phytotechnics (entomology) from UFRRJ. She was a laboratory technician hired by the USDA (United States Department of Agriculture), Agricultural Research Service (ARS), working at the Center for Agricultural Medical and Veterinary Entomology-CMAVE (Gainesville, FL, USA) (paid work permit granted by the US Citizenship and Immigration Services), having the opportunity to also work in the Quarantine Laboratory of the DPI (Division Plant Industry) of the Florida Department of Agriculture and Consumer Services (Gainesville, FL, USA). Professor at the Programa de Pós-graduação em Agricultura Orgânica (PPGAO) at UFRRJ, since 2010. She is permanent professor in the Postgraduate Program in Plant Science (PPGF) at UFRRJ, since January/2019 and coordinator of the Integrated Center for Integrated Pest Management (CIMP), DEnF/ICBS/UFRRJ, since 2009. She has experience in agronomy, with an emphasis on agricultural entomology, working mainly on the following topics: agroecological pest management; conservative biological control and augmentation of agricultural pests; selectivity of defensive alternatives to predatory ladybugs; organic agriculture, agroecology, ecology, and management of fruit flies (Diptera: Tephritidae) in the state of Rio de Janeiro; and ecology, behavior, and termite control in urban and rural areas in the state of Rio de Janeiro.

Heloisa Alves de Lima holds a degree in physical and biological sciences (graduation and bachelor of ecology) from the Federal University of Rio de Janeiro (1980 and 1981), a master's degree in biological sciences (botany) from the Federal University of Rio de Janeiro (1986), and a Ph.D. in biological sciences (botany) from the Federal University of Rio de Janeiro (2002). She is a professor at the National Museum/Federal University of Rio de Janeiro since 1995. She has experience in botany, with emphasis on reproductive biology of angiosperms. She is a professor in the Postgraduate Program in Biological Sciences (Botany), National

Museum/Federal University of Rio de Janeiro since 2003, where she teaches disciplines, develops research projects, and supervises undergraduate, master's, and doctoral students.

Vera Lúcia de Moraes Huszar is an aquatic ecologist and full professor in the Department of Botany, National Museum, Federal University of Rio de Janeiro (MN-UFRJ). Her Ph.D. is in ecology, and she is interested in phytoplankton ecology in inland waters, searching for a better understanding of the structuring of plankton communities, seasonality, diversity, macroecology, and recovery of aquatic systems, mainly in reservoirs, flood plain lakes, coastal lagoons, and estuaries. She is also a master's and doctoral advisor in the Graduate Program in Botany at the Museu Nacional (MN-UFRJ).

Patrick de Oliveira holds a degree in biological sciences from the State University of Rio de Janeiro (UERJ), is a specialist in science and biology teaching from the Federal University of Rio de Janeiro (UFRJ), and holds master's degree in education, management, and dissemination in biosciences from Federal University of Rio de Janeiro (UFRJ). He has completed his internship in the Laboratory of Reproductive Biology of Angiosperms – Museu Nacional/Federal University of Rio de Janeiro (2009– 2011). He has experience in botany and biology teaching. He has been a teacher in public and private networks (since 2011).

Lygia Dolores Ribeiro de Santiago-Fernandes is an associate professor in the Botany Department at the National Museum/Federal University of Rio de Janeiro. During the last 30 years, she has supervised many undergraduate, master's, and doctoral students. Fernandes holds a degree in biological science from Santa Ursula University (1980) and a specialization in genetics from the Federal University of Rio de Janeiro (1982). In 1987, she earned her master's degree in biological sciences (botany) from the Federal University of Rio de Janeiro, followed by her Ph.D. in biological sciences (genetics) from the same University in 1996. Lygia was the representative of the National Museum at CEPG (Counsel of Education for Graduates) for 6 consecutive years and occupied the position of associate director of education at the National Museum/Federal University of Rio de Janeiro from 2014 to 2020.

Luci de Senna-Valle graduated in natural history from the State University of Rio de Janeiro, and holds a master's degree in biological sciences (botany) from the Federal University of Rio de Janeiro and Ph.D. in plant biotechnology from the Federal University of Rio de Janeiro. She is currently holder of the Federal University of Rio de Janeiro (2015). She has published 72 articles, 6 book chapters, and 1 book. She has supervised 75 undergraduates, and master's and doctoral students. She has experience in the field of botany, with an emphasis on ethnobotany, working mainly on the following topics: traditional communities, historical ethnobotany, medicinal plants, and taxonomy.

Gabriel Henrique Gomes de Souza Freitas Teixeira graduated in biological sciences from Instituto Federal do Rio de Janeiro and completed his master's degree in biological sciences (botany) at the National Museum/Federal University of Rio de Janeiro. He is a doctoral student of biological sciences (botany) at the Museu Nacional/UFRJ, developing his thesis in melissopalynology. He has experience in botany, with an emphasis on palynology and biodiversity.

Raquel Maria Batista Souza de Souza holds a degree in biological sciences from Castelo Branco University (2010), a master's in plant biology from the State University of Rio de Janeiro (2014), and a Ph.D. in botany from the National Museum/UFRJ (2018). She has experience in the field of botany, with an emphasis on systematics, palynology, and asteraceae taxonomy.

Thiago Sampaio de Souza is an agronomist who graduated from the Universidade Federal Rural do Rio de Janeiro (UFRRJ), having been a scholarship holder at the Botanical Garden of UFRRJ and scientific initiation scholarship holder of CNPq of the Integrated Center for Pest Management at UFRRJ. He is a resident in agronomy in the company Olearys Tecnologia e Ciência S/A, in the field of phytotechnics with emphasis on pest, disease, and irrigation control using precision agriculture (Triângulo Mineiro and Goiás). He holds a Master of Science in plant health and applied biotechnology in the field of applied plant health (UFRRJ). Thiago is a specialist in medical entomology from the Oswaldo Cruz Foundation (IOC/FIOCRUZ). He is currently a doctoral student in the postgraduate program in phytotechnics/UFRRJ, in the agroecology concentration area. He is interested in insect behavior, ecological pest, and disease management. He has experience in agronomy, with an emphasis on augmentative and conservative biological control of agricultural pests, selectivity of alternative pesticides to predatory ladybugs, cultivation of vegetables and large crops, urban pests, and botanical identification.

Luciana Carolina Oliveira Sepúlveda do Nascimento holds a degree in biological sciences from Castelo Branco University (2009), is an aquaculture technician from the Federal Institute of Education, Science and Technology of São Paulo/IFSP (2019), is a postgraduate in environmental education from São Brás College (2019), and holds a degree in didactics and methodology of science teaching from Educacional Faculty of Lapa FAEL (2021) and biology teaching from Cruzeiro do Sul University (2021). She completed her internship in the Laboratory of Reproductive Biology of Angiosperms – Museu Nacional/Federal University of Rio de Janeiro (2007–2009). She has experience in botany, biology teaching, and zoology. She has been working as a professor at the Secretary of State for Education of São Paulo (since 2013). She is a professor (since 2020) at São Sebastião College and Anglo Módulo College.

Juliana Barreto Oliveira dos Santos holds a bachelor's degree in biological sciences from Gama Filho University, and a master's and Ph.D. degree in botany from National Museum/UFRJ. She has vast experience in botany taxonomies of desmids and phytoplankton ecology, and emphasizes on functional diversity and metacommunities. Currently, she is an assistant professor at Celso Lisboa University and substitute professor at UNIRIO. In the past and until now, she worked as an environmental consultant monitoring freshwater ecosystems for distribution and human use.

Luiz Fernando Dias Duarte is a full professor in the graduate program in social anthropology at Museu Nacional, Federal University of Rio de Janeiro (UFRJ), Brazil. Luiz has a Ph.D. in social sciences from UFRJ and is pursuing postdoctoral studies at Ecole de Hautes Etudes en Sciences Sociales, Paris. Luiz is a visiting professor at the universities of Paris-Nanterre, Liège, Buenos Aires, UNAM – México, Brasília, Misiones, Córdoba/AR, and Rio Grande do Norte. Luiz is a senior researcher of the Brazilian Council for Technical and Scientific Development (CNPq), former director of the Museu Nacional/UFRJ, former member of the National Trust Commission (IPHAN), and full member of the Brazilian Academy of Sciences.

Roberto Lourenço Esteves graduated in biological sciences from the State University of Rio de Janeiro (1975), and holds a Master of Biological Sciences (Botany) from the Federal University of Rio de Janeiro (1993) and a Ph.D. in plant biology from the State University of Campinas (2001) in the Asteraceae family (Eupatorieae). He is currently an associate professor at the State University of Rio de Janeiro in the Department of Plant Biology at UERJ. He has supervised students through the Plant Biology Program (UERJ) and the Postgraduate Program in Biological Sciences (Botany) at the Museu Nacional/UFRJ. He also develops studies in palynology.

Vinicius José Fernandes is an agricultural engineer who graduated from Universidade Federal Rural do Rio de Janeiro (UFRRJ). He holds a Master of Science in the postgraduate course in plant health and applied biotechnology (PGFBA) from the Universidade Federal Rural do Rio de Janeiro (UFRRJ), with a concentration in applied entomology. Has experience in the area of behavior and termite control in urban and rural areas and also in urban and rural pest control. He is currently a doctoral candidate in the postgraduate program in Ppytotechnics at UFRRJ, in the agroecology concentration area. He is interested in insect behavior, ecological pest management, biological control, and urban pests.

Indhira Viana Freire is graduate in biological sciences from the Centro Universitário Celso Liboa, and Master of Biological Sciences (Botany) from the National Museum-UFRJ, where she developed the dissertation on phytoplankton ecology in a highly eutrophicated coastal lagoon and evaluated the mitigation of cyanobacterial blooms through the application of geoengineering.

Vania Gonçalves-Esteves is titular professor in the Department of Botany at Universidade Federal do Rio de Janeiro, Museu Nacional, and is a research fellow at Conselho Nacional de Desenvolvimento Científico e Tecnológico (CNPq). She received a graduate degree in biological sciences (botany) from Universidade do Estado da Guanabara (1974); a master's in botany (palynology) from the Uiniversidade Federal do Rio de Janeiro, Museu Nacional (1981); and a Ph.D. in from the Universidade de São Paulo (1994). Her research involves palynology of the current flora in several Brazilian ecosystems.

Alexandre Verçosa Greco holds a degree in biological sciences from State University of Rio de Janeiro (UERJ) (2011) and a specialization in environmental education and sustainability from the International University Center (2018). He had taken an internship in the Laboratory of Reproductive Biology of Angiosperms, Museu Nacional/Federal University of Rio de Janeiro (2008–2011). He is currently a master's student in the graduate program in history of science and health at Casa de Oswaldo Cruz/Fundação Oswaldo Cruz. He has experience in botany and environmental education in schools and conservation units. He is a primary school professor at the Secretary of State for Education of Rio de Janeiro.

Isabella Veríssimo Nader Haddad holds a degree in biological sciences from the Veiga de Almeida University (2007), a master's degree in biological sciences (botany) from the Federal University of Rio de Janeiro (2011), and a Ph.D. in biological sciences (botany) from the Federal University of Rio de Janeiro (2017). She is a collaborating researcher at the National Museum/Federal University of Rio de Janeiro since 2019. She has experience in botany, with emphasis on Anatomy, working mainly on the development and anatomy of reproductive structures focusing on sterility. She supervised nine undergraduate students.

Luciene Campos São Leão holds a degree in biological sciences (environment biology) from the City University Center (2008), a master's degree in biological sciences (botany) from the Federal University of Rio de Janeiro (2012), and a Ph.D. in biological sciences (botany) from the Federal University of Rio de Janeiro (2017). She has experience in botany and floral anatomy. She is a volunteer researcher in the Botany Department, National Museum, Federal University of Rio de Janeiro (since 2017); professor at FAETEC (Technical Support Foundation School), since 2019; and tutor at the Center for Distance Higher Education of the State of Rio de Janeiro, since 2021.

Alice Teodorio Lixa is an agricultural engineer from the Universidade Federal Rural do Rio de Janeiro. She holds a Master of Science degree after completion of the postgraduate course in plant health and applied biotechnology, with a concentration area in applied entomology at the same university, a period in which she was a scholarship holder of the National Center for Research in Agrobiology (EMBRAPA Agrobiology), where she had the opportunity to develop research in the areas of biological pest control, organic agriculture, agroecology, and sustainable development. She has just obtained a Doctor of Science degree in the postgraduate program in phytotechnics at UFRRJ, with a concentration in agroecology.

Elysiane Barros Marinho holds an undergraduate degree in biological sciences (2008), a bachelor's degree in marine biology (2009), a specialization in education (2010) from the State University of Rio de Janeiro – FFP/UERJ, master's degree in education (2013) from the Federal University of the State of Rio de Janeiro (UNIRIO), a master's degree in botany (2013) and doctorate in botany (2017) from the Federal University of Rio de Janeiro – Museu Nacional/UFRJ. She has experience in the field of education, with an emphasis on educational practices with a science, technology, society, and environment (CTSA) bias. She teaches at Fluminense Federal University the following subjects: Natural-Material Sciences and Methods (Pedagogy) and Research and Teaching Practices I and II (Biological Sciences). She has experience in botany, with an emphasis on palynology, acting on the following subjects: palynology, palynotaxonomy, pollen evolution, palynotaxonomy of neotropical podostemaceae (dissertation), and palinotaxonomy of Sapindaceae (thesis).

Thiago Ávila Medeiros holds a degree in biological sciences from Gama Filho University (2007), specialization in environmental management and planning from Estácio University (2007), and master's degree in science teaching from the Federal Institute of Education, Science and Technology of Rio de Janeiro (2014). He has experience in botany and biology teaching. He is a professor in the Higher Education Faculty at Faculdades Integradas Vital Brasil and the Brazilian Conservatory of Music – University Center; professor and coordinator at Faculdades São José; tutor at the Center for Distance Higher Education of the State of Rio de Janeiro; and professor at the Secretary of State for Education of Rio de Janeiro.

Claudia Barbieri Ferreira Mendonça is a professor in the Department of Botany of the Universidade Federal do Rio de Janeiro, Museu Nacional, and is a research fellow at Conselho Nacional de Desenvolvimento Científico e Tecnológico (CNPq). She graduated in biological sciences, and has a master's degree and Ph.D. in biological sciences (botany) from Universidade Federal do Rio de Janeiro, Museu Nacional (1999/2006). Her research involves palynology of the current flora in several Brazilian ecosystems.

Mariângela Menezes has a long career in teaching in countries of Latin America, such as Mexico, Colombia, Argentina, and Brazil. Mariângela is a CAPES postdoctoral fellow, Department of Phycology, University of Copenhagen. Completed her Ph.D. in biological science (botany-cryptogams) at São Paulo University. Her research areas are morphology, taxonomy, molecular phylogeny of flagellated photosynthetic algae, and botanical collections. Currently, Mariângela is a full professor at the Federal University of Rio de Janeiro, National Museum, Department of Botany, where she perform teaching and research activities in undergraduate Biological Sciences and graduate of Botany courses, and she is curator of the Cryptogams Collection from the Herbarium of the National Museum (R).

Ana Carolina Mezzonato-Pires graduated in biological sciences from Universidade Federal de Juiz de Fora, Minas Gerais, Brazil, in in the modalities of Licenciatura (2010) and Bachelor (2011), and holds a master's degree (2013), Ph.D. (2017), and postdoctorate PNPD/CAPES (2019) from the postgraduate program in biological sciences (Botany) from the Museu Nacional/ Universidade Federal do Rio de Janeiro. She works in the postgraduate program at the Museu Nacional in the disciplines: eudicotyledon diversity and systematics and evolution of pollen characters in angiosperms. She is a substitute professor at Universidade Federal de Juiz de Fora, Minas Gerais (UFJF). She has experience in botany with emphasis on taxonomy, palynology, biodiversity, conservation, evolution, and management of botanical collections. Ana is a specialist in the Passifloraceae family s.s.

Marina Muniz Moreira holds a degree in biological sciences (botany) from the Federal University of Rio de Janeiro (2011), a master's degree in biological sciences (botany) from the Federal University of Rio de Janeiro (2014), and a Ph.D. in ecology from the Federal University of Rio de Janeiro (2018). She has experience in ecology and botany. She is a volunteer researcher in the Botany Department, National Museum, Federal University of Rio de Janeiro, since 2018.

Leila Nunes Morgado graduated in animal science from Universidade Federal Rural do Rio de Janeiro, and holds a master's degree in agronomy (entomology) from Universidade Federal de Lavras, a Ph.D. in science (ecology) from the State Universidade do Rio de Janeiro, a postdoc from the Department of Entomology at Universidade Federal de Lavras, postdoctorate in palynology from the Department of Botany/Museu Nacional/Universidade Federal do Rio de Janeiro, and postdoctorate in biology at the Universidade dos Azores (Portugal). She has experience in agronomy, animal science, and biology, with an emphasis on entomology, ecology, ethology, beekeeping, and meliponiculture, acting on the following topics: palynology applied to the study of interaction, apoidea ecology in tropical ecosystems, pollination ecology, and animal behavior.

Letícia Barbosa Quesado holds a bachelor's degree in biological sciences from UNIRIO, specialization in science teaching from UERJ, and a master's degree and Ph.D. in ecology from UFRJ and UFRN, respectively. She has experience in ecology in limnology with functional approaches of phytoplankton, with macroecological and intra and interspecific perspectives. She also has experience in education, with biology teaching and environmental education, working with ecological interactions in textbooks, School Agenda 21, prevention and confronting accidents, and natural disasters.

Luiz José Soares Pinto graduated in biological sciences from the State University of Rio de Janeiro, and holds a master's degree and a Ph.D. in biological sciences (botany) from the Federal University of Rio de Janeiro. He is currently a teacher in the State Education Network of Rio de Janeiro at David Capistrano State School. He has published 19 articles in specialized journals and more than 50 works in events, most of them on botany. He has published one book, *Medicinal Plants: Identification, Properties and Cultivation Manual*, and wrote six book chapters. He has supervised six undergraduates and pursues in research that has emphasis in ethnobotany and vegetable taxonomy.

Maria Carolina Sotero holds a degree in biological sciences from the Federal Rural University of Pernambuco (2008), a master's degree in development and environment from the Federal University of Pernambuco (2013), and a Ph.D. in ethnobiology and nature conservation from the Federal Rural University of Pernambuco (2020). She has been a teacher of basic education in the State of Pernambuco since 2010 and at the Municipality of Vitória de Santo Antão since 2016. She works in the coordination and training of teachers and teaches in science classes. She conducts research with a focus on ethnobiology and science education.

Renata Barreto Tostes graduated in biological sciences from Federal University of Juiz de Fora, and obtained her Master of Botany from the Federal University of Viçosa and Doctor of Plant Biology from the National Museum/Federal University of Rio de Janeiro. Since 2017, she has been a professor at the State University of Minas Gerais, where she has been developing research in the area of ethnobotany. During this time, she has already supervised several undergraduate research students and final monographs.

Part I
Angiosperms Reproductive Aspects

This part will address the reproductive aspects of the angiosperms as one of the various areas of Botany that will be discussed in the chapters of *Aspects of Brazilian Floristic Diversity: From Botany to Traditional Communities*. Here, certain aspects will be presented that will allow us to have a deeper comprehension of this theme, having as a guide the experiences of research carried out in the Brazilian territory, which will be the first step of our journey through this book.

So, in Part I, we will have the opportunity to understand phenology as a multidisciplinary science that can transit from individual to a population range for searching the recurrent biological events. In this perspective, flowering phenology in long-term studies can be very useful for better understanding and assessing the diversity and availability of floral resources of a community in the current scenario of global climate change.

Also, we will see that dioecy, the occurrence of separate male and female individuals in a population, can be evaluated by studying the dimorphic sexual system, the pollination system, the attributes related to visitor attraction, and the floral visitors of dioecious species to better understand flora structure and functioning in a specific vegetation type.

Finally, going into another approach, the floral reproductive development as a gene-controlled process that involves organogenesis of reproductive whorls and male and female gametophytes development will be discussed. This subject will be addressed through a critical analysis of literature data combining the elements of structure and functionality during gametophytes development and by considering the entire programmed cell death processes, commonly recruited to cause sterility in unisexual flowers. The intention of this data analysis will be to show the significance of these events in the reproductive development of neotropical plants.

Chapter 1
Flowering Phenology in a *Restinga* Community: 7 Years of Study

Ana Tereza Araújo Rodarte, Cristine Rodrigues Benevides, Marina Muniz Moreira, Alexandre Verçosa Greco, Luciene Campos São Leão, Patrick de Oliveira, Thiago Ávila Medeiros, and Heloisa Alves de Lima

Abstract Since changes in climate can generate phenological shifts and temporal mismatches between plants and their pollinators, long-term studies of flowering phenology have become more common in the scenario of global climate change. Although in tropical environments, flowering cycles are diverse, irregular, and complex, the existing phenology studies have evaluated mainly tree species over short periods. We characterized, over 7 years, flowering events of a restinga (sandy coastal plain) plant community in southeastern Brazil, including 829 individuals, 78 species, and 36 families, with diverse life forms. In restinga, the general flowering strategy is annual, regular, with intermediate duration, although some species show continuous, sub-annual, or supra-annual strategies. Plants of various life forms flower continuously or sub-annually, whereas only trees flower annually. We recorded flowers throughout all the study years, but the highest rates of flowering activity and intensity occurred in the warmer and wetter season (October to March). Nectar, oil, pollen, and resin were available to pollinators throughout the year. We found significant positive correlations between the indexes of activity and intensity and the mean temperature and day length, but not precipitation. Our results suggest a high predictability of flowering periods and availability of floral resources for pollinators throughout the year.

Keywords Atlantic Forest · Climatic change · Floral resource

A. T. Araújo Rodarte (✉) · C. R. Benevides · M. M. Moreira · A. V. Greco · L. C. S. Leão · P. de Oliveira · T. Á. Medeiros · H. A. de Lima
Departamento de Botânica, Laboratório de Biologia Reprodutiva de Angiospermas, Museu Nacional, Universidade Federal do Rio de Janeiro (MN/UFRJ), Rio de Janeiro, RJ, Brazil
e-mail: atrodarte@gmail.com

M. F. T. Medeiros, B. de Sá Haiad (eds.), *Aspects of Brazilian Floristic Diversity*,
https://doi.org/10.1007/978-3-031-07453-0_1

1.1 Introduction

Phenological studies on flowering in plant communities are important for understanding plant reproduction and the spatiotemporal organization of floral resources available to animals (Newstrom et al. 1994a). The distribution of flowering events of the plant species in a community is selected over time by abiotic and biotic factors (van Schaik et al. 1993). Tropical environments are highly challenging for phenological studies because of the wide range of interactions and the environmental heterogeneity in these regions (Ramírez 2002). In temperate ecosystems and dry tropical forests, climatic seasonality directly influences plant phenology (Morellato et al. 2013; van Schaik et al. 1993). In contrast, most tropical forests have less pronounced climatic seasons; the highly diverse plant species with different life forms can provide blooms at any time of the year, with varying frequencies, regularities, and synchronisms, resulting in diverse and complex phenological patterns (Morellato et al. 2013; van Schaik et al. 1993). Therefore, tropical areas provide reliably and continuously available plant resources that support a rich spectrum of forage animals (Morellato et al. 2016).

Phenological studies in tropical environments focus mainly on tree species of forest vegetation and over short periods of time, mainly 1 or 3 years (Morellato 2008; Morellato et al. 2016), making it difficult to understand the factors that shape flowering patterns in plants with different life forms (Newstrom et al. 1994b). Long-term phenological data are rare but are beginning to gain significance in light of the importance of understanding phenological patterns in communities and, more recently, as a tool for understanding plant sensitivity to global climate change (Morellato 2008; Morellato et al. 2016; Dunham et al. 2018). Changes in period, duration, and amplitude of flowering events caused by climate changes have been reported (Primack et al. 2004; Morellato et al. 2016; Dunham et al. 2018) and may result in imbalances in the interactions between plants and their pollinators (Morellato et al. 2016).

The *restingas* are part of the Atlantic vegetation complex. They are distributed along the coastal plain formed by marine sediments of Quaternary origin. *Restinga* flora arose mainly from the Atlantic Forest (Scarano 2002, 2008). This environment is stressful due to the sandy soil with low water retention, low air humidity, and strong sea wind action (Rizzini 1979; Scarano 2002). Scarano (2002) proposed that epiphyte and hemi-epiphyte plants of the Atlantic Forest would have been mainly responsible for the colonization and diversification of Atlantic Forest marginal environments. Canopy plants are undemanding of resources from the substrate and, once migrating, would have been able to settle as terrestrial plants on sandy soils, creating conditions for the establishment of other species (Scarano 2002, 2008). This facilitation process would be the main explanation for the high diversity found in an environment with such low environmental potential (Scarano 2002, 2008). *Restinga* vegetation has been considered stable and little sensitive to climate fluctuations (Scheel-Ybert 2000, 2002). Paleoenvironmental studies have shown that the *restinga* vegetation of the southeastern coast of Rio de Janeiro state (RJ) did not vary in diversity through the second half of the Holocene (5500–1400 years BP)

(Scheel-Ybert 1999, 2000, 2001). The *restinga* at Maricá, RJ, does not have well-defined climatic seasons and did not have a dry season during the years 1989–2000 (Mantovani and Iglesias 2001); however, there are relatively frequent records of water deficit in the last 10 years, in July and August (INMET/RJ). In general, the mean monthly rainfall decreases significantly during the winter, but months with high rainfall in the winter and low rainfall in the summer are often recorded, characterizing an irregular rainfall distribution pattern (Mantovani and Iglesias 2001).

This study characterized the flowering phenophase of the species occurring in a *restinga* vegetation community at Maricá over 7 years. We analyzed the frequency, duration, regularity, and the percentages of activity (Bencke and Morellato 2002b) and intensity (Fournier 1974) of the flowering period of the species, relating them to the possible influences of climate variables. The following questions guided our work: (1) What are the flowering phenological strategies in this community? (2) Are there abiotic variables with significant potential to trigger flowering in the area? (3) Do flowering phenophases follow seasonal rhythms in this community? (4) Is the distribution of floral resources such as pollen, nectar, oil, and resin predictable through the year?

1.2 Material and Methods

1.2.1 Study Site

We conducted the study in the Maricá *restinga*, an environmentally protected area located on the border of the districts of Barra de Maricá and Itaipuaçu, municipality of Maricá, Rio de Janeiro state, Brazil (22°57′45″S to 42°53′33″W and 22°57′52″S to 42°53′48″W) (Fig. 1.1a).

The Maricá *restinga* is composed of two sandy ridges (inner and outer) formed between 3500 and 5500 years ago, respectively, in episodes of the last marine transgressions (Perrin 1984) (Fig. 1.1b). The study site contains a shrubby-arboreal formation that encompasses three typical *restinga* physiognomies: (1) flooded open shrubby area, located in periodically flooded places (Fig. 1.1b); (2) non-flooded open shrubby area, with thickets interspersed with bare spaces (Fig. 1.1b–d); and (3) non-flooded closed arboreal area, with dense vegetation (Fig. 1.1b, d).

1.2.2 Climate

The climate is Aw tropical humid (Köppen 1948), with rainy summers and dry winters (Mendonça and Danni-Oliveira 2007). Climatic data for the Maricá *restinga* from 2004 to 2010 were provided by the *Instituto Nacional de Meteorologia* for Maricá Station, located inside the protected area (22°055"S to 42°049"W). We calculated day length according to Pereira et al. (2001) and Varejão-Silva (2000)

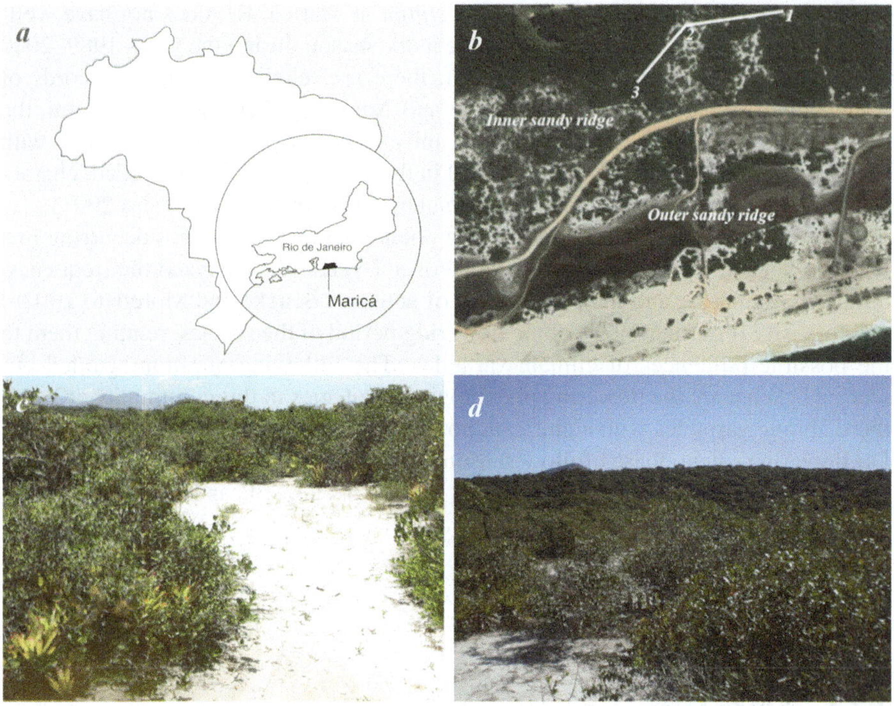

Fig. 1.1 The Maricá *restinga*. (**a**) Location of Maricá, Rio de Janeiro state, Brazil. (**b**) Aerial view of the study area, showing the transect across the inner sandy ridge and the three physiognomies (1, 2, and 3) sampled. (**c**) View of area 2 (non-flooded open shrubby area). (**d**) View of area 2 (non-flooded open shrubby area) in foreground and of area 3 (non-flooded and closed arboreal area) in background

(Fig. 1.2a). We considered two seasons: warmer/wetter (October to March) and colder/drier (April to September) (Figs. 1.2 and 1.3). The mean annual temperature ranged from 22.9 °C (2004) to 24.6 °C (2005). July was the coldest month in all years (Fig. 1.2b), with mean temperatures ranging from 19.5 °C (2010) to 21.8 °C (2005). In the warmer/wetter months (October to March), the mean temperature ranged from 25.6 °C (January 2004) to 28.5 °C (February 2010).

Total annual rainfall ranged from 1197.8 mm (2007) to 1435.8 mm (2005) (Fig. 1.2c). We recorded water deficits in the years 2004 (August, 15.1 mm; September, 31.2 mm; October, 38.2 mm), 2005 (August, 5.7 mm), 2007 (March, 33 mm; June, 33.8 mm; August, 30.6 mm; September, 16.6 mm), 2008 (August, 32 mm), and 2010 (August, 36.3 mm; September, 33.3 mm) (Fig. 1.3). Higher rainfall volumes were recorded from October to March of 2007–2008 and 2009–2010. Atypical rain peaks occurred in June 2004 and 2010 (Fig. 1.3). Although we have considered these two seasons (warmer/wetter and colder/drier), the temperature varied much less than the rainfall, which varied widely from year to year (Fig. 1.2b, c).

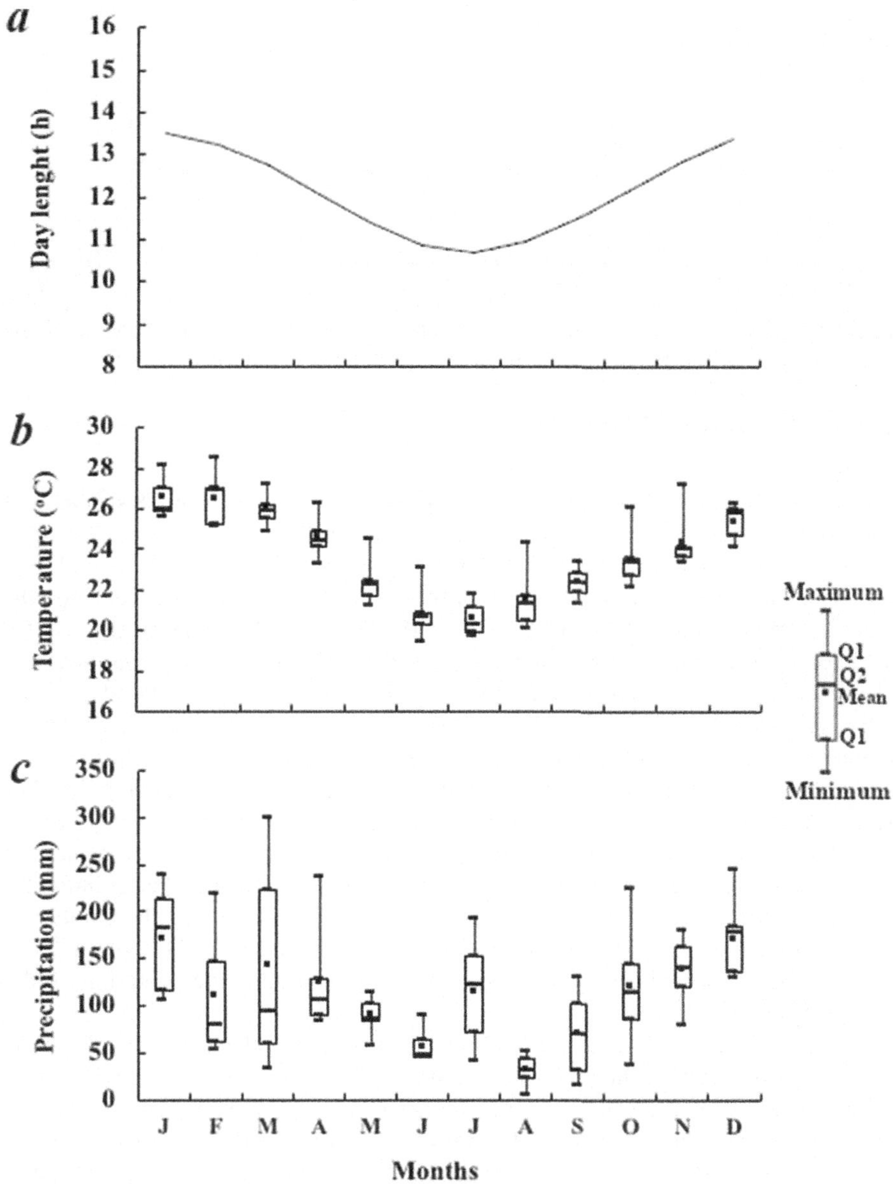

Fig. 1.2 (**a**) Day length (h), (**b**) box plots of monthly temperature (°C), (**c**) precipitation (mm) data for 2004–2010 in the Maricá *restinga*, Rio de Janeiro state, southeastern Brazil. (Source: INMET/RJ)

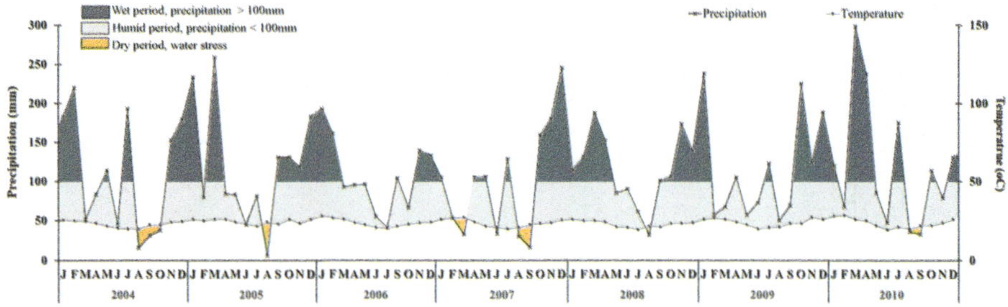

Fig. 1.3 Climatic diagram showing wet, humid, and dry periods during the study years (2004–2010) in the Maricá *restinga*, Rio de Janeiro state, Brazil. (Source: INMET/RJ)

1.2.3 Flowering Phenology

To conduct the phenological study, we marked 829 individuals from 78 species and 36 families. These individuals were inspected weekly from 2004 to 2010. We made these observations along a transect covering 5000 m^2 (500 m long × 10 m wide) on the inner sandy ridge, oriented east-west parallel to the coast and between coordinates 22°057′45″S to 42°053′33″W and 22°057′52″S to 42°053′48″W (Fig. 1.1b). This transect covered the following physiognomies: (1) flooded open shrubby area (1600 m^2), (2) non-flooded open shrubby area (1300 m^2), and (3) non-flooded closed arboreal area (2100 m^2) (Fig. 1.1b). For species more than 1 m tall (here termed upper stratum), we marked up to 15 individuals with aluminum plates. Species with fewer than five individuals were disregarded (Fournier and Charpantier 1975). For species less than 1 m tall (here termed lower stratum), the sampling unit was 1 m^2 (n = 150 squares). Three 1 m^2 square frames were used to cover the vegetation at every 10 m along the transect. Where possible, we counted the number of individuals within the squares. For Poaceae and Cyperaceae, we estimated the number of individuals by the occupied proportion of the square, considering 100% of the occupied area as 50 individuals. Voucher specimens are housed in the herbarium of the Museu Nacional (R).

To quantify the presence/absence of the flowering phenophase, we calculated the percentage of activity, which indicates the proportion of species that are manifesting a given phenological event in a given month (Bencke and Morellato 2002a). To quantify the flowering intensity of the individuals, we adopted the method proposed by Fournier (1974) and modified by Bianchini et al. (2006), which uses a semiquantitative scale, from 0 to 3, to express the intensity of flowers per plant: 0 (absence of flowers); 1 (low intensity), 1–33% flowering branches; 2 (moderate intensity), 34–66% flowering branches; and 3 (high intensity), more than 66% flowering branches. Each individual was evaluated weekly, assigned a score from 0 to 3, and the mode of the assigned values was used to estimate the intensity of the event in each species, over the various months of follow-up.

To evaluate the flowering intensity in the community, we used the percentage of intensity (Fournier 1974), given by the formula:

$$\%\text{Intensity} = \left(\Sigma \,\text{monthly Fournier categories of each species} / 3 \times N\right) / 100 \qquad (1.1)$$

where N is the total number of species and 3 is the highest category.

We classified the flowering patterns according to Newstrom et al. (1994a, b) based on the frequency (*continuous*, flowering throughout the year; *annual*, one flowering event per year; *sub-annual*, more than one flowering event per year; and *supra-annual*, flowering events with an interval longer than 1 year), duration (*brief*, up to 1 flowering month; *intermediate*, 2–4 flowering months; *long*, more than 4 flowering months), and regularity (*regular*, flowering event always occurs in the same period; *irregular*, unpredictable flowering events).

For each species, we recorded the life form, sexual system, and floral resources available to pollinators. The life form was defined according to Flora do Brasil (2021), considering the prevalent life form of the species in the community. To determine the available floral resources, we used four classes: (1) pollen (species that offer only pollen), (2) nectar (species that offer nectar, even if together with other resources), (3) oil (species with elaiophores), (4) resin (resinous species), and (5) anemophilous (species with no resource). To assess the availability of resources over the years, we plotted the percentage of intensity of the species in flower in each month, according to the resource offered.

1.2.4 Statistical Analysis

We correlated the flowering of the species to climatic factors (total monthly precipitation, mean monthly temperature, and day length). Correlations were determined using the Spearman correlation index in StatSoft Inc (2005).

1.3 Results

1.3.1 The Study Community: General Characterization

In the delimited area of the community, the 78 plant species marked and analyzed belong to 71 genera and 36 families (Table 1.1). Bromeliaceae, with seven species, was the most important family in number of species, followed by Myrtaceae and Fabaceae (six species each); Rubiaceae, Euphorbiaceae, and Apocynaceae (four species each); and Orchidaceae and Malpighiaceae (three species each). The three sampled physiognomies had different species compositions, with different dominances. The flooded open shrubby physiognomy contained species as tall as 5 m; *Pleroma gaudichaudianum* (Fig. 1.4a), *Myrcia ilheosensis* (Fig. 1.4b), and *Tapirira guianensis* (Fig. 1.4c) were the most important species in the upper stratum, and species of Poaceae (Fig. 1.4d) and Cyperaceae were the most important in the lower

Fig. 1.4 Species from the inner sandy ridge of the Maricá *restinga*, Rio de Janeiro state, Brazil. (**a**) *Pleroma gaudichaudianum* (Melastomataceae), (**b**) *Myrcia ilheosensis* (Myrtaceae), (**c**) *Tapirira guianensis* (Anacardiaceae), (**d**) Poaceae, (**e**) *Myrsine parvifolia* (Primulaceae), (**f**) *Chamaecrista ramosa* (Fabaceae), (**g**) *Clusia lanceolata* (staminate flower), (**h**) *Clusia lanceolata* (pistillate flower), (**i**) *Neoregelia cruenta*. Scales: 2 cm (**a**), 1 cm (**b–i**)

stratum. In the non-flooded open shrubby physiognomy, *Neomitranthes obscura* and *Myrsine parvifolia* (Fig. 1.4e) were prominent in bush clumps, while *Chamaecrista ramosa* (Fig. 1.4f) and *Cuphea flava* predominated in bare spaces. In the non-flooded closed arboreal physiognomy, individuals from 1 m to 6 m tall occurred; *Clusia lanceolata* (Fig. 1.4g, h) and *Neomitranthes obscura* were the most important species in the upper stratum and *Neoregelia cruenta* (Fig. 1.4i) and *Aechmea nudicaulis* (Fig. 1.5b) in the lower stratum.

Regarding the life form, most species were shrubs (30.8%) or herbs (28.2%), followed by subshrubs (15.4%), vines (14.1%), and trees (11.5%). In relation to the sexual system, hermaphroditism (74%) prevailed, followed by dioecism (17%) and monoecism (9%). In general, the species have attractive flowers that provide a variety of floral resources for biotic pollen vectors (Figs. 1.5 and 1.6). Nectar is provided by 52 species from 27 families (67% of the species), the most important being Bromeliaceae, Apocynaceae, Fabaceae, and Rubiaceae (Fig. 1.5 and Table 1.1). Pollen is provided by 14 species from 7 families (18% of the species), the most important being Myrtaceae (Figs. 1.4b and 1.6a) (except *Myrrhinium atropurpureum*, whose petals are the main resource; Fig. 1.6b), Araceae, Fabaceae, and

Table 1.1 List of species, including the respective number of individuals analyzed, life form, sexual system, floral resource, frequency, regularity, duration, and flowering period (with Fournier scale) over 7 years (2004–2010) in the Maricá *restinga*, Rio de Janeiro state, Brazil

Family	Species	*n* individuals	Life form	Resource	Sexual system	Frequency	Regularity	Duration	Years	J	F	M	A	M	J	J	A	S	O	N	D
Anacardiaceae	*Tapirira guianensis* Aubl.	15	Tree	Nectar	D	Annual	Irregular	Inter	**2004**	1	1	1	1						1	2	2
									2005	2	2	1	1								
									2006										1	2	1
									2007	1	2								1	2	1
									2008											2	3
									2009		1										1
									2010									1	2	2	3
Apocynaceae	*Aspidosperma pyricollum* Müll.Arg.	15	Tree	Nectar	H	Annual	Regular	Inter	**2004**		1	1								1	3
									2005	1										1	1
									2006	1										2	3
									2007	1										2	3
									2008	1										1	
									2009												1
									2010										1	3	2
	Ditassa banksii R.Br. ex Schult.	15	Climber	Nectar	H	Annual	Regular	Inter	**2004**											2	3
									2005	1									2	3	3
									2006	2									3	2	3
									2007	1	1								3	2	3
									2008	1										3	1
									2009	2											1
									2010	3								1	3	2	1

(continued)

Table 1.1 (continued)

Family	Species	*n* individuals	Life form	Resource	Sexual system	Frequency	Regularity	Duration	Years	J	F	M	A	M	J	J	A	S	O	N	D
	Mandevilla guanabarica Casar. ex M.F.Sales et al.	15	Climber	Nectar	H	Continual	Regular	Long	**2004**	3	3	3	2	2	2	2	2	2	2	3	3
									2005	3	3	3	2	2	2	2	1	2	2	3	2
									2006	3	3	2	2	2	2	1	1	1	1	2	3
									2007	3	2	2	2	2	1	1	1	2	2	3	2
									2008	3	2	2	2	2	2	2	1	2	2	2	3
									2009	3	3	3	3	3	3	2	1	2	2	2	3
									2010	3	3	2	3	1	1	3	1	2	2	2	2
	Peplonia asteria (Vell.) Fontella & E.A.Schwarz	15	Climber	Nectar	H	Continual	Irregular	Long	**2004**	2	3	2	3	2	2	2	3	3	3	2	2
									2005	2	3	2	3	2	2	2	2	3	2	3	2
									2006	2	3	2	2	2	2	2	2	1	3	2	2
									2007	2					1	2	2	1	3	2	2
									2008	2					1	2	2			3	2
									2009												
									2010	2	2	3	2	2	1	1	1	2	1	2	2
Araceae	*Anthurium harrisii* (Graham) G.Don	15	Herb	Pollen	M	Annual	Irregular	Inter	**2004**	3											2
									2005	3											2
									2006	2	2								1	2	
									2007	3											2
									2008	3											2
									2009		3							2	2		
									2010	2	2	3	1	1	1	3	1			2	

	Thaumatophyllum corcovadense (Kunth) Sakur., Calazans & Mayo	9	Herb	Pollen	M	Annual	Irregular	Inter	**2004**	3									2		
									2005	3									2		
									2006										2	2	1
									2007	3									2	2	1
									2008	3											
									2009										3		
									2010									1	2		
Arecaceae	*Allagoptera arenaria* (Gomes) Kuntze	7	Subshrub	Pollen	M	Subannual	Irregular	Inter	**2004**				1	2					2		
									2005				1	2							
									2006	2									3	1	3
									2007	2									3	1	3
									2008	2											
									2009	2				1			1	2	2	1	
									2010					1		1	1	2	1		
Asteraceae	*Lepidaploa subsquarrosa* (DC) H. Rob.	10	Subshrub	Nectar	H	Annual	Irregular	Inter	**2004**	2											2
									2005	2											
									2006	2											2
									2007	2											2
									2008	2											
									2009								2				1
									2010					3	2	1		3			
	Mikania hoehnei B.L.Rob.	14	Climber	Nectar	H	Annual	Regular	Inter	**2004**				3	3	2	1					
									2005				3	3	2	1					
									2006				3	2	2						
									2007				3	2	2						
									2008				3	2	2						
									2009						1						
									2010				3	3							

(continued)

Table 1.1 (continued)

Family	Species	*n* individuals	Life form	Resource	Sexual system	Frequency	Regularity	Duration	Years	J	F	M	A	M	J	J	A	S	O	N	D
Bignoniaceae	*Jacaranda jasminoides* (Thunb.) Sandwith	6	Shrub	Nectar	H	Subannual	Irregular	Inter	**2004**	2	2										
									2005		1	2							2		
									2006			3									
									2007			3									
									2008			3							2		
									2009										1		2
									2010		1										
	Lundia corymbifera (Vahl) Sandwith	15	Climber	Nectar	H	Annual	Regular	Inter	**2004**				3	3	2						
									2005				3	3	2						
									2006				3	2							
									2007				3	2							
									2008				3	2							
									2009					2							
									2010				3	1							
Bromeliaceae	*Aechmea nudicaulis* (L.) Griseb.		Herb	Nectar	H	Subannual	Irregular	Long	**2004**	3	2	2	3	3			2	2	2	3	
									2005	3	2	2	3	3			2	2	2	3	2
									2006	2	2	3	2	2	2	2	3	2	2	3	2
									2007	3	2			2	1	2	3	2	2	3	2
									2008	3	2			2	1	2	3	2	2	3	2
									2009	1	1			2	1	2	2				3
									2010	2	1		3	3	2	3	2	1	2	1	1

	Billbergia amoena (Lodd.) Lindl.		Herb	Nectar	H	Annual	Regular	Inter	**2004**				3	1	2						
									2005				3	3	2	1					
									2006				3	1	2						
									2007				2	3	1						
									2008				2	3	1						
									2009				2	3	1						
									2010				1	2	2	1					
	Neoregelia cruenta (R.Graham) L.B.Sm.		Herb	Nectar	H	Annual	Irregular	Long	**2004**		3	2	3	2	1						
									2005		3	2	3	2	1						
									2006		3	3	3								
									2007	2	2	3									
									2008	2	2	3	2	1						2	2
									2009												
									2010			2	3	2	1					1	2
	Tillandsia gardneri Lindl.		Herb	Nectar	H	Supra-annual	Regular	Inter	**2004**								2	1			
									2005												
									2006												
									2007							2	1				
									2008												
									2009												
									2010							2					
	Tillandsia stricta Sol.		Herb	Nectar	H	Annual	Regular	Inter	**2004**	3	2	1									1
									2005	3	3	2									3
									2006	2	3	2									1
									2007	3	3	2									3
									2008	3	3	2									2
									2009	2	3										
									2010	3	2									1	2

(continued)

Table 1.1 (continued)

Family	Species	*n* individuals	Life form	Resource	Sexual system	Frequency	Regularity	Duration	Years	J	F	M	A	M	J	J	A	S	O	N	D
	Vriesea neoglutinosa Mez		Herb	Nectar	H	Annual	Regular	Long	**2004**							1	1	1			
									2005							2	2	3	3	2	
									2006								1	2	2	1	
									2007								1	2	2	1	
									2008								1	2	2	3	
									2009												
									2010							2			2		
	Vriesea procera (Mart. ex Schult. & Schult.f.) Wittm.		Herb	Nectar	H	Annual	Regular	Long	**2004**	3	2	3	2	2	1						3
									2005	3	2	3	2	2	2						3
									2006	2	3	3	1	1	1						
									2007	2	3	2	2								
									2008	2	3	2	1								3
									2009	2	2										
									2010	3	3	3	2	1							2
Burseraceae	*Protium brasiliense* (Spreng.) Engl.	10	Shrub	Nectar	D	Annual	Regular	Inter	**2004**								3	2	3		
									2005								2	3	2		
									2006							3	3	3	2		
									2007							3	3	3	2		
									2008							3	3	3	2		
									2009									2	2		
									2010							1	2	3	2		

Cactaceae	*Cereus fernambucensis* Lem.	8	Herb	Nectar	H	Subannual	Irregular	Inter	**2004**		3							2	3	2	
									2005		3								3	2	
									2006		3	2	1					3	2	1	
									2007		3	2	1					3	2	1	
									2008		3	2	1								
									2009	2											
									2010	3									3		
	Pilosocereus arrabidae (Lem.) Byles & Rowley	15	Shrub	Nectar	H	Subannual	Irregular	Inter	**2004**	2	3	2	1		1		2		2	2	3
									2005	2	3	2	1		1			2	1		3
									2006		3	3				2	3			1	1
									2007		3	3				2	3			1	1
									2008		3	3				2	3	2	1		3
									2009		3	3				2		1	2	2	
									2010	2		2	1	1			1	1	2	2	1
Celastraceae	*Monteverdia obtusifolia* (Mart.) Biral	15	Shrub	Nectar	D	Annual	Irregular	Long	**2004**							1	2	3	2	1	1
									2005	2	1						2	3	2	2	1
									2006	2		1				1	2	3	2	1	
									2007						1	1	2	3	2	1	
									2008						1	1	2			1	
									2009								1	2	3	1	1
									2010				1			2	2	2	2	1	

(continued)

Table 1.1 (continued)

Family	Species	*n* individuals	Life form	Resource	Sexual system	Frequency	Regularity	Duration	Years	J	F	M	A	M	J	J	A	S	O	N	D
Clusiaceae	*Clusia lanceolata* Cambess.	15	Tree	Resin	D	Annual	Regular	Long	**2004**		2	2	3	2	2	2	2	2	1		
									2005	2	2	2	3	3	2	2	1	1			
									2006	1	2	3	3	2	2	3	2	2	1	1	
									2007		1	2	2	1	2	3	2	2	1	1	
									2008		1	2	2	1	2	3	2	1			
									2009		1	1	2	1	2	3	2	1		1	
									2010		1		1	2	3	3	2	1	1		
	Garcinia brasiliensis Mart.	15	Tree	Nectar	D	Annual	Irregular	Inter	**2004**						2	3	3	1			1
									2005	1			1					3	2		
									2006								3	2			
									2007												
									2008									2			
									2009									2		2	
									2010	1							2	2	1		
Com-melinaceae	*Commelina erecta* L.	8	Herb	Nectar	H	Annual	Irregular	Long	**2004**	3	3	2	1						2	3	3
									2005	3	3	2	1						2	3	3
									2006	2	3	2	1						1	2	2
									2007	1	2	2	2	2					1	2	2
									2008	1	2	2	2	2				1	2	3	3
									2009	3				2	3	2					3
									2010	1	1	2	3	3	1	1	1		1	2	2

Cyperaceae	*Cyperus* L.		Herb	AN	H	Annual	Regular	Inter	**2004**	2	1										3
									2005	2	1										2
									2006	3	3	2	1								
									2007	3	3	2	2	1							
									2008	3	3	2	2	1							
									2009	2	2			1	2						2
									2010	2			3								
	Rhynchospora Vahl		Herb	AN	H	Annual	Irregular	Inter	**2004**	3	2										2
									2005	3	2										2
									2006	2	1										
									2007	3	2										2
									2008	3	2										
									2009					2	1						
									2010				3	2							1
Ericaceae	*Agarista revoluta* (Spreng.) J.D. Hook. ex Nied.	15	Tree	Nectar	H	Annual	Regular	Inter	**2004**							2	3	1	1		
									2005							2	3				
									2006						2	3	2	2			
									2007							3	2				
									2008							3	2	1			
									2009								3	2			
									2010						1	2	1	2			
	Gaylussacia brasiliensis (Spreng.) Meisn.	9	Subshrub	Nectar	H	Subannual	Irregular	Inter	**2004**	2	3	3	2	2	2	3	3	1	2	3	
									2005	2	3	3	2	2	2	3	2				
									2006	1		1	1			2	1			1	2
									2007	3		1	1			2	1			1	2
									2008	3		1	1			2	1				
									2009	3							3	1		2	
									2010					2	1	1	1		1	2	2

(continued)

Table 1.1 (continued)

Family	Species	*n* individuals	Life form	Resource	Sexual system	Frequency	Regularity	Duration	Years	J	F	M	A	M	J	J	A	S	O	N	D
Erythroxylaceae	*Erythroxylum ovalifolium* Peyr.	15	Shrub	Nectar	H	Annual	Regular	Inter	**2004**											3	1
									2005	2										3	2
									2006	1									1	3	2
									2007										1	3	2
									2008										3		
									2009										1	1	1
									2010										2	3	1
	Erythroxylum subsessile (Mart.) O.E.Schulz	10	Subshrub	Nectar	H	Subannual	Irregular	Inter	**2004**		3		1		3				2	3	
									2005		3	1	1		3				2		
									2006		2	3		3				2	1		
									2007		2	1						2	1		
									2008		2	1							2		
									2009				3								
									2010	2	1			3					3		
Euphorbiaceae	*Alchornea triplinervia* (Spreng.) Müll. Arg.	15	Shrub	AN	D	Annual	Irregular	Long	**2004**	3	3	2	3	3	2	3	3	1	1		1
									2005	2	3	2	1	2	3				2	2	
									2006	3	3	3	1	1	1				3	2	2
									2007	3			1	3					3	2	2
									2008	3	2	2	1	3					2	2	
									2009	3	2	2	1	3	1	2	1	3	2	2	3
									2010	1			1	1	1	1	2	2	1	2	1

	Astraea klotzschii Didr.	15	Subshrub	Nectar	M	Continual	Regular	Long	**2004**	3	2	3	1	1	1	1	1	1	2	3	1
									2005	3	2	3	1	1	1	1	1	1	2	3	1
									2006	2	2	2	1	1	2	2	3	2	1	2	2
									2007	3	3	3	2	2	2	2	3	2	1	2	2
									2008	3	2	3	1	1	1	1	1	1	2	3	1
									2009	3	2	2	1	1	1	1	1	1	2	2	1
									2010	3	2	2	2	2	2	1	2	2	2	3	2
	Dalechampia micromeria Baill.	7	Climber	Resin	M	Continual	Irregular	Long	**2004**	3	1	1	1	2					3	2	3
									2005	3	1	1	1	2					3	2	3
									2006	3	2	2	2	2	1		2	2	2	3	3
									2007	2	2	2	2	1		1	2	2	2	3	3
									2008	2	2	2	2	1		1	2	1	2	2	3
									2009	2	2	2	2					1	2	2	2
									2010	2	2	2	1	1	1	1	1	1	2	1	2
	Microstachys corniculata (Vahl) Griseb.	15	Subshrub	Pollen	M	Continual	Regular	Long	**2004**	2	3	2	2	2	1	1	3	1	2	3	3
									2005	2	3	2	2	2	1	1	3	1	3	3	3
									2006	3	3	3	2	2	2	2	3	1	3	3	3
									2007	3	3	3	3	2	2	2	3	1	3	3	3
									2008	3	3	3	2	2	2	2	3	1	3	3	3
									2009	3	2	2	1	1	1	1	1	1	1	2	2
									2010	3	2	3	1	3	3	2	2	2	3	2	2

(continued)

Table 1.1 (continued)

Family	Species	*n* individuals	Life form	Resource	Sexual system	Frequency	Regularity	Duration	Years	J	F	M	A	M	J	J	A	S	O	N	D
Fabaceae	*Andira legalis* (Vell.) Toledo	8	Tree	Nectar	H	Annual	Regular	Brief	**2004**									2			
									2005									3			
									2006									3			
									2007									3			
									2008												
									2009									3			
									2010									1			
	Chamaecrista ramosa (Vogel) H.S.Irwin & Barneby	15	Subshrub	Pollen	H	Continual	Regular	Long	**2004**	3	3	3	2	2	2	2	2	2	2	3	3
									2005	3	3	3	2	2	2	2	2	2	2	3	3
									2006	3	3	3	2	1	2	3	2	2	2	3	3
									2007	3	3	2	2	1	1	2	2	2	2	2	3
									2008	3	3	2	2	1	1	2	2	2	2	2	1
									2009	1	3	3	2	2	1	1	1	1	2	2	2
									2010	3	1	1	3	1	1	2	2	2	2	3	2
	Clitoria laurifolia Poir.	12	Climber	Nectar	H	Annual	Irregular	Long	**2004**	3	3	2	1				2	2	1	2	2
									2005	3	3	2	1						1	2	2
									2006	3	3	2								1	2
									2007	3	3	2								1	2
									2008	3	3	2							1	2	2
									2009	2	2			1	2	3	1			2	2
									2010	1	2	1	3	2	1	1				2	3

	Ormosia arborea (Vell.) Harms	15	Tree	Nectar	H	Annual	Regular	Inter	**2004**	3											1
									2005	3	2										
									2006		1	2									
									2007		2										
									2008		2										
									2009		1										
									2010			1	1								
	Stylosanthes guianensis (Aubl.) Sw.	15	Subshrub	Nectar	H	Continual	Regular	Long	**2004**	2	2	2	1	3	2	2	1	1	2	3	3
									2005	2	2	2	1	3	2	2	3	1	3	3	3
									2006	3	3	3	3	2			3	3	3	3	
									2007	2	3	3	3	2	2	3	3	3	3	3	3
									2008	2	3	3	3	1	1		3	1	3	3	3
									2009			2	2	2	1				1	2	2
									2010	3	3	3	1	1	1	3	3	2			1
	Swartzia apetala Raddi	6	Shrub	Pollen	H	Annual	Regular	Inter	**2004**	1	1	1									
									2005	1	3										
									2006	1											
									2007	2	1										
									2008	2	1										
									2009	2											
									2010												
Iridaceae	*Neomarica candida* (Hassl.) Sprague	8	Herb	Nectar	H	Annual	Regular	Inter	**2004**										3		
									2005										3		
									2006									1			
									2007										3		
									2008										3		
									2009										2	2	
									2010									1	2		

(continued)

Table 1.1 (continued)

Family	Species	*n* individuals	Life form	Resource	Sexual system	Frequency	Regularity	Duration	Years	J	F	M	A	M	J	J	A	S	O	N	D
Lamiaceae	*Vitex polygama* Cham.	8	Shrub	Nectar	H	Annual	Irregular	Long	**2004**	1	1	1	1	1						2	2
									2005	3	2	2						2	3	2	2
									2006	2		1		2	3	2	1			1	3
									2007	3	2	1	1	1		2	1			1	3
									2008	3	2	1	1	1			1	2	3	2	2
									2009	3	2	1	1	1					3	2	
									2010					1	1				1	2	1
Lauraceae	*Ocotea notata* (Nees & Mart.) Mez	15	Tree	Nectar	D	Annual	Regular	Long	**2004**	2	3	2								2	2
									2005	2	3	3	2							1	2
									2006	3	3	3	1						2	1	2
									2007	2	3	1							2	1	2
									2008	2	3	1									1
									2009	2	3	1								1	2
									2010	2	1	1	1						1	1	1
Loranthaceae	*Psittacanthus dichroos* (Mart.) Mart.	8	Herb	Nectar	H	Annual	Regular	Inter	**2004**										3	2	
									2005									1	3	2	
									2006									2	3	1	
									2007											3	1
									2008									2	3		
									2009										3	2	
									2010											2	2

	Struthanthus taubatensis Eichler	10	Herb	Nectar	H	Annual	Regular	Inter	**2004**		1	1									
									2005		1	1									
									2006	2	1	1									
									2007		1	2	1	1							
									2008		1	2	1	1							
									2009		2	2									
									2010	1	2										
Lythraceae	*Cuphea flava* Spreng.	15	Subshrub	Nectar	H	Continual	Regular	Long	**2004**	3	3	2	1	2	1	1	1	2	1	3	3
									2005	3	3	2	1	2	1	1	1	2	3	2	3
									2006	3	3	3	2	1	1	1	1	2	3	3	3
									2007	1	3	3	2	1	1	1	1	2	3	3	3
									2008	1	3	3	2	1	1	1	1	2	3	2	3
									2009	2	2	1								1	2
									2010	2	1	2	1	1	1				1	3	3
Malpighiaceae	*Byrsonima sericea* DC.	15	Shrub	Oil	H	Annual	Regular	Long	**2004**	2	3	2	1							3	2
									2005	1	1	1	1							2	3
									2006	2	1	1	1						2	3	2
									2007	1	3	1							2	3	2
									2008	1	3	1								2	2
									2009	1	1							1	1	2	2
									2010	1								2	3	2	2
	Peixotoa hispidula A.Juss.	15	Climber	Oil	H	Continual	Irregular	Long	**2004**	3	2	2	2	2	2	1	1	2	2	3	2
									2005	3	2	2	2	2	2	2	2	2	3	2	2
									2006	3	2	3	2	1	1	2	2	1	2	2	2
									2007	3	2	3	2			2	2	1	2	2	2
									2008	3	2	3	2			2	2	2	3	2	2
									2009	2	1	1								1	2
									2010	3	1	1	1	1		1			1	2	1

(continued)

Table 1.1 (continued)

Family	Species	*n* individuals	Life form	Resource	Sexual system	Frequency	Regularity	Duration	Years	J	F	M	A	M	J	J	A	S	O	N	D
	Stigmaphyllon paralias A.Juss.	6	Subshrub	Oil	H	Annual	Irregular	Long	**2004**	2	3	2	1	1					3	2	2
									2005	2	3	2	1	1						2	2
									2006	2	3	2	1	1	1			1	2	2	2
									2007	2	3	2	2	1	1			1	2	2	2
									2008	2	3	2	2	1	1					2	2
									2009	2	2	2								2	3
									2010	3	2	3	1						2	2	2
Melastomataceae	*Marcetia taxifolia* (A.St.-Hil.) DC.	6	Subshrub	Pollen	H	Subannual	Irregular	Inter	**2004**		2	3		2	3		2		3	1	
									2005		2	3		2	3			2	3		3
									2006		2			3	3	1		3		3	3
									2007		2			2		1	2	3		3	3
									2008		2			2		1	2	2	3		3
									2009	1		3					3	3	2	2	
									2010	1	1	1		2	1	1	1			2	
	Pleroma gaudichaudianum (DC.) A. Gray	15	Shrub	Pollen	H	Annual	Irregular	Long	**2004**	3	2	2							1	2	3
									2005	2	2	2	2	1					1	2	3
									2006	3	1	1								1	2
									2007	2										1	2
									2008	2										1	3
									2009	3	2	1							1	1	2
									2010	1				1		1	1		1	2	3

Myrtaceae	*Myrcia ilheosensis* Kiaersk.	15	Shrub	Pollen	H	Annual	Regular	Inter	**2004**		3	2									
									2005	1	3	1									
									2006	3	2										
									2007	3	3										
									2008	3	3										1
									2009	1	3										1
									2010	2	1									1	2
	Myrcia loranthifolia (DC.) G.P.Burton & E.Lucas	6	Tree	Pollen	H	Annual	Regular	Inter	**2004**		3										
									2005	2	3	1									
									2006		3										
									2007		3	2									
									2008		3	2									
									2009		3	2									
									2010	3	3	1								1	1
	Myrcia vittoriana Kiaersk.	12	Shrub	Pollen	H	Annual	Regular	Inter	**2004**	1									1	1	1
									2005												2
									2006	3											2
									2007	2											2
									2008												2
									2009												1
									2010	1										1	1

(continued)

Table 1.1 (continued)

Family	Species	*n* individuals	Life form	Resource	Sexual system	Frequency	Regularity	Duration	Years	J	F	M	A	M	J	J	A	S	O	N	D
	Myrciaria floribunda (H.West ex Willd.) O.Berg	15	Shrub	Pollen	H	Annual	Regular	Inter	**2004**		1	1								1	1
									2005	2	3	3									
									2006												
									2007		2	1									
									2008		2	1									
									2009												1
									2010	1	1	1									1
	Myrrhinium atropurpureum Schott	15	Shrub	Petal	H	Annual	Regular	Long	**2004**							1	3	2	2	1	
									2005								2	3	3	2	1
									2006							3	3	2	1		
									2007							3	3	2	1		
									2008							3	3				
									2009								1	1	1	1	
									2010						1	1	1	1	1	1	1
	Neomitranthes obscura (DC.) N.Silveira	15	Shrub	Pollen	H	Annual	Regular	Inter	**2004**					3	2	1					
									2005					3	1	1					
									2006												
									2007						3						
									2008						3						
									2009				1	2							
									2010				1	1							

Nyctaginaceae	*Guapira opposita* (Vell.) Reitz	15	Shrub	Nectar	D	Annual	Irregular	Long	**2004**							1	3	2	1	1	1
									2005			1	1	2	2	3	2				
									2006							3	3	2	1	1	
									2007					2	3	3					
									2008					2	3	3					
									2009					.	3	3					
									2010							3	3	2	2	1	1
	Guapira pernambucensis (Casar.) Lundell	15	Shrub	Nectar	D	Annual	Irregular	Long	**2004**	1	3	2	1	1	1	1	1		1	2	2
									2005	2	3	3	2	3					2	2	3
									2006	3	3	2	2	1	1	1			1	2	3
									2007	1	2	2	3	1							1
									2008	1	2	2	3	1						2	3
									2009	3	3									3	3
									2010	3	2	2	3	1					2	2	2
Ochnaceae	*Ouratea cuspidata* (A.St.-Hil.) Engl.	15	Shrub	Pollen	H	Continual	Regular	Long	**2004**	2	2	2	1	1	1	2	2	2	1	2	2
									2005	2	3	2	3	2	1	1	2	2	2	2	2
									2006	3	1	3	2	1	2	2	2	2	1	2	2
									2007	3	1	3	2	1	1	2	2	2	1	2	2
									2008	3	1	3	2	1	2	2	2				
									2009	2	1	3	2	1	2	2	1	2	3	2	2
									2010				1	2	1	2	2	2	2	2	1

(continued)

Table 1.1 (continued)

Family	Species	*n* individuals	Life form	Resource	Sexual system	Frequency	Regularity	Duration	Years	J	F	M	A	M	J	J	A	S	O	N	D
Olacaceae	*Ximenia americana* L.	11	Shrub	Nectar	H	Supra-annual	Irregular	Inter	**2004**												
									2005												
									2006												
									2007		3	2									
									2008												
									2009												
									2010	2	1										
Orchidaceae	*Epidendrum denticulatum* Barb.Rodr.	6	Herb	Nectar	H	Continual	Regular	Long	**2004**	3	2	2	3	2	2	2	1	1	2	2	3
									2005	3	2	2	3	2	2	2	3	2	2	2	3
									2006	3	2	3	2	2	1	2	1	2	2	2	2
									2007	3	2	2	2	1	1	1	1	2	2	2	2
									2008	3	2	2	2	1	1	2	1	2	2	2	3
									2009	3	3	2	1	1	1	1	1	1	1	2	2
									2010	2	2	1	2	1	2	1	2	1	1	2	2
	Epidendrum orchidiflorum (Salzm.) Lindl.	6	Herb	Nectar	H	Continual	Regular	Long	**2004**	3	2	2	3	2	2	2	2	1	2	2	3
									2005	3	2	2	3	2	2	2	3	2	2	2	3
									2006	3	2	3	3	2	1	3	2	3	3	2	2
									2007	3	2	2	2	2	2	3	2	3	3	2	2
									2008	3	2	2	2	2	2	3	2	2	2	2	3
									2009	3	3	2	1	1	1	1	1	1	1	2	3
									2010	1	3	2	2	1	2	2	2	2	2	2	2
	Vanilla chamissonis Klotzsch	15	Herb	Nectar	H	Supra-annual	-	Brief	**2004**												
									2005												
									2006												
									2007												
									2008								3				
									2009												
									2010												

Passifloraceae	*Passiflora silvestris* Vell.	15	Climber	Nectar	H	Subannual	Irregular	Long	**2004**	2	2	1	1						1	2	2
									2005	2	2	1								1	2
									2006	3	2	2	1		1	2		2		3	2
									2007	2	3					2		2		3	2
									2008	2	3					2					
									2009	2										1	2
									2010	3	3	2			1				1	3	1
Poaceae	*Paspalum maritimum* Trin.		Herb	AN	H	Annual	Irregular	Inter	**2004**	2	2										
									2005	2	2										
									2006		3	1	1								
									2007	3	3	1									
									2008	3	3	1									
									2009					2	2						
									2010				3	2							
	Schizachyrium sanguineum (Retz.) Alston		Herb	AN	H	Annual	Irregular	Inter	**2004**	2	2										
									2005	2	2										
									2006		2	1									
									2007	3	3	1									
									2008	3	3	1									
									2009					2	2						
									2010				3	2							
Polygonaceae	*Coccoloba arborescens* (Vell.) R.A.Howard	15	Climber	Nectar	D	Annual	Regular	Inter	**2004**							3	2	1	1		
									2005							3	2	1	1		
									2006							3	3	.			
									2007							3	3	1			
									2008							3	3	2	1		
									2009							1	2	1			
									2010							3	2				

(continued)

Table 1.1 (continued)

Family	Species	*n* individuals	Life form	Resource	Sexual system	Frequency	Regularity	Duration	Years	J	F	M	A	M	J	J	A	S	O	N	D
Primulaceae	*Myrsine parvifolia* A.DC.	15	Shrub	AN	D	Annual	Regular	Inter	**2004**		1	1									
									2005	2	3	2	1								
									2006		3	2	1								
									2007	1	3										
									2008	1	3	2									
									2009		1										
									2010	3	1	2									
Rubiaceae	*Amaioua intermedia* Mart. ex Schult. & Schult.f.	11	Shrub	Nectar	D	Supra-annual	Irregular	Inter	**2004**												2
									2005												
									2006						`1						
									2007						3	2					
									2008	1									1	2	1
									2009												
									2010												1
	Chiococca alba (L.) Hitchc.	12	Shrub	Nectar	H	Subannual	Irregular	Inter	**2004**	1	3	2							2	1	
									2005	1	3	2									
									2006		2	1	2			1				2	3
									2007		3	2				1				2	3
									2008		3	2				1				3	3
									2009	2	2			2	2			2	2		
									2010	2	3	2			2				3	3	2

	Denscantia cymosa (Spreng.) E.L.Cabral & Bacigalupo	10	Herb	Nectar	H	Annual	Irregular	Long	**2004**	3	2	2							1	2	2
									2005	3	2	2							1	2	2
									2006	3	2	2	1	2	3	2	2	2	2	3	3
									2007	3	2	2							1	2	2
									2008	3	2	2							1	2	2
									2009	2	2			2	3	1			2	2	2
									2010	2	2	2	2	2	1	1	2		1	2	2
	Tocoyena formosa (Cham. & Schltdl.) K.Schum.	6	Shrub	Nectar	H	Annual	Regular	Inter	**2004**											2	2
									2005	2											2
									2006											2	3
									2007	3	1									2	3
									2008	3											3
									2009	1									2	3	2
									2010	1									3	3	1
Sapindaceae	*Allophylus puberulus* (Cambess.) Radlk.	8	Shrub	Nectar	D	Annual	Regular	Inter	**2004**							1	3	2	1		
									2005					1	2	2	1				
									2006							3	2	3			
									2007							3	2	3			
									2008							3	2				
									2009								2	2	1	1	
									2010								2	1	2	1	

(continued)

Table 1.1 (continued)

Family	Species	*n* individuals	Life form	Resource	Sexual system	Frequency	Regularity	Duration	Years	J	F	M	A	M	J	J	A	S	O	N	D
	Paullinia weinmanniifolia Mart.	15	Climber	Nectar	M	Annual	Regular	Inter	**2004**			3	3	3	1						
									2005			3	3	3	1						
									2006		1		2	1							
									2007		1		2	1							
									2008		1		2	1							
									2009						2						
									2010				3	2	3						
Verbenaceae	*Stachytarpheta canescens* Kunth	6	Subshrub	Nectar	H	Annual	Irregular	Long	**2004**	2	1	2	2	2	1	1			2	2	2
									2005	2	1	2	2	2	1	1			2	2	2
									2006	2	1	2	2	1					2	1	2
									2007	2	1	2	2	1					2	1	2
									2008	2	1	2	2	1					2	1	1
									2009	1	1										2
									2010	1	2	2	1	2	2	1	1	2	2	1	2

AN anemophilous, *H* hermaphroditic, *D* dioecious, *M* monoecious, *Inter* intermediate

Fig. 1.5 Species from the inner sandy ridge of the Maricá *restinga*, Rio de Janeiro state, Brazil. (**a**) *Ditassa banksii* (Apocynaceae), (**b**) *Aechmea nudicaulis* (Bromeliaceae), (**c**) *Tillandsia gardneri* (Bromeliaceae), (**d**) *Lundia cordata* (Bignoniaceae), (**e**) *Pilosocereus arrabidae* (Cactaceae), (**f**) *Garcinia brasiliensis* (Clusiaceae), (**g**) *Agarista revoluta* (Ericaceae), (**h**) *Tocoyena bullata*. Scales: 1 cm

Fig. 1.6 Species from the inner sandy ridge of the Maricá *restinga*, Rio de Janeiro state, Brazil, and their floral resources: pollen (a), petal (b), oil (c), resin (d). (**a**) *Myrciaria floribunda* (Myrtaceae), (**b**) *Myrrhinium atropurpureum* (Myrtaceae), (**c**) *Peixotoa hispidula* (Malpighiaceae), (**d**) *Dalechampia micromeria* (Euphorbiaceae). Scales: 2 cm (**a**, **b**), 1 cm (**c**), 0.5 cm (**d**)

Melastomataceae (Fig. 1.4a and Table 1.1). Oil is provided by only three (4% of the species) members of Malpighiaceae (*Stigmaphyllon paralias*, *Byrsonima sericea*, and *Peixotoa hispidula*; Fig. 1.6c and Table 1.1), while resin is offered by only two species (3% of the species), *Dalechampia micromeria* (Euphorbiaceae) (Fig. 1.6d) and *Clusia lanceolata* (Clusiaceae) (Fig. 1.4g, h and Table 1.1). Six species (8% of the species) from four families (Cyperaceae, Euphorbiaceae, Primulaceae, and Poaceae) do not provide any resources, as they are anemophilous (Fig. 1.4d, e and Table 1.1).

1.3.2 Flowering Phenology

The plants of the Maricá *restinga* community bore flowers throughout the study period (Fig. 1.7b). We recorded cyclical and regular increases in the percentages of activity and intensity of the flowering phenophase in the warmer/wetter periods (October through March) and decreases in these percentages in the colder/drier periods (April through September) (Fig. 1.7a, b).

The maximum percentages of activity (between 45% and 65%) coincided with the maximum percentages of intensity (between 32% and 49%) and characterized the flowering peaks in the community (i.e., months with a large proportion of species exhibiting very intense flowering) (Fig. 1.7b, c). In general, the flowering peaks were repeated through the years, in January and/or February (Fig. 1.7b). However, in 2010, two flowering peaks occurred, one in January and the other in October through December (Fig. 1.7b). The minimum percentages of activity (between 19% and 38%) also coincided with the minimum percentages of intensity (between 11% and 19%) and characterized months with a small proportion of species showing mostly low flowering intensity in the area (Fig. 1.7b, c). In general, periods of low flowering were also repeated yearly, in June or July (Fig. 1.7b), although low flowering also occurred in April 2009.

In practically every year and month, species flowering in moderate intensity (2 on the Fournier scale) predominated (Fig. 1.7c). The highest intensities of flowering (scale 3) were observed in the warmer/wetter season (October to March), especially in January and February, from 2004 to 2008. Fewer species with high flowering intensity were observed in the colder/drier period (April to September) (Fig. 1.7c), when the number of species flowering at low intensity (Fournier scale 1) increased. Differently from previous years, in 2009 and 2010, more species flowered with low intensity, and the number of species flowering with high intensity was surprisingly small (Fig. 1.7c).

We recorded significant positive correlations between flowering (intensity and activity percentages) and the mean temperature and day length (Table 1.2). Flowering was positively and significantly correlated with total precipitation only in 2006 and 2008 (Table 1.2).

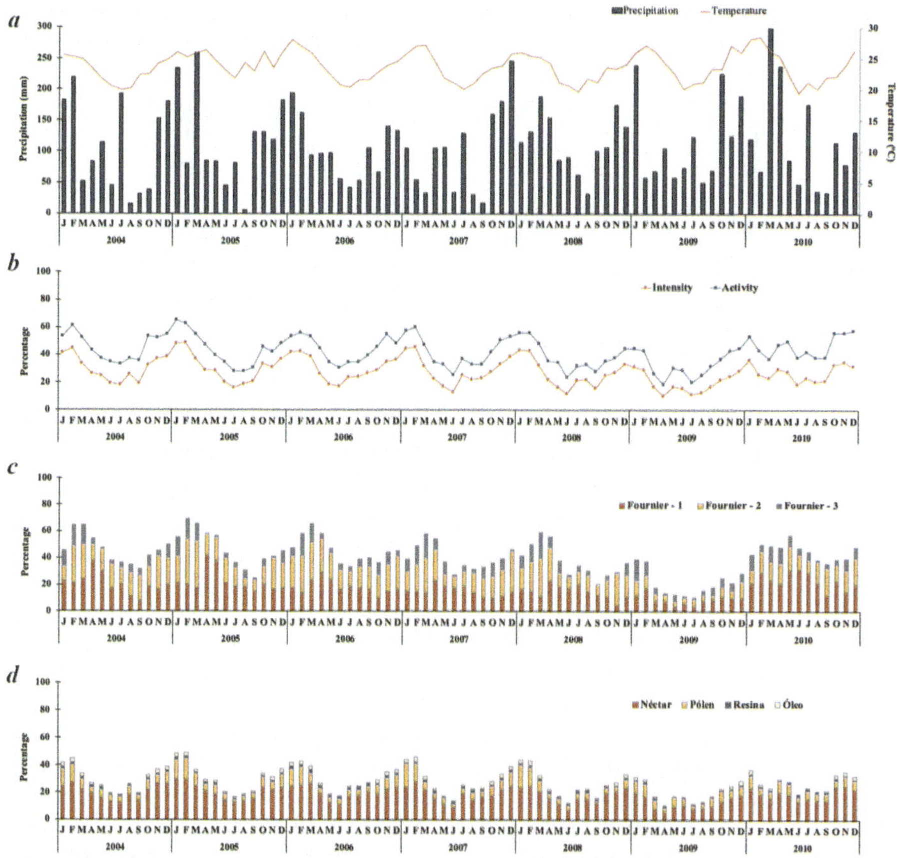

Fig. 1.7 Climatic data and flowering events during the study period (2004–2010) in the Maricá *restinga*, Rio de Janeiro state, Brazil. (**a**) Monthly data for mean temperature (°C) and total precipitation (mm) (Source: INMET/RJ). (**b**) Flowering intensity and activity percentages for the community. (**c**) Monthly proportion of species flowering, according to Fournier semiquantitative intensity scale: 0 (absence of flowers), 1 (low flowering intensity), 2 (moderate flowering intensity), 3 (high flowering intensity). (**d**) Monthly availability of floral resources, based on the proportion of flowering species with nectar, pollen, oil, and resin flowers

Although the community as a whole had some species in flower year-round, most individual species showed seasonal flowering events in an annual frequency (67%), followed by continuous (15%), sub-annual (13%), and supra-annual (5%) (Table 1.1).

The annual flowering pattern was recorded for all trees and for a large proportion of the shrubs, herbs, and climbing species. The continuous flowering pattern was observed in subshrubs, climbers, and herbaceous species, but rarely in shrubs (*Ouratea cuspidata*) and never in trees. Continuous emission of flowers predominated in species from open areas between bushes (*Chamaecrista ramosa* and

Table 1.2 Spearman correlation indices (rs) obtained in the correlation analysis between the climatic factors (mean temperature, total precipitation, and day length) and the percentages of Fournier and activity of flowering (78 species) of the Maricá *restinga*, Rio de Janeiro state

Year	Percentages	Temperature	Precipitation	Day length
2004	F	0.90*	0.45	0.91*
	A	0.81*	0.36	0.86*
2005	F	0.67*	0.53	0.93*
	A	0.70*	0.57	0.86*
2006	F	0.88*	0.71*	0.90*
	A	0.86*	0.73*	0.86*
2007	F	0.73*	0.35	087*
	A	0.77*	0.44	0.87*
2008	F	0.86*	0.62*	0.88*
	A	0.87*	0.67*	0.89*
2009	F	0.70*	0.35	0.87*
	A	0.61*	0.36	0.80*
2010	F	0.64*	0.42	0.73*
	A	0.46	0.26	0.58*
All years	F	0.90*	0.45	0.91*
	A	0.81*	0.36	0.86*

F percentages of Fournier, *A* percentages of activity of flowering
*Significant values for $p < 0.05$

Cuphea flava) and edges of bushes (*Epidendrum denticulatum*, *Mandevilla guanabarica*, and *Peplonia asteria*) in the open, non-floodable shrub physiognomy. In some years, species that tended to flower continuously reduced or even stopped flowering for a few weeks or months, as occurred with *Ouratea cuspidata* (2008 and 2010), *Peplonia asteria* (2007–2009), *Dalechampia micromeria* (2004–2006 and 2009), *Cuphea flava* (2009–2010), *Stylosanthes guianensis* (2006 and 2008–2010), and *Peixotoa hispidula* (2007–2010). The sub-annual and supra-annual flowering patterns were least common, but occurred in species with varied life forms, except for trees. Among the species with sub-annual flowering, *Marcetia taxifolia* (Melastomataceae) had the largest number of floral episodes per year (four events) (Table 1.1).

The regular flowering pattern predominated (54%), and the irregular (45%) pattern occurred in all the species with sub-annual frequency of flowering. We did not classify regularity for *Vanilla chamissonis*, as this species flowered only once during the study (Table 1.1). Three species alternated between sub-annual and annual flowering over the 7 years (*Gaylussacia brasiliensis*, *Erythroxylum subsessile*, and *Chiococca alba*) (Table 1.1).

Restinga species tended to have a long (41%) or intermediate (56%) flowering duration. Brief flowerings occurred in only 3% of species (*Andira legalis*, *Neomarica candida*, and *Vanilla chamissonis*) (Table 1.1).

1.3.3 Availability of Floral Resources During the Year

Nectar, pollen, oil, and resin were available to anthophile visitors year-round (Fig. 1.7d). Nectar and pollen were abundant and continuously available (Fig. 1.7d), while oil and resin were less abundant and occasionally absent, mainly in the colder/drier periods (oil, May and June; resin, May, August, September, and November) (Fig. 1.7d).

Unlike nectar and pollen, which were offered by many different species flowering concomitantly, the annual supply of oil and resin was provided by a small number of species that alternated periods or peaks of flowering. For example, when *Clusia lanceolata* ceased flowering in the beginning and middle of the warmer/wetter season (October–December), floral resin was still supplied by continuous emission from *Dalechampia micromeria* inflorescences. Similarly, when *Byrsonima sericea* ceased flowering at the beginning of the colder/drier season, floral oil was still available from *Peixotoa hispidula* and *Stigmaphyllon paralias*, which tended to exhibit more extensive or continuous blooms over part of the colder/drier period. Differently from other years, in the 2009–2010 cold/wet season, *Peixotoa hispidula* underwent a sharp reduction in flowering activity, probably also reducing the amount of floral oil available.

1.4 Discussion

The phenological flowering pattern in the Maricá *restinga* shows a seasonal flowering rhythm at both the community and species level. This conclusion is supported by the occurrence of regular peaks (in warmer/wetter seasons) and depressions (in colder/drier seasons) in the activity and intensity curves over the 7 years of the study. We also observed a high proportion of species flowering in an annual (66.7%) and regular (55%) pattern. This flowering pattern has also been recorded in areas with less-pronounced seasonal climatic changes in other phytophysiognomies of the Atlantic Forest Domain in southeastern Brazil (Morellato et al. 2000; Talora and Morellato 2000; Straggemeier and Morellato 2011). Taken together, these studies reinforce the idea that the flowering phenophase in the Atlantic Forest follows a general pattern (Morellato et al. 2000; Talora and Morellato 2000; Straggemeier and Morellato 2011).

The flowering events recorded in the Maricá *restinga* showed significant and positive correlations with mean temperature and day length, suggesting that these factors determine the seasonality of flowering at this site. In contrast to expectation for a harsh environment such as the *restinga* (characterized by high salinity, unstable sandy soil with low water retention, strong winds, direct sunlight, and high temperatures; Scarano 2002), we did not observe a strong relationship between phenological events and rainfall. For the Marambaia *restinga*, which is also located in Rio de Janeiro state, Carvalho et al. (2015) also found a low correlation between the phenological flowering events of Sapindaceae species and rainfall. To explain this lack of correlation, the authors suggested that the Marambaia region does not have sufficiently low precipitation levels to induce water stress during the dry season (Mattos 2005) and that the summer rains replenish swamps, swampy forests, seasonal ponds, and seasonally flooded forests, regularly providing groundwater for non-flooded areas. The Maricá region has no regular prolonged dry season that might expose the vegetation to periods of water limitation (Mantovani and Iglesias 2001). In this *restinga*, dew and the shallow water table (Dau 1960) can also serve as water sources, which may contribute to the absence of a correlation between flowering events and rainfall (Gomes et al. 2008). Studies in other phytophysiognomies of the Atlantic Forest where the climate is less seasonal than in Maricá have also found no correlation between phenology and rainfall and have shown that day length is highly important for flowering periodicity (Morellato et al. 2000; Straggemeier and Morellato 2011). Therefore, our results reinforce the hypothesis that day length is important for the seasonal distribution of flowering events through the year in areas with an irregular rainfall regime and without a severe and well-defined dry season.

Among the species monitored in the Maricá *restinga*, 92% provided some floral resources during the flowering period, suggesting that these plants are highly dependent on biotic pollen vectors for sexual reproduction. Four types of floral resources (nectar, pollen, oil, and resin) occur. Notably, *Myrrhinium atropurpureum* offers sweet petals as a floral resource (Roitman et al. 1997). As in other assessments of the proportion and temporal distribution of floral resources in plant communities, in the Maricá *restinga*, nectar is the most frequent resource, followed by pollen, oil, and resin; however, resin was not found in some of these studies (Atlantic Forest, Silva et al. 1997; *restinga*, Viana 1999; Cerrado, Silberbauer-Gottsberger and Gottsberger 1988 and Oliveira and Gibbs 2000; Caatinga, Rodarte 2003 and Machado and Lopes 2004). This was expected since nectar is used by a wide variety of anthophilous visitors, both diurnal and nocturnal (Simpson and Neff 1983; Endress 1994; Agostini et al. 2014). Pollen is a rich source of proteins and amino acids together with numerous other compounds of potential nutritional value; it is used mainly by bees but also by beetles and mammals (Endress 1994; Agostini et al. 2014). Floral oil is collected by solitary female bees and together with pollen grains serves as food for the larvae (Endress 1994; Agostini et al. 2014). Resin is a non-nutritive resource, also collected only by female bees for coating and asepsis of the nests (Simpson and Neff 1983; Endress 1994; Agostini et al. 2014).

Although we observed a concentration of species blooming in the warmer/wetter period, the Maricá *restinga* has some species blooming throughout the year, and floral resources are always available for pollinators. Among the species with a flowering peak restricted to the colder/drier season are *Lundia corymbifera*, *Billbergia amoena* (both pollinated by hummingbirds), *Agarista revoluta* (pollinated mainly by nocturnal insects), and *Protium brasiliensis*, *Guapira opposita*, *Coccoloba arborescens*, and *Neomithrantes obscura* (all pollinated by various insects). These species are annual, regular, with intermediate to long and intense flowering, and attract diurnal (Rodarte 2008; Benevides 2011, 2015a; Moreira et al. 2017, 2019) and nocturnal visitors (Moreira et al. 2017, 2019).

The large number of continuously flowering species recorded here may be an effect of sampling plants with different life forms, as most phenology studies that have evaluated mainly tree species found a very small number of species with continuous flowering (Newstrom et al. 1994a, b; Engel and Martins 2005). Species with continuous flowering predominated among herbs, vines, and subshrubs, with a tendency for higher flowering intensity in the warmer/wetter season. Particularly important are *Dalechampia micromeria*, which supplies some Euglossini and Meliponini bees with floral resin when *Clusia lanceolata* stops flowering at the end of the year (Rodarte 2008), and *Peixotoa hispidula*, an important source of floral oil, used mainly by Centridini bees (Vieira 1995) when *Byrsonima sericea* is not flowering.

The temporal distribution and the intensity of the flowering periods in a community affect the quantity and quality of floral resources available to pollinators and are important community structural factors (Newstrom et al. 1994b; Olesen et al. 2008; Rosas-Guerrero et al. 2014). The regularity of events over the years provides predictability in the supply of resources for these pollen vectors (Morellato et al. 2016), promoting consistent, diversified, and often specialized plant-pollinator interactions, as observed in most species in this study, particularly for species pollinated by bees (Rodarte 2008).

The climate is one of the main factors involved in the control and regulation of plant phenological events (leaf emission and loss, flowering, and fruiting). In recent years, studies of phenology have gained in importance in the face of the threat of climate change, which is potentially capable of disrupting interactions such as those between plants and pollinators (Morellato et al. 2016). Anthracological studies using carbonized woody remains from *sambaquis* (prehistoric kitchen middens) on the southeast coast of Rio de Janeiro state found a high floristic diversity in the *restinga* areas, correspondence between the taxa recorded and currently existing plants, and virtually no vegetation variation during the entire second half of the Holocene, approximately 5500 to 1400 years BP (Scheel-Ybert 1999, 2000, 2001, 2004). Based on these results, Scheel-Ybert (2001, 2004) suggested that *restinga* vegetation appears to be stable and not exceptionally sensitive to climatic fluctuations. In parallel, the regular flowering rhythms and the consequent predictable availability of floral resources found here for *restinga* vegetation over 7 years also suggest that interactions between plants and pollinators in this environment can be quite stable and resilient. Certain phenological characteristics such as annual and long-lasting flowering (more than 4 months), together with diversified plant-pollinator interactions, from generalists to specialists (Rodarte 2008), reinforce the idea of resilient interactions, leading to sexual reproductive success. The high capacity for vegetative propagation that is commonly documented in *restinga* plants (Cirne and Scarano 2001; Faria et al. 2006; Benevides et al. 2015b) is also an important reproductive strategy and contributes to plant resilience in this ecosystem. The *restinga* vegetation comprises a pool of species with high ecological plasticity that migrated from the Atlantic Forest and colonized the geologically younger and resource-poor *restingas* (Scarano 2002). Despite the high ecological plasticity, the colonization and succession in natural *restinga* areas characterizes this vegetation as unique and vulnerable (Scarano 2002).

Recent studies based on models that compared current climate conditions to future climate scenarios, projected to 2050 to assess potential changes in taxonomic and functional diversities of woody *restinga* species, revealed that climatic changes can critically alter the biodiversity in *restinga* communities, mainly through loss of species and homogenization of the vegetation (Inague et al. 2021). The present study of flowering phenology in a *restinga* area showed how the temporal distribution of flowering events, availability of floral resources, pollination, and reproductive success of plant species are interdependent events that can become mismatched, resulting in irreparable losses of ecosystem services. Therefore, anthropic actions and climate changes together would increase the vulnerability of these coastal communities and require urgent conservation measures such as restoration of degraded areas, increases in the number of *restinga* areas, and strengthening of legal protections of *restingas*, in order to mitigate the expected impacts.

Acknowledgments We thank Dr. Janet W. Reid (JWR Associates) for the revision of the English text and Dr. Maria Célia Bezerra Pinheiro and Dr. Dorothy Sue Dunn de Araújo for thesis orientation. This study is part of the Ph.D. thesis and visitant professor project of Dr. Ana Tereza Araújo Rodarte, which were carried out in the Postgraduate Program in Biological Sciences (Botany) of the Museu Nacional, Universidade Federal do Rio de Janeiro (UFRJ), Rio de Janeiro, Brazil, and were supported by CAPES (Coordenação de Aperfeiçoamento de Pessoal de Nível Superior).

References

Agostini K, Lopes AV, Machado IC (2014) Recursos florais. In: Reich AR, Agostini K, Oliveira PE, Machado IC (eds) Biologia da polinização. Projeto Cultural, Rio de Janeiro, pp 130–150

Bencke CSC, Morellato LPC (2002a) Comparação de dois métodos de avaliação da fenologia de plantas, sua interpretação e representação. Rev Bras Bot 25:269–275

Bencke CSC, Morellato LPC (2002b) Estudo comparativo de nove espécies arbóreas em três tipos de floresta atlântica no sudeste do Brasil. Rev Bras Bot 25:237–248

Benevides CR (2011) Biologia floral e reprodutiva de espécies dioicas na restinga de Maricá, RJ. Ph.D. Thesis, Universidade Federal do Rio de Janeiro/Museu Nacional, Rio de Janeiro

Benevides CR, Rodarte ATA, Lima HA (2015a) Strategies of flowering in dioecious plants in tropical coastal vegetation, Brazil. Braz J Bot 38:585–595

Benevides CR, Rodarte ATA, Lima HA (2015b) Vegetative propagation as a successful reproductive strategy in woody dioecious species in tropical coastal vegetation, southeast Brazil. Braz J Bot 38:579–584

Bianchini E, Pimenta JA, Santos FAM (2006) Fenologia de *Chrysophyllum gonocarpum* (Mart. & Eichler) Engl. (Sapotaceae) em floresta semidecídua do sul do Brasil. Rev Bras Bot 29:595–602

Carvalho ALG, Somner GV, Aleen J (2015) Is the phenology of all restinga species the same? A taxonomically-focused study of Sapindaceae in a highly threatened coastal environment. Flora 215:92–101

Cirne P, Scarano FR (2001) Resprouting and growth dynamics after fire of the clonal shrub *Andira legalis* (Leguminosae) in a sandy coastal plain in south-eastern Brazil. J Ecol 89:351–357

Dau L (1960) Microclimas das restingas do sudeste do Brasil. I. Restinga interna de Cabo Frio. Arq Mus Nac Rio de Janeiro 50:79–133

Dunham AE, Razafindratsima OH, Rakotonirina P et al (2018) Fruiting phenology is linked to rainfall variability in a tropical rain forest. Biotropica 50:396–404

Endress PK (1994) Diversity and evolutionary biology of tropical flowers. Cambridge University Press, Cambridge

Engel V, Martins FR (2005) Reproductive phenology of Atlantic forest tree species in Brazil: an eleven year study. Trop Ecol 46:1–16

Faria APG, Matallana G, Wendt T et al (2006) Low fruit set in the abundant dioecious tree *Clusia hilariana* (Clusiaceae) in a Brazilian restinga. Flora 201:606–611

Flora do Brasil (2021) http://floradobrasil.jbrj.gov.br/. Accessed May 2021

Fournier LA (1974) Un método cuantitativo para la medición de características fenológicas en árboles. Turrialba 24:422–423

Fournier LA, Charpantier C (1975) El tamaño de la muestra y la frecuencia de las observaciones en el estudio de las características fenológicas de los árboles tropicales. Turrialba 25:45–48

Gomes R, Pinheiro MCB, Lima HA (2008) Fenologia reprodutiva de quatro espécies de Sapotaceae na restinga de Maricá, RJ. Rev Bras Bot 31(4):679–687

Inague GM, Zwiener VP, Marques MCM (2021) Climate change threatens the woody plant taxonomic and functional diversities of the Restinga vegetation in Brazil. Perspect Ecol Conserv 19:53–60

Köppen W (1948) Climatología: con un estudio de los climas de la tierra. Fondo de Cultura Económica, México, D.F.

Machado IC, Lopes AV (2004) Floral traits and pollination systems in the Caatinga, a Brazilian tropical dry forest. Ann Bot 94:365–376

Mantovani A, Iglesias RR (2001) Bromélias terrestres na restinga de Barra de Maricá, Rio de Janeiro: influência sobre o microclima, o solo e a estocagem de nutrientes em ambientes de borda de moitas. Leandra 16:17–37

Mattos CCLV (2005) Caracterização climática da Restinga da Marambaia, R.J. In: Menezes LFT, Peixoto AL, Araújo DSD (eds) História Natural da Marambaia. EDUR, Seropédica, pp 55–66

Mendonça F, Danni-Oliveira IM (2007) Climatologia: noções básicas e climas do Brasil. Oficina de Textos, São Paulo

Moreira MM, Miranda AS, Lima HA (2017) *Agarista revoluta* (Ericaceae): a generalist plant with self-compatible and self-incompatible individuals. Flora 234:7–14

Moreira MM, Miranda AS, Sá-Haiad B et al (2019) Diurnal versus nocturnal pollinators and the effect of anthesis onset on the reproductive success of Agarista revoluta (Ericaceae). Plant Syst Evol 305:375–384

Morellato LPC (2008) Fenologia de plantas e os efeitos das mudanças climáticas. In: Buckeridge MS (ed) Biologia e mudanças climáticas no Brasil. RiMa, São Carlos, pp 181–191

Morellato LPC, Talora DC, Takahasi A et al (2000) Phenology of Atlantic rain forest trees: a comparative study. Biotropica 32:811–823

Morellato LPC, Camargo MGG, Gressler E (2013) A review of plant phenology in South and Central America. In: Schwartz MD (ed) Phenology: an integrative environmental science. Springer, Dordrecht, pp 91–113

Morellato LPC, Alberton B, Alvarado ST et al (2016) Linking plant phenology to conservation biology. Biol Conserv 195:60–72

Newstrom LE, Frankie GW, Baker HG et al (1994a) Diversity of long-term flowering patterns. In: McDade LA, Bawa KS, Hespenheide HA, Hartshorn GS (eds) Ecology and natural history of a neotropical rain forest. Chicago University Press, Chicago, pp 142–160

Newstrom LE, Frankie GW, Baker HG (1994b) A new classification for plant phenology based on flowering patterns in lowland tropical rain forest trees at la selva, Costa Rica. Biotropica 26:141–159

Olesen JM, Bascompte J, Elberling H et al (2008) Temporal dynamics in a pollination network. Ecology 89:1573–1582

Oliveira PE, Gibbs PE (2000) Reproductive biology of woody plants in a cerrado community of Central Brazil. Flora 195:311–329

Pereira AR, Angelocci LR, Sentelhas PC (2001) Agrometeorologia: fundamentos e aplicações práticas. Editora Agropecuária, Guaíba.

Perrin P (1984) Evolução da costa fluminense entre as pontas de Itacoatiara e Negra: preenchimentos e restingas. In Restingas: origem, estrutura, processos (Lacerda LD, Araujo DSD, Cerqueira R, Turcq B orgs.) Universidade federal Fluminense/CEUFF, Niterói

Primack D, Imbres C, Primack RB et al (2004) Herbarium specimens demonstrate earlier flowering times in response to warming in Boston. Am J Bot 91:1260–1264

Ramírez N (2002) Reproductive phenology, life-forms, and habitats of the Venezuelan Central Plain. Am J Bot 89:836–842

Rizzini CT (1979) Tratado de fitogeografia do Brasil. HUICITEC/EDUSP, São Paulo

Rodarte ATA (2003) Flora fanerogâmica das dunas interiores da margem esquerda do médio rio São Francisco, Ibiraba, Barra, Bahia, com ênfase na flora apícola. M.Sc. Dissertation, Instituto de Biologia da Universidade Federal da Bahia, Salvador

Rodarte ATA (2008) Caracterização espacial, temporal e biologia floral das espécies de restinga, com ênfase nos recursos florais. Ph.D. Thesis, Universidade Federal do Rio de Janeiro/Museu Nacional, Rio de Janeiro

Roitman GG, Montaldo NH, Medan D (1997) Pollination biology of *Myrrhinium atropurpureum* (Myrtaceae): sweet, fleshy petals attract frugivorous birds. Biotropica 29(2):162–168

Rosas-Guerrero V, Aguilar R, Martén-Rodríguez S et al (2014) A quantitative review of pollination syndromes: do floral traits predict effective pollinators? Ecol Lett 17:388–400

Scarano FR (2002) Structure, function and floristic relationships of plant communities in stressful habitats marginal to the Brazilian Atlantic rainforest. Ann Bot 90:517–524

Scarano FR (2008) Mudanças climáticas globais: até que ponto a ecologia como ciência pode ajudar na mitigação? In: Buckeridge MS (ed) Biologia e mudanças climáticas no Brasil. RiMa, São Carlos, pp 117–130

Scheel-Ybert R (1999) Paleoambiente e paleoetnologia de populações sambaquieiras do sudeste do Estado do Rio de Janeiro. Rev Mus Arqueol Etnol 9:43–59

Scheel-Ybert R (2000) Vegetation stability in the Southeastern Brazilian coastal area from 5500 to 1400 14C yr BP deduced from charcoal analysis. Rev Palaeobot Palynol 110:111–138

Scheel-Ybert R (2001) Man and vegetation in southeastern Brazil during the Late Holocene. J Archaeol Sci 28(5):471–480

Scheel-Ybert R (2002) Vegetation stability in the Brazilian littoral during the late holocene: anthracological evidence. Pesqui em Geociênc 28(2):315–323

Scheel-Ybert R (2004) Teoria e métodos em antracologia. 1. Considerações teóricas e perspectivas. Arq Mus Nac Rio de Janeiro 62(1):3–14

Silberbauer-Gottsberger I, Gottsberger G (1988) A polinização de plantas do Cerrado. Rev Bras Biol 48:651–663

Silva AG, Guedes-Bruni RR, Lima MPM (1997) Sistemas sexuais e recursos florais do componente arbustivo-arbóreo em mata preservada na reserva ecológica de Macaé de Cima. In: Lima MPM, Guedes-Bruni R R (eds.) Serra de Macaé de Cima: diversidade florística e conservação em mata Atlântica. Rio de Janeiro. Jardim Botânico do Rio de Janeiro

Simpson BB, Neff JL (1983) Evolution and diversity of floral rewards. In: Jones CE, Little RL (eds) Handbook of experimental pollination biology. Scientific and Academic Editions, New York, pp 142–159

StatSoft Inc (2005) STATISTICA (data analysis software systems), version 7.1. www.Statsoft.com

Straggemeier VG, Morellato LPC (2011) Reproductive phenology of coastal plain Atlantic forest vegetation: comparisons from seashore to foothills. Int J Biometeorol 55:843–854

Talora DC, Morellato LPC (2000) Fenologia de espécies arbóreas em floresta de planície litorânea do sudeste do Brasil. Rev Bras Bot 23:13–26

van Schaik CP, Terborgh JW, Wright SJ (1993) The phenology of tropical forests: adaptive significance and consequences for primary consumers. Annu Rev Ecol Evol Syst 24:353–377

Varejão-Silva MA (2000) Meteorologia e climatologia. Instituto Nacional de Meteorologia (INMET), Brasília

Viana BF (1999) Biodiversidade da apifauna e flora apícola das dunas litorâneas da APA das lagoas e Dunas de Abaeté, Salvador, Bahia: Composição, fenologia e suas interações. Thesis, Universidade de São Paulo

Vieira ACM (1995) Biologia da polinização e sistema de reprodução de *Peixotoa hispidula* Juss. (Malpighiaceae). Dissertation, Universidade Federal do Rio de Janeiro/Museu Nacional

Chapter 2
Dioecy: The Dimorphic Sexual System and Pollination in *Restinga* Vegetation

Cristine Rodrigues Benevides, Marina Muniz Moreira, Ana Tereza Araújo Rodarte, Angela Arruda e Albuquerque, Emanoela Mano Muniz da Silva, Luciana Carolina Oliveira Sepúlveda do Nascimento, and Heloisa Alves de Lima

Abstract Dioecy, the occurrence of separate male and female individuals in a population, relies on biotic or abiotic pollen vectors for sexual reproduction. We evaluated the dimorphic sexual system, pollination, attractive floral attributes, and floral visitors of ten dioecious species in *restinga* vegetation. Although most species showed full dioecy, we found one case of cryptic dioecy, in *Garcinia brasiliensis*, and one case of gynodioecy, in *Monteverdia obtusifolia*. Fruit set was dependent on insect pollination in the majority of species, but *Myrsine parvifolia* is wind-pollinated, and *Guapira opposita* is both wind- and insect-pollinated (ambophily). The entomophilous species fell into two groups: one, with specialist pollination, had attractive and multiovulate flowers, as well as floral resources such as resin, nectar, and pollen collected by a limited group of insects (stingless bees and night beetles); the other group, with generalist insect pollination, produced many inconspicuous nectar flowers with few ovules, visited by diverse groups of insects. Among the visitors the stingless bee *Trigona spinipes* visited almost all the species. Plant species with dimorphic sexual systems are important and diverse in the *restinga* vegetation, with different expressions of sexual dimorphism as well as diverse pollination systems, including anemophily, ambophily, and specialist and generalist entomophily.

Keywords Ambophily · Anemophily · Cryptic dioecy · Generalist pollination · Gynodioecy · Specialist pollination

---

C. R. Benevides (✉) · M. M. Moreira · A. T. A. Rodarte · A. Arruda e Albuquerque · E. M. M. da Silva · L. C. O. S. do Nascimento · H. A. de Lima
Departamento de Botânica, Laboratório de Biologia Reprodutiva de Angiospermas, Museu Nacional, Universidade Federal do Rio de Janeiro (MN/UFRJ), Rio de Janeiro, RJ, Brazil
e-mail: cristinebenevides@gmail.com

M. F. T. Medeiros, B. de Sá Haiad (eds.), *Aspects of Brazilian Floristic Diversity*,
https://doi.org/10.1007/978-3-031-07453-0_2

2.1 Introduction

Most species of flowering plants are hermaphroditic, producing flowers with both female (carpels) and male (stamens) sexual organs. Stamens and carpels can occur in diverse combinations at the levels of flower, individual, and population, leading to a great diversity of monomorphic (only one morph that contributes to the next generation, such as female and male parents) or dimorphic (two distinct sexual morphs that function primarily as either female or male parents) sexual systems (Barrett 2002).

Dioecy is a dimorphic sexual system; plants in these populations can be strictly unisexual (dioecy), or with one morph hermaphroditic and the other female (gynodioecy) or male (androdioecy), resulting in dependence on pollen vectors for pollination (Bawa and Opler 1975; Vamosi and Otto 2002; Barrett 2002, 2010). Other dimorphic systems exist, with various combinations of functionally pistillate, staminate, and perfect flowers, at the level of individuals and populations (Barrett 2010). The great diversity of cases makes it difficult to classify plants based only on morphology (Lloyd 1979; Barrett 1998, 2002; Delph and Wolf 2005).

Dioecy frequently evolves from hermaphroditism (Barrett 2002, 2010). Recent studies have suggested that the evolution of this sexual system results from an interplay of several forces, which include the benefit of cross-fertilization, preventing inbreeding depression, and an increased ability to adjust investment in male and female functions independently, maximizing the success of both (Spigler and Ashman 2012; Sinclair et al. 2013; Käfer et al. 2017).

While dioecy is rare in plants, occurring in only 5–6% of angiosperms (Renner and Ricklefs 1995; Renner 2014), it occurs in about half of all angiosperm families, including basal and derived groups (Barrett 2002). This scattered phylogenetic distribution of dioecy was long taken to support the idea that dioecy is an evolutionary dead end, leading to reduced diversification or increased extinction (Vamosi and Otto 2002). Nevertheless, more recent phylogenetic and experimental studies have shown that dioecy can be lost and regained repeatedly (Renner 2014; Käfer et al. 2017; Barrett 2021; Crossard et al. 2021) and that there is no evidence for reduced diversification in dioecious angiosperms (Käfer et al. 2017). Selective forces such as pollen limitation and reproductive assurance seem to be involved in the transition from dioecy to hermaphroditism and are being investigated (Pannel 2000; Crossman and Charlesworth 2013).

Dioecy has been associated with many ecological attributes (Bawa and Opler 1975; Bawa 1980; Renner and Ricklefs 1995; Vamosi et al. 2003). Tropical floras generally contain more dioecious species than temperate ones, because dioecy prevails in trees and shrubs with fleshy fruits (Bawa and Opler 1975; Bawa 1980), which are more common in the tropics (Vamosi et al. 2003). In tropical forests, frequencies of dioecy range from 3% to 29.2% (Bawa 1980; Flores and Schemske 1984; Bawa et al. 1985; Bullock 1985; Vamosi 2006; Queenborough et al. 2009; Vary et al. 2011). Generalist pollination by diverse insects has been recorded for many dioecious species whose flowers are small and inconspicuous, greenish yellow, and with easily accessible nectar (Bawa 1980). A recent review of plants pollinated by diverse insects (DI) confirmed the results of Bawa (1980) and found a high proportion (31%) of dioecy in DI plant species (Moreira and Freitas 2000). In addition to entomophilous pollination, studies have shown that wind pollination is also correlated with dioecy and that anemophily evolved more frequently from ancestral lineages with unisexual flowers pollinated by animals (Friedman and Barrett 2008; Barrett 2010; Renner 2014). Ambophily (insect and wind pollination) has also been reported for dioecious species (Bullock 1994; Culley et al. 2002; Friedman and Barrett 2008, 2009) and would be a stable or intermediate stage in the evolution of animal pollination for wind pollination (Bullock 1994; Culley et al. 2002).

The *restinga* is a recent environment with harsh conditions (such as the low water-retention capacity of the sandy soils, low air relative humidity, and strong action of marine winds) (Rizzini 1979; Scarano 2002). *Restinga* vegetation is composed mainly of plants that migrated from neighboring areas of the Atlantic Forest (Scarano 2002). In this ecosystem, many dioecious species have been recorded, mainly among the dominant woody species (Ormond et al. 1991; Matallana et al. 2005). The success of dioecy in *restinga* vegetation can be explained by the idea that resource allocation for sexual functions in different morphs can be optimized, increasing the success of male and female individuals in a stressful and/or oligotrophic environment (Matallana et al. 2005; Delph and Wolf 2005). Entomophily is the most commonly recorded pollination system for dioecious *restinga* species, with Clusiaceae the most-studied group (Correia et al. 1993, 1999; Lopes and Machado 1998; Silva 2005; Faria et al. 2006; Cesário 2007). However, despite the increasing number of studies on dioecious species from *restingas* (Lenzi and Orth 2004; Cesário and Gaglianone 2008; Albuquerque et al. 2013; Leal et al. 2012, 2013; Benevides et al. 2013, 2015a, b), we still know little about the sexual expression and pollination system of dimorphic species with unspecialized flowers.

Here, we characterized the sexual and pollination systems of sexually dimorphic species from the *restinga* vegetation, integrating morphological, functional, and experimental approaches.

2.2 Material and Methods

2.2.1 Study Area and Species

We carried out this study in the Maricá *restinga*, an environmentally protected area of 8.3 km^2 located in the municipality of Maricá, Rio de Janeiro state, Brazil (22°57′45″S to 42°53′33″W and 22°57′52″S to 42°53′48″W). The Maricá *restinga* is composed of two sandy ridges (inner and outer) formed between 3500 and 5500 years ago, respectively, in episodes of the last marine transgressions (Perrin 1984). The climate is Aw tropical humid (Köppen 1948). The air temperature in

Table 2.1 Species with sexual dimorphism studied in the *restinga* of Maricá, Rio de Janeiro

Family	Species	N individuals
Anacardiaceae	*Tapirira guianensis* Aubl.	20
Burseraceae	*Protium brasiliense* (Spreng.) Engl.	22
Celastraceae	*Monteverdia obtusifolia* (Mart.) Biral	36
Clusiaceae	*Clusia criuva* Cambess.	20
	Clusia lanceolata Cambess.	20
	Garcinia brasiliensis Mart.	20
Lauraceae	*Ocotea notata* Mez.	21
Myrsinaceae	*Myrsine parvifolia* A. DC.	30
Nyctaginaceae	*Guapira opposita* (Vell.) Reitz.	20
Polygonaceae	*Coccoloba arborescens* (Vell.) R.A.Howard	22

winter varies between 8.6 and 15.2 °C and in summer between 15 and 37.7 °C. The mean annual rainfall is 1130 mm and is more intense in the summer, although some rainfall occurs in all months (Mantovani and Iglesias 2001).

We selected ten species with sexual dimorphism (Table 2.1) belonging to eight families, along the inner sandy ridge of the Maricá *restinga*. We followed these species through weekly excursions lasting 1–3 days, for 4 years (2007–2010).

2.2.2 Sexual System and Floral Traits

To assess the sexual system for each species, we analyzed the presence of reproductive structures in flowers of each individual as well as the fruit formation. Using a stereoscopic microscope, we checked for the presence of pollen and ovules. According to the floral morphology, flowers were classified as pistillate, staminate, or perfect and individual plants as female, male, or hermaphroditic. In consecutive reproductive seasons, flowers from the same individuals were observed in order to check for possible changes in sexuality.

We characterized floral traits of pistillate, staminate, and perfect flowers in relation to the presence of scent, color, anthesis (day or night), and floral resources (Faegri and Van Der Pijl 1979). To assess the presence of scent, we collected and deposited flowers in clean vials for 15 min, in order to concentrate the odor, and then sniffed above the vial mouth (Dafni et al. 2005). The floral size was evaluated from flowers collected from five individuals of each sexual morph and fixed in 70% ethanol. In the laboratory, measurements were performed with the aid of an analog steel vernier caliper (0.05 × 150 mm) and a stereoscopic microscope. Based on these measurements, we classified the flowers in two categories: large (>1 cm) or small (≤1 cm).

To complement the study of floral morphology in species with inconspicuous flowers, we made observations in a JEOL JSM-6390LV scanning electron microscope at the Universidade Federal do Rio de Janeiro/Museu Nacional. Flowers were fixed in 70% ethanol, dehydrated in an ethyl series, taken to the CO_2 critical point in a Leica EM CPD030 critical point dryer, attached to an aluminum support covered with double-sided carbon tape, and metallized with gold in a Denton Vacuum Desk IV cold sputter system (Silveira 1998).

In flowers that showed pollen grains, we quantified the pollen viability in three to five flowers per individual ($n = 5$ individuals), using acetic carmine (Alexander 1980). We also checked for the presence of ovules in the same flowers used to calculate pollen viability.

To determine the anthesis period, we marked flowers of both sexual morphs at the beginning of anthesis and followed them until they withered. We considered that flowers were in anthesis when the anthers were dehiscent, releasing pollen grains, or the stigmatic area was receptive. Stigmatic receptivity was assessed using the 3% H_2O_2 method (Kearns and Inouye 1993). We classified stigmatic areas according to Heslop-Harrison and Shivanna (1977).

We counted the number of buds per inflorescence (*C. arborescens*, *G. opposita*, *O. notata*, and *T. guianensis*) or per branch (*M. obtusifolia* and *P. brasiliensis*) in male and female morphs of each species, using five individuals of each morph. The use of branches for counting buds of *M. obtusifolia* and *P. brasiliensis* was necessary, since the inflorescences in these two species are congested and buds extend along a large part of the branch. Therefore, we counted the number of buds present within 10 cm of the apex of the flowering branches. We compared the mean number of buds among the morphs using Student's *t*-test (StatSoft Inc 2005).

2.2.3 Reproductive and Pollination Systems

We conducted manual pollination experiments using buds bagged in pre-anthesis. We conducted the following treatments: for *cross-pollination*, we pollinated pistillate and perfect flowers (in the case of *M. obtusifolia*) with pollen from other individuals; for *autonomous apomixis*, we bagged only pistillate flowers; for *anemophily*, we bagged pistillate flowers with fine-meshed bags that allow pollen grains to enter by wind action but exclude possible floral visitors; and for *natural pollination*, we randomly marked pistillate flowers and exposed them to natural pollination. In the species with very small and congested flowers, arranged along the inflorescences (*T. guianensis*) or branches (*M. obtusifolia*, *P. brasiliensis*), we estimated the proportion of fruits formed after natural pollination by counting the number of buds in 10 cm of branches or inflorescences; later, during the fruiting period, we counted the number of fruits in the same 10 cm of branches or inflorescences. In all experiments, we observed the flowers until they fell or set fruit. For each treatment we used a minimum of 15 flowers and then calculated the fruit production.

We randomly collected ripe fruits that were naturally pollinated (10–30 fruits from at least 3 female individuals) to evaluate the size and number of viable seeds per fruit. For gynodioecious species, we also collected and evaluated the fruits of hermaphroditic individuals. A steel vernier caliper (0.05 × 150 mm) was used for all

Table 2.2 Hours of observation of floral visitors in each sexual morph in the *restinga* of Maricá, Rio de Janeiro

Species	Male morph	Female morph	Hours of observation	N individuals
Coccoloba arborescens	7	7	14	2
Guapira opposita	10	10	20	4
Monteverdia obtusifolia	18	18	36	4
Ocotea notata	11	11	22	6
Protium brasiliensis	11	11	22	4
Tapirira guianensis	7	7	14	2
Total	64	64	128	22

measurements. We considered well-formed seeds to be viable and seeds that were withered and without embryos to be aborted.

To assess seed germination percentages, we placed viable seeds in Petri dishes, lined with filter paper and moistened with distilled water, immediately after the field collections. We observed the seeds daily in order to record the emergence of the main root and to calculate the germination percentage (percentage of germinated seeds in relation to the number of seeds able to germinate) (Gonçalves et al. 2008).

For six species, we observed focal plants to record the floral visitors (Dafni et al. 2005). We performed the observations on each sexual morph, from 08:00 to 15:00 h, for a total of 128 h of observation for all species in the study (Table 2.2). For the other species (*C. criuva*, *C. lanceolata*, *G. brasiliensis*, and *M. parvifolia*), data on floral visitors were obtained from the literature (Correia et al. 1993, 1999; Silva 2005; Albuquerque et al. 2013). Specimens of insects were collected for later identification by specialists.

Floral traits, observations of floral visitors, and manual reproductive experiments were used to classify the pollination systems as ambophily, anemophily, or specialist or generalist entomophily.

2.3 Results

2.3.1 Sexual Systems and Floral Traits

The ten species studied are shrubs or trees. Nine of them have a dioecious sexual system, with female plants bearing pistillate flowers and male plants bearing staminate flowers. Only *M. obtusifolia* bears both female and hermaphroditic (perfect) flowers, being classified as gynodioecious.

In general, staminate flowers bear pistillodes, while pistillate flowers bear staminodes, except in *C. lanceolata*, *C. criuva*, and *G. brasiliensis* whose staminate flowers have no traces of the female sex, although the pistillate flowers have staminodes (Fig. 2.1a–f). In *C. lanceolata* flowers, staminodes are the resin-secreting structures in both pistillate and staminate flowers. Stamens of pistillate and staminate flowers

Fig. 2.1 Dioecious plants of the *restinga* of Maricá, Rio de Janeiro. *Clusia criuva*, (**a**) staminate flowers, (**b**) pistillate flowers; *Clusia lanceolata*, (**c**) staminate flowers, (**d**) pistillate flowers; *Garcinia brasiliensis*, (**e**) staminate flowers, (**f**) pistillate flowers; (**g**) monoecious plant of *Garcinia brasiliensis*; (**h**) male plant of *Guapira opposita* with fruits. Scales 1 cm (**a–g**)

of *G. brasiliensis* did not differ in length, and pistillate flowers seemed to be perfect but lacked pollen grains, i.e., they were functionally female; this species was considered to be cryptic dioecious. In the pistillate flowers of *G. opposita*, *C. arborescens*, and *M. obtusifolia*, empty staminodes have dehiscent anthers (Figs. 2.2, 2.3, 2.4, 2.5, 2.6, 2.7 and 2.8), but in the other species, the staminodes are indehiscent.

Female plants of all species produced fruits; only one male plant of *C. arborescens* and one male plant of *G. opposita* (Fig. 2.1h) showed some fruits in a single reproductive event. In the gynodioecious *M. obtusifolia*, hermaphroditic plants always produced fruits but fewer than in female plants. In *G. brasiliensis*, three

Fig. 2.2 *Coccoloba arborescens* in the *restinga* of Maricá, Rio de Janeiro. Photographs of (**a**) staminate flowers and (**b**) pistillate flowers. Scanning electron micrographs of (**c**) staminate flower, (**d**) pistillate flower, (**e**) pistillode of staminate flower, and (**f**) stigma surface of pistillate flower

Fig. 2.3 *Guapira opposita* in the *restinga* of Maricá, Rio de Janeiro. Photographs of (**a**) staminate flowers and (**b**) pistillate flowers. Scanning electron micrographs of (**c**) staminate flower, (**d**) pistillate flower, (**e**) dehiscent anther of staminate flower, (**f**) stigma surface of pistillode in staminate flower, and (**g**) stigma surface of pistillate flower

monoecious individuals were observed, two with a higher frequency of staminate flowers (one fruit was observed) and one with a higher frequency of pistillate flowers (Fig. 2.1g).

All species have light-colored flowers, with radial symmetry, little or no depth effect, diurnal anthesis (except *C. criuva*), and fragrance (except *M. parvifolia*) (Table 2.3 and Figs. 2.1, 2.2, 2.3, 2.4, 2.5, 2.6, 2.7 and 2.8). In most species the fragrances are mild, but in *C. criuva* the flowers are intensely fragrant, especially at night. Small and inconspicuous flowers predominated, but among the Clusiaceae the flowers are large and conspicuous: in *C. criuva* they are completely white (Fig. 2.1a, b), in *C. lanceolata* they are white with a reddish central area (Fig. 2.1c, d), and in *G. brasiliensis* they have numerous attractive stamens (Fig. 2.1e, f and

Fig. 2.4 *Monteverdia obtusifolia* in the *restinga* of Maricá, Rio de Janeiro. Photographs of (**a**) perfect flowers and (**b**) pistillate flowers. Scanning electron micrographs of (**c**) perfect flower, (**d**) pistillate flower, (**e**) stigma surface of perfect flower, (**f**) stigma surface of pistillate flower, (**g**) anthers of perfect flower with pollen grains, and (**h**) empty anthers of pistillate flower

Fig. 2.5 *Myrsine parvifolia* in the *restinga* of Maricá, Rio de Janeiro. Photographs of (**a**) staminate flowers and (**b**) pistillate flowers. Scanning electron micrographs of (**c**) anthers with pollen grains and pistillode of staminate flower, (**d**) stigma and staminode of pistillate flower, (**e**) detail of pollen grain, and (**f**) anthers of pistillate flower. Scales 5 mm (**a, b**)

Table 2.4). Floral resources consist of resin (in *C. lanceolata*), pollen (in *C. criuva*), or nectar, always exposed and easily accessible (in the other studied species). Nectaries are usually disc-shaped, conspicuous, and occupying practically the entire floral receptacle (Figs. 2.2c–e, 2.4c, d, 2.7a, b, 2.8g, h). In *G. brasiliensis* the nectary is disc-shaped, surrounding the gynoecium in pistillate flowers and globose in staminate flowers (Fig. 2.1e, f). In *T. guianensis* (Fig. 2.8g), stomata were observed in the nectary. *Ocotea notata* has six globose nectaries located between the internal and external stamens (Fig. 2.6c, d).

Fig. 2.6 *Ocotea notata* in the *restinga* of Maricá, Rio de Janeiro. Photographs of (**a**) staminate flowers and (**b**) pistillate flowers. Scanning electron micrographs of (**c**) staminate flower, (**d**) pistillate flower, (**e**) stigma surface of pistillode in staminate flower, (**f**) stigma surface of pistillate flower, (**g**) valved anthers of staminate flower, and (**h**) valved anthers of pistillate flower; the valve is indehiscent

Fig. 2.7 *Protium brasiliense* of the *restinga* of Maricá, Rio de Janeiro. Photographs of (**a**) staminate flowers and (**b**) pistillate flowers. Scanning electron micrographs of (**c**) staminate flower, (**d**) pistillate flower, (**e**) stigma surface of pistillode in staminate flower, (**f**) stigma surface of pistillate flower, and (**g**) indehiscent anthers of pistillate flowers

The stigmatic surfaces are in general papillate and wet when receptive, except in *C. lanceolata*, *M. obtusifolia*, and *M. parvifolia*, which have stigmas of the non-papillate type (Table 2.3). However, sinuosity on the stigmatic surface of *M. obtusifolia* could be observed (Fig. 2.4). *Guapira opposita* has plumose stigmas (Fig. 2.3). All stigmas are receptive at the time of floral opening, except in the perfect flowers of *M. obtusifolia*, which are protandrous. The number of ovules in general is low, with several uniovulate species (*C. arborescens*, *G. opposita*, *M. parvifolia*, *O. notata*, and *T. guianensis*) or species with up to 10 ovules (*G. brasiliensis*, *M. obtusifolia*, and *P. brasiliense*); only *Clusia* species are pluriovulate (> 25 ovules) (Table 2.3).

Fig. 2.8 *Tapirira guianensis* of the *restinga* of Maricá, Rio de Janeiro. Photographs of (**a**) staminate flowers and (**b**) pistillate flowers. Scanning electron micrographs of (**c**) staminate flower, (**d**) pistillate flower, (**e**) stigma surface of pistillode in staminate flower, (**f**) stigma surface of pistillate flower, (**g**) pistillode and nectary of staminate flower, and (**h**) indehiscent anthers of pistillate flowers

Table 2.3 Floral biology data for dioecious species in the *restinga* of Maricá, Rio de Janeiro

Species	Anthesis	Floral shape	Floral resource	N of ovules	Stigma	Dehiscence	% Pollen viability (N)
Clusia criuva[1]	N	Di	P	25 to 45	Pa	Lo	99% (1000)
Clusia lanceolata[2]	D	Di	R	31 to 43	NPa	Po	97% (2500)
Coccoloba arborescens	D	Di	Ne	1	Pa	Lo	96% (3000)
Garcinia brasiliensis[3]	D	Di	Ne	2 to 6	Pa	Lo	97% (8300)
Guapira opposita	D	Bf	Ne	1	Pl	Lo	94% (1500)
Monteverdia obtusifolia	D	Di	Ne	4	NPa	Lo	70% (13500)
Myrsine parvifolia[4]	D	Di	P	1	NPa	Lo	97% (300)
Protium brasiliense	D	Di	Ne	8 to 10	Pa	Lo	98% (1000)
Ocotea notata	D	Di	Ne	1	Pa	V	98% (1000)
Tapirira guianensis	D	Di	Ne	1	Pa	Lo	96% (2500)

N night, *D* day, *Di* dish, *Bf* bell-funnel, *P* pollen, *R* resin, *Ne* nectar, *Pa* papillate, *NPa* non-papillate, *Pl* plumose, *Lo* longitudinal, *Po* pores, *V* valves, [1]Correia et al. (1993), [2]Correia et al. (1999), [3]Silva (2005), [4]Albuquerque et al. (2013)

In general, pollen grains are released through anthers with longitudinal dehiscence. In *C. lanceolata*, the pollen grains are released by "pores" of the stamens that compose the synander, and in *O. notata*, the anthers are valved (Fig. 2.6g, h). *Monteverdia obtusifolia* had a mean pollen viability of 70%; all other species had more than 94% viable pollen grains (Table 2.3).

Male plants produced more buds than female plants in *C. lanceolata*, *C. arborescens*, *G. brasiliensis*, *O. notata*, and *T. guianensis*. The members of Clusiaceae showed the lowest mean number of buds per inflorescence (from 1.3 in the female plant of *C. lanceolata* to 8.6 in the male plant of *C. criuva*). The other species produced more buds per inflorescence (from 78 in the female plant of *O. notata* to 1139 in the male plant of *T. guianensis*) (Table 2.4).

The anthesis period ranged from 12 h for *C. arborescens* flowers to 8 days for perfect *M. obtusifolia* flowers and for pistillate *T. guianensis* flowers. In most species, pistillate flowers remained in anthesis longer than staminate flowers, or the periods of anthesis were similar in the two sexual morphs. Small, inconspicuous flowers tended to last longer than large, attractive flowers, except for *C. arborescens* and *G. opposita*, whose inconspicuous flowers had the shortest period of anthesis (Table 2.4).

Table 2.4 Number of buds, floral size, and floral longevity in staminate and pistillate flowers of dioecious species in the *restinga* of Maricá, Rio de Janeiro

Species	Morph	Buds (N)	Floral size	Floral longevity
Clusia criuva	♀	7.5 ± 2.5 (30)[a]	Large	24 h
	♂	8.6 ± 2.1 (30)[a]	Large	24 h
Clusia lanceolata	♀	1.3 ± 0.5 (25)[a]	Large	30 h
	♂	2.1 ± 0.6 (25)[b]	Large	48 h
Coccoloba arborescens	♀	248 ± 46 (5)[a]	Small	12 h
	♂	538 ± 17 (25)[b]	Small	12 h
Garcinia brasiliensis	♀	2.4 ± 1.3 (35)[a]	Large	36 h
	♂	5.2 ± 2.7 (10)[b]	Large	24 h
Guapira opposita	♀	103 ± 33 (23)[a]	Small	24 h
	♂	98 ± 36 (32)[a]	Small	18 h
Monteverdia obtusifolia	♀	181 ± 62 (8)[a]	Small	6 days
	⚥	220 ± 90 (15)[a]	Small	8 days
Myrsine parvifolia[1]	♀	60.5 (25)[a]	Small	3 days
	♂	54 (25)[a]	Small	3 days
Ocotea notata	♀	78 ± 28 (9)[a]	Small	3 days
	♂	167 ± 38 (12)[b]	Small	2 days
Protium brasiliense	♀	364 ± 256 (8)[a]	Small	3 days
	♂	519 ± 269 (6)[a]	Small	2 days
Tapirira guianensis	♀	101 ± 27 (10)[a]	Small	8 days
	♂	1139 ± 469 (12)[b]	Small	3 days

N number of inflorescences or branches sampled. Different letters (a, b) between pistilate and staminate floral morphs indicate that they differ statistically at the 95% significance level. [1]Albuquerque et al. (2013)

2.3.2 Reproductive and Pollination Systems

Reproductive and pollination experiments (Table 2.5) showed that bagging pistillate flowers of all species prevented fruit formation, indicating that they are not autonomous apomicts and depend on biotic and/or abiotic pollen vectors to form fruits. *Myrsine parvifolia* (11.7%) and *G. opposita* (3.6%) set fruits after the flowers were bagged with bags that excluded pollinators, but not the pollen dispersed by wind, indicating the importance of wind for the reproductive success of these species. Fruit production from natural pollination ranged from 5.6% for *O. notata* to 85% for *G. brasiliensis*. For six of the ten species, manual cross-pollination increased the natural fruit production, suggesting failures in natural pollination (pollen limitation).

The fruits are fleshy and colored at maturity in almost all species, except *C. lanceolata*, *C. criuva*, *M. obtusifolia*, and *P. brasiliensis*, which have dehiscent fruit and exposed seeds with attractive arils. The majority of the species have only one seed per fruit (*C. arborescens*, *G. opposita*, *M. parvifolia*, *O. notata*, and *T. guianensis*), or fewer than two viable seeds per fruit on average (*G. brasiliensis*, *M. obtusifolia*, and *P. brasiliense*) (Table 2.6). The fruits of *C. criuva* and *C. lanceolata*

Table 2.5 Reproductive and pollination experiments: anemophily, cross-pollination, natural pollination, and autonomous apomixis performed with dioecious species in the *restinga* of Maricá, Rio de Janeiro

Species	% Anemophily (N)	% Cross-pollination (N)	% Natural pollination (N)	% Autonomous apomixis (N)
Clusia criuva	0 (15)	50 (18)	39 (224)	0 (50)
Clusia lanceolata	0 (15)	100 (25)	77 (100)	0 (96)
Coccoloba arborescens	0 (131)	14 (134)	17 (372)	0 (90)
Garcinia brasiliensis	0 (15)	85 (40)	85 (158)	0 (30)
Guapira opposita	3.6 (2000)	39 (236)	19.5 (1583)	0 (203)
Monteverdia obtusifolia ♀	0 (200)	19 (79)	41 (694)	0 (150)
M. obtusifolia ⚥	0 (150)	3 (30)	2 (2759)	0 (215)
Myrsine parvifolia[a]	11.7 (1303)	56.1 (73)	16.2 (975)	0 (100)
Ocotea notata	0 (1221)	0 (100)	5.7 (535)	0 (193)
Protium brasiliense	0 (143)	17 (47)	16 (603)	0 (100)
Tapirira guianensis	0 (150)	58 (50)	28 (101)	0 (150)

[a]Albuquerque et al. (2013)

Table 2.6 Number of viable seeds per fruit and germination percentage of dioecious species from the *restinga* of Maricá, Rio de Janeiro

Species	Seeds per fruit (N fruits)	Percentage of germination (N seeds)
Clusia criuva	19.8 ± 4.86 (26)	99% (140)
Clusia lanceolata[a]	30 ± 6.96 (10)	100% (48)
Coccoloba arborescens	1 (25)	2% (50)
Garcinia brasiliensis[b]	1.87 (79)	90.45% (21)
Guapira opposita	1 (25)	100% (30)
Monteverdia obtusifolia	1.49 ± 0.6 (30)	100% (25)
Myrsine parvifolia[c]	1 (25)	44.44% (18)
Ocotea notata	1 (25)	96% (25)
Protium brasiliense	1.32 ± 0.48 (25)	72% (25)
Tapirira guianensis	1 (25)	100% (25)

[a]Correia et al. (1999), [b]Silva (2005), [c]Albuquerque et al. (2013)

contained the largest numbers of viable seeds per fruit, 19.8 and 30, respectively. In general, the seeds showed high germination rates (> 70%), with the exception of *C. arborescens* (2%) and *M. parvifolia* (44.44%) (Table 2.6).

All flowers were visited exclusively by insects (Coleoptera, Diptera, Hymenoptera, and Lepidoptera), except the pistillate flowers of *M. parvifolia*, which were not visited, and therefore the species constitutes a case of exclusively

Table 2.7 Floral visitors observed in the dioecious species in the *restinga* of Maricá, Rio de Janeiro

Flower visitors		Cc	Cl	Ca	Gb	Go	Mo	On	Pb	Tg
Coleoptera	*Astylus lineatus* (Fabricius, 1775)	x					x			
	Cantharidae sp.						x	x	x	x
	Coleoptera sp2								x	
	Dinaltica bahiaensis (Jacoby, 1902)	x								
	Elateridae	x								
	Oedemeridae	x								
Diptera	*Agoravinia rufiventris*						x	x		
	Archytas sp.			x						
	Chamaemyiidae sp1			x		x	x	x		
	Chamaemyiidae sp2			x			x			
	Cochliomyia macellaria (Fabricius)			x			x			
	Chrysomya albiceps (Wiedemann)			x		x				
	Chrysomya megacephala (Fabricius)			x						
	Cochliomyia hominivorax (Coquerel)			x						
	Eristalinus taeniops (Wiedemann)			x		x				
	Eristalis sp1			x				x		
	Eristalis sp2			x						
	Ocyptamus sp1						x			
	Ocyptamus sp2						x			
	Ornidia sp.							x		
	Sarcophagidae sp.						x	x		
	Tachinidae sp.			x						
	Tachina sp.							x		
	Ylasoia sp.									x
	Diptera sp1					x	x			
	Diptera sp2						x			
	Diptera sp3						x			
	Diptera sp4							x		
	Diptera sp5						x			
	Diptera sp6							x		
	Diptera sp7							x		
	Diptera sp8						x			
	Diptera sp9									x
	Diptera sp10									x
	Diptera sp11									x

(continued)

Table 2.7 (continued)

Flower visitors		Cc	Cl	Ca	Gb	Go	Mo	On	Pb	Tg
Hymenoptera	*Apis mellifera* L.		x		x	x	x		x	x
	Dialictus sp.						x	x		
	Euglossa cordata L.		x							
	Pseudaugochlora gramínea (F.)		x							
	Trigona spinipes (Fabricius, 1793)		x	x	x	x	x	x	x	x
	Xylocopa sp.				x					
	Vespidae sp1								x	
	Vespidae sp2						x		x	
	Vespidae sp3						x			
	Vespidae sp4									x
Lepidoptera	*Agraulis vanillae maculosa* (Stichel, 1907)			x						
	Iapsis talaya (Hewitson, 1868)			x						
	Pseudolycaena marsyas (Linnaeus, 1758)			x						x
	Timochares trifasciata trifasciata (Hewitson, 1868)							x		
	Urbanus proteus proteus L.				x					
	Lepidoptera sp1					x				
	Lepidoptera sp2							x		
	Lepidoptera sp3									x

Cc Clusia criuva, Cl Clusia lanceolata, Ca Coccoloba arborescens, Gb Garcinia brasiliensis, Go Guapira opposita, Mo Monteverdia obtusifolia, On Ocotea notata, Pb Protium brasiliense, Tg Tapirira guianensis

anemophilous pollination. Fruit production from anemophily experiments and observation of floral visitors in both morphs of *G. opposita* confirmed ambophily for this species. Although no observations were made during the night, there is a record of pollination of *C. criuva* flowers by the nocturnal beetle *Dinaltica bahiaensis* (Jacoby, 1902) (Correia et al. 1993). The other species were visited by diurnal insects (Table 2.7). *Trigona spinipes* (Fabr., 1775) was the most common floral visitor to all the species, except that it was not present in flowers of *C. criuva* (species with nocturnal anthesis); this stingless bee was the only insect observed in the staminate flowers of *M. parvifolia*, collecting pollen (Table 2.7).

2.4 Discussion

Full dioecy was the predominant dimorphic sexual system in these ten species of the Maricá *restinga*. We also found a case of cryptic dioecy and one of gynodioecy. Most plants classified as dioecious had strictly unisexual individuals in this population and did not show variations in their sexuality during the reproductive events monitored for 4 years. We observed rare events of fruit formation in male plants, one each of *Coccoloba arborescens*, *Guapira opposita*, and *Garcinia brasiliensis*. Recently, Crossard et al. (2021) experimentally demonstrated the dissolution of dioecy in *Mercurialis annua* L. (Euphorbiaceae), an annual herb, through the production of staminate flowers in isolated female individuals, after three generations. This experiment validated a mechanism for rapid reversal to functional hermaphroditism during periods of low population density, repeated colonization, or range expansion (Crossard et al. 2021). Therefore, the sporadic events of sexual plasticity such as those observed in trees, e.g., *C. arborescens*, *G. opposita*, and *G. brasiliensis* (Leal et al. 2013), should not be disregarded, as they may reflect possible transitions in the sexual systems and mating behavior of plants.

Cryptic dioecy (Mayer and Charlesworth 1991) characterized the population of *G. brasiliensis* in the Maricá *restinga*, due to the occurrence of male plants as well as plants with morphologically perfect flowers that are in fact pistillate, given the absence of pollen (Silva 2005; Leal et al. 2012, 2013). A hypothesis to explain the maintenance of developed but sterile stamens in pistillate flowers of cryptic dioecious species involves selective pressure to maintain the similarity of floral morphs, assuring recognition by pollinators (Mayer and Charlesworth 1991; Cane 1993; Davis 1997), and genetic constraints for reducing the size of stamens in pistillate flowers without affecting the stamens of staminate flowers (Davis 1997), or the species is undergoing an evolutionary process toward full dioecy (Mayer and Charlesworth 1991). The last two hypotheses are more likely, since the flowers of *G. brasiliensis* are nectariferous, and pistillate flowers with stamens experimentally removed and submitted to natural pollination showed high fruit set (Silva 2005).

Monteverdia obtusifolia was the only case of gynodioecy (i.e., female and hermaphroditic plants in the population) recorded. Female plants produced only pistillate flowers in all flowering events, giving them a large advantage in fruiting percentage over the hermaphroditic plants (Benevides et al. 2013). According to Benevides (2011) and Benevides et al. (2013), hermaphroditic plants of *M. obtusifolia* showed continuous variation in certain floral and reproductive traits (e.g., sizes of the style and stigmatic surface, pollen viability, and fruit set), leading to a continuous gender variation among them (in some plants the female functions are more efficient and in others the male functions are more efficient) and hence to different levels of fruit set. The high frequency of dioecy in tropical areas encompasses diverse environments such as forests (Bawa 1980; Bawa et al. 1985; Bullock 1985; Silva et al. 1997; Vamosi 2006; Queenborough et al. 2009; Vary et al. 2011), the Cerrado (Oliveira and Gibbs 2000), and *restingas* (Ormond et al. 1991; Matallana et al. 2005); but see Machado et al. (2006) for the Caatinga. Although the effects of environmental variation on the frequency of different sexual systems (dioecy, monoecy, and hermaphroditism), especially the effects of water and resource availability, are still under discussion (Wang et al. 2019), some studies have indicated that the levels of oligotrophy may influence gender specialization (Delph and Wolf 2005; Ashman 2006). Differences related to the availability of water and nutrients in the forest (moist, nutrient-rich soil) and in the *restinga* (dry, nutrient-poor soil) in Brazil do not seem to have influenced the dioecy rates of plants in these contrasting environments, which generally range between 10% and 14% (Ormond et al. 1991; Silva et al. 1997; Matallana et al. 2005), but may be related to the gender specialization of dimorphic sexual system. Our results seem to support the hypothesis that gender specialization is favored when resources are limited or the environment is harsh, since nine of the ten *restinga* species showed full dioecy. This idea is reinforced by the fact that some species that are dioecious in the *restinga* can exhibit monomorphic sexual systems in forest environments; for example, *Tapirira guianensis* sporadically bear perfect flowers in female plants in a gallery forest, Minas Gerais (Lenza and Oliveira 2005), and hermaphroditic (perfect flowers), andromonoecious (perfect and staminate flowers) and male (staminate flowers) plants exist in populations of the Atlantic Forest, Espírito Santo (Santana et al. 2009). Cavalcante (1996) recorded that plants of *Garcinia brasiliensis* in the Amazon Forest have a monoecious sexual system (staminate and pistillate flowers). Study of the gender specialization of dioecious *restinga* species with distributions encompassing other contrasting environments, in relation to the availability of water and nutrients, would be quite enlightening.

Most dioecious species in the Maricá *restinga* depend on the visits by insects for fruit production. *Guapira opposita* and *M. parvifolia* were the only species that produced fruits in inflorescences isolated from access by insect pollinators, but not from pollen transported by wind. In *G. opposita* the pollen is transferred primarily by insects, with wind playing a supplemental role, whereas *M. parvifolia* depends exclusively on wind to set fruits. Therefore, three pollination systems were recorded among the dioecious species in this *restinga* vegetation: ambophily, anemophily, and entomophily.

Guapira opposita has some floral traits associated with the anemophilous pollination syndrome (Faegri and Van Der Pijl 1979; Culley et al. 2002): the staminate flowers have well-exposed stamens, with small, dry, slightly sculpted pollen grains (Souza et al. 2010) released in a powdery cloud (personal observation); the pistillate flowers are uniovulate, with a well-branched exposed stigma. Nevertheless, the flowers release odor and offer nectar to many visitors. The floral morphology, the insect visits to the two sexual morphs, and the results of bag experiments to test anemophily are consistent in suggesting that *G. opposita* is an ambophilic species in the *restinga*. The hypothesis of facultative anemophilous pollination for *Guapira* species, based solely on floral characters, was proposed by Bullock (1994) in Central American forests and by Oliveira (1996) for two species in the Cerrado. We suggest that the pollination of *Guapira* species should be investigated, including the hypothesis of ambophily and assessing possible variations in temporal or spatial importance of wind and insects for mixed-pollination species.

On the other hand, flowers of *M. parvifolia* are open, dish-shaped, odorless, and devoid of nectar. Floral traits such as small dry pollen grains released in a powdery cloud and the sessile stigma well exposed, with a large receptive area, the presence of pollen grains dispersed in the air, and the absence of visits by insects to the female flowers confirmed a strictly anemophilous system for the species (Albuquerque et al. 2013). Anemophily is recognized as a derivative condition among flowering plants, arising independently in numerous families from ancestors pollinated by animals (Ackerman 2000). Phylogenetic analysis indicates that wind pollination evolved more frequently in clades with characteristics compatible with wind pollination (Friedman and Barrett 2008, 2009; Renner 2014). At least in *Myrsine*, the absence of nectar, fruits with one seed, conspicuous stigmas, and free anthers were already present or evolved along with dioecy, providing new evolutionary possibilities in the genus, including anemophily (Otegui and Cocucci 1999).

Regarding entomophilous pollination systems, the dioecious *restinga* species can be grouped in two categories, according to the degree of pollination specialization. In the specialist entomophilous pollination group, species with this pollination system have inflorescences with few, large, and showy flowers, with multiovulate ovaries, and fruits with many seeds, including the three species of Clusiaceae. Specialist pollination in Clusiaceae is related mainly to the supply of restricted collection resources, such as resin, in *C. lanceolata*, and pollen made available to the pollinator at night, in *C. criuva* (Correia et al. 1993, 1999). In the generalist entomophilous pollination group, species with this pollination system have inflorescences with many congested flowers; the flowers are inconspicuous, small, white to greenish yellow, dish-shaped, nectariferous, mildly fragrant, and uniovulate or with few ovules. These floral attributes are the same as described for some dioecious species from tropical rainforests that are pollinated by "small insects," mainly small bees (Bawa and Olper 1975; Bawa 1980), which is in agreement with the floral visitors observed here.

The two entomophilous pollination groups described above differed greatly in terms of female reproductive success. The group with specialist pollination showed fruit set between 39% and 85%, while the group with generalist pollination showed relatively lower levels, ranging from 5.6% to 41%. The high fruit/flower ratio obtained for the species in the first group is in agreement with the results obtained by Sutherland and Delph (1984) for dioecious species in general (73.8% on average) and supports the hypothesis that dioecy increases the ability to independently adjust investment in male and female functions, maximizing the success of both (Wilson 1983; Sutherland and Delph 1984). This hypothesis, however, does not apply to our results for the group of dioecious plants with generalist pollination, where the reproductive success was very low, including the entomophilous and ambophilous systems. In this group, fruiting percentages were very similar, particularly in *C. arborescens* (17%), *G. opposita* (19.5%), and *P. brasiliense* (16%). *Ocotea notata* had the lowest fruiting capacity (5.6%), whereas *T. guianensis* and *M. obtusifolia* had the highest (28 and 41%, respectively). In general, low reproductive success can be interpreted as a result of pollen limitation (Knight et al. 2005). Factors commonly associated with pollen limitation are low frequency of pollinators, insufficient deposition of pollen on the stigma, or inappropriate pollen deposition (Ashman et al. 2004). Although species with generalist pollination are visited by many insects, the large investment in flower number observed in these species probably results in a low frequency of visits per flower. Several investigators (Knight et al. 2005; Harder and Aizen 2010) have considered that dioecy is a sexual system susceptible to pollen limitation, as the obligate cross-pollination between the two different sexual morphs limits the chances of successful pollination.

On the other hand, in general, the flowers of the generalist entomophilous species lasted longer than in the specialist entomophilous species. Pistillate flowers of *M. obtusifolia* and *T. guianensis* had an anthesis period of 6 and 8 days, respectively, and also showed the highest reproductive success of the generalist entomophilous group. Long periods of anthesis in this group may be associated mainly with an increased chance of pollen arrival, minimizing pollen limitation. According to Knight et al. (2005), flower longevity is negatively related to the degree of pollen limitation, and our results reinforced this hypothesis. Longer floral longevity has also been interpreted as a possible adaptive solution for species with inconspicuous flowers or with a smaller floral display compared to other simultaneously flowering species, to increase reproductive success when pollen limitation is due to a low frequency of visits (Ashman et al. 2004).

2.5 Concluding Remarks

The plant community in the Maricá *restinga* includes 15 species with sexual dimorphism, of which the 10 studied here (67%) are important, with a frequency of 36.03% in the upper stratum and a VI (importance value) of 73.63% (Rodarte 2008). Our results confirm most of the expected patterns for dioecy in tropical vegetation that are extensively reported in the literature, such as woodiness, predominance of fleshy fruits, and pollination systems involving wind and insects. This small group of species shows diverse reproductive traits, including floral attributes (attractive and inconspicuous flowers), floral resources (resin, pollen, and nectar), pollination systems (entomophily, ambophily, and anemophily), and fruit set (5.6–85%). This study showed for the first time that entomophily in dioecious *restinga* species is not associated exclusively with the generalist entomophilous system. Some of the dioecious species that we studied have a specialist entomophilous system involving two different groups of pollinators, bees and beetles. However, a more thorough analysis of the effectiveness and frequency of floral visitors in the pollination of species classified as generalist entomophilous may reveal the existence of other pollination systems with different degrees of generalization.

Acknowledgments We thank Dr. Janet W. Reid (JWR Associates) for the revision of the English text. We also express our gratitude to Dr. Cátia Antunes de Mello Patiu, Dr. Márcia Souto Couri, Dr. Favizia Freitas de Oliveira, and MSc. Alexandre Soares for the taxonomic identification of insects and to Sérgio Nascimento for photographs a and b in Figures 2, 3, 5, and 6. This study is part of the Ph.D. thesis of Cristine Rodrigues Benevides, which was carried out in the Postgraduate Program in Biological Sciences (Botany) of the Museu Nacional, Universidade Federal do Rio de Janeiro (UFRJ), Rio de Janeiro, Brazil, and was supported by CAPES (Coordenação de Aperfeiçoamento de Pessoal de Nível Superior).

References

Ackerman JD (2000) Abiotic pollen and pollination: ecology, functional and evolutionary perspectives. Plant Syst Evol 222:167–185

Albuquerque AA, Lima HA, Gonçalves-Esteves V et al (2013) *Myrsine parvifolia* (Primulaceae) in sandy coastal plains marginal to Atlantic rainforest: a case of pollination by wind or by both wind and insects? Braz J Bot 36:65–73

Alexander MP (1980) A versatile stain for pollen, fungi, yeast and bacteria. Stain Technol 55:13–18

Ashman TL (2006) The evolution of separate sexes: a focus on the ecological context. In: Harder LD, Barrett SCH (eds) Ecology and evolution of flowers. Oxford University Press, Oxford, pp 204–222

Ashman TL, Knight TM, Steets JA et al (2004) Pollen limitation of plant reproduction: ecological and evolutionary causes and consequences. Ecology 85:2408–2421

Barrett SCH (1998) The evolution of mating strategies in flowering plants. Trends Plant Sci 3:335–341

Barrett SCH (2002) The evolution of plant sexual diversity. Nat Rev Genet 3:274–284

Barrett SCH (2010) Understanding plant reproductive diversity. Phylos Trans R Soc B Biol Sci 365:99–109

Barrett SCH (2021) Plant sex: best to be bisexual when mates are scarce. Curr Biol 31:282–309

Bawa KS (1980) Evolution of dioecy in flowering plants. Annu Rev Ecol Syst 11:15–39

Bawa KS, Opler PA (1975) Dioecism in tropical forest trees. Evolution 29:167–174

Bawa KS, Perry DR, Beach JH (1985) Reproductive biology of tropical lowland rain forest trees. I. Sexual systems and incompatibility mechanisms. Am J Bot 72:331–345

Benevides CR (2011) Biologia floral e reprodutiva de espécies dioicas na restinga de Maricá, RJ. Thesis, Universidade Federal do Rio de Janeiro/Museu Nacional

Benevides CR, Haddad IVN, Barreira NP et al (2013) *Maytenus obtusifolia* Mart. (Celastraceae): a tropical woody species in a transitional evolutionary stage of the gynodioecy– dioecy pathway. Plant Syst Evol 299:1693–1707

Benevides CR, Rodarte ATA, Lima HA (2015a) Strategies of flowering in dioecious plants in tropical coastal vegetation, Brazil. Braz J Bot 38:585–595

Benevides CR, Rodarte ATA, Lima HA (2015b) Vegetative propagation as a successful reproductive strategy in woody dioecious species in tropical coastal vegetation, Southeast Brazil. Braz J Bot 38:579–584

Bullock SH (1985) Breeding systems in the flora of a tropical deciduous forest in Mexico. Biotropica 17:287–301

Bullock SH (1994) Wind pollination of Neotropical Dioecious Trees. Biotropica 26:172–179

Cane JH (1993) Reproductive role of sterile pollen in cryptically dioecious species of flowering plants. Curr Sci 65:223–225

Cavalcante PB (1996) Frutas comestíveis da Amazônia. Coleção Adolpho Ducke, 6th edn. CNPq/ Museu Paraense Emílio Goeldi, Belém

Cesário LF (2007) Recompensas florais e visitantes de duas espécies simpátricas de *Clusia hilariana* Schltd. e *Clusia spirito-sanctensis* Mariz & Weinberg (Clusiaceae), em área de restinga. Dissertation, Universidade Estadual do Norte Fluminense

Cesário LF, Gaglianone MC (2008) Biologia floral e fenologia reprodutiva de *Schinus terebinthifolius* Raddi (Anacardiaceae) em Restinga do Norte Fluminense. Acta Bot Bras 22:828–833

Correia MCR, Ormond WT, Pinheiro MCB et al (1993) Estudo da biologia floral de *Clusia criuva Camb.* – Um caso de mimetismo. Bradea 24:209–219

Correia MCR, Ormond WT, Pinheiro MCB et al (1999) Biologia da reprodução de *Clusia lanceolata* Camb. Hoehnea 26:61–73

Crossard GG, Gerchen JF, Li X et al (2021) The rapid dissolution of dioecy by experimental evolution. Curr Biol 3:1277–1283

Crossman A, Charlesworth D (2013) Breakdown of dioecy: models where males acquire cosexual functions. Evolution 62:426–440

Culley TM, Weller SG, Sakai AK (2002) The evolution of wind pollination in Angiosperms. Trends Ecol Evol 17:361–369
Dafni A, Keavan PG, Husband BC (2005) Practical pollination biology. Enviroquest, Cambridge
Davis SL (1997) Stamens are not essential as an attractant for pollinators in female of cryptically dioecious *Thalictrum pubescens* Pursch. (Ranunculaceae). Sex Plant Reprod 10:93–299
Delph LF, Wolf DE (2005) Evolutionary consequences of gender plasticity in genetically dimorphic breeding systems. New Phytol 166:119–128
Faegri K, Van Der Pijl L (1979) The principles of pollination ecology, 3rd edn. Pergamon Press, Oxford
Faria APG, Matallana G, Wendt T et al (2006) Low fruit set in the abundant dioecious tree *Clusia hilariana* (Clusiaceae) in a Brazilian restinga. Flora 201:606–611
Flores S, Schemske DW (1984) Dioecy and monecy in the flora of Puerto Rico and the Virgin Islands: ecological correlates. Biotropica 16:132–139
Friedman J, Barrett SCH (2008) A phylogenetic analysis of the evolution of wind pollination in angiosperms. Int J Plant Sci 169:49–58
Friedman J, Barrett SCH (2009) Wind of change: new insights on the ecology and evolution of pollination and mating in wind-pollinated plants. Ann Bot 103:1515–1527
Gonçalves IP, Gama MC, Correia MCR et al (2008) Caracterização dos frutos, sementes e germinação de quatro espécies de leguminosas da Restinga de Maricá, Rio de Janeiro. Rodriguésia 59:497–512
Harder LD, Aizen MA (2010) Floral adaptation and diversification under pollen limitation. Phylos Trans R Soc B Biol Sci 365:529–543
Heslop-Harrison Y, Shivanna KR (1977) The receptive surface of the angiosperm stigma. Ann Bot 41:1233–1258
Käfer J, GAB M, Panel JR (2017) On the rarity of dioecy in flowering plants. Mol Ecol 26:1225–1241
Kearns CA, Inouye D (1993) Techniques for pollination biologists University Press of Colorado, Colorado
Knight TM, Steets JA, Vamosi JC et al (2005) Pollen limitation of plant reproduction: pattern and process. Annu Rev Ecol Evol Syst 36:467–497
Köppen W (1948) Climatología: con un estudio de los climas de la tierra. Fondo de Cultura Económica, México, D.F
Leal DO, Malafaia C, Cesar R et al (2012) Floral structure of *Garcinia brasiliensis* in relation to flower biology and evolution. Int J Plant Sci 173:172–183
Leal DO, Benevides CR, Silva RCP et al (2013) *Garcinia brasiliensis*: insights into reproductive phenology and sexual system in a Neotropical environment. Plant Syst Evol 299:1577–1585
Lenza E, Oliveira PE (2005) Biologia reprodutiva de *Tapirira guianensis* Aubl. (Anacardiaceae), uma espécie dioica em mata de galeria do Triângulo Mineiro, Brasil. Rev Bras Bot 28:179–190
Lenzi M, Orth AI (2004) Fenologia reprodutiva, morfologia e biologia floral de *Schinus terebinthifolius* Raddi (Anacardiaceae), em restinga da Ilha de Santa Catarina, Brasil. Biotemas 17:67–89
Lloyd DG (1979) Parental strategies of angiosperms. N Z J Bot 17:595–606
Lopes AV, Machado IC (1998) Floral biology and reproductive ecology of *Clusia nemorosa* (*Clusiaceae*) in northeastern Brazil. Plant Syst Evol 213:71–90
Machado IC, Lopes AV, Sazima M (2006) Plant sexual systems and a review of the breeding system studies in the Caatinga, a Brazilian tropical dry forest. Ann Bot 97(2):277–287
Mantovani A, Iglesias RR (2001) Bromélias terrestres na restinga de Barra de Maricá, Rio de Janeiro: influência sobre o microclima, o solo e a estocagem de nutrientes em ambientes de borda de moitas. Leandra 16:17–37
Matallana G, Wendt T, Araújo DSD et al (2005) High abundance of dioecious plants in a tropical coastal vegetation. Am J Bot 92:1513–1519
Mayer SS, Charlesworth D (1991) Cryptic dioecy in flowering plants. Tree 6:320–325
Moreira MM, Freitas L (2000) Review of the pollination system by small diverse insects. Neotrop Entomol 49:472–481
Oliveira PE (1996) Dioecy in the Cerrado vegetation of Central Brazil. Flora 191:235–243

Oliveira PE, Gibbs PE (2000) Reproductive biology of woody plants in a Cerrado community of Central Brazil. Flora 195(4):311–329

Ormond WT, Pinheiro MCB, Lima HÁ et al (1991) Sexualidade das plantas da restinga de Maricá, RJ. Bol Mus Nac (N S Botânica) 87:1–24

Otegui M, Cocucci A (1999) Flower morphology and biology of *Myrsine laetevirens*, structural and evolutionary implications of anemophily in Myrsinaceae. Nord J Bot 19:71–85

Pannel JR (2000) A hypothesis for the evolution of androdioecy: the joint influence of reproductive assurance and local mate competition in a metapopulation. Evol Ecol 14:195–211

Perrin P (1984) Evolução da costa fluminense entre as Pontas de Itacoatiara e Negra: Preenchimento e restinga. In: Lacerda LD, Araújo DSD, Cerqueira R, Turcq B (eds) Restingas: origem, estrutura, processos. Universidade Federal Fluminense/CEUFF, Niterói, pp 65–74

Queenborough SA, Mazer SJ, Vamosi SM et al (2009) Seed mass, abundance, and breeding system among tropical forest species: do dioecious species exhibit compensatory reproduction or abundances? J Ecol 97:555–566

Renner S (2014) The relative and absolute frequencies of Angiosperm sexual systems: Dioecy, monoecy, gynodioecy, and an updated online database. Am J Bot 101:1588–1596

Renner SS, Ricklefs RE (1995) Dioecy and correlates in the flowering plants. Am J Bot 82:596–606

Rizzini CT (1979) Tratado de fitogeografia do Brasil. HUICITEC/EDUSP, São Paulo

Rodarte ATA (2008) Caracterização espacial, temporal e biologia floral das espécies de restinga, com ênfase nos recursos florais. Ph.D. Thesis, Universidade Federal do Rio de Janeiro/Museu Nacional

Santana WMS, Silva-Mann R, Ferreira RA et al (2009) Morfologia de flores, frutos e sementes de pau-pombo (*Tapirira guianensis* Aubl. – Anacardiaceae) na região de São Cristóvão, SE, Brasil. Sci For 37:47–54

Scarano FR (2002) Structure, function, and floristic relationships of plant communities in stressful habitats marginal to the Brazilian Atlantic rain forest. Ann Bot 90:517–524

Silva RCP (2005) Estudos reprodutivos em *Garcinia brasiliensis* Mart. na restinga de Maricá, RJ. Dissertation, Universidade Federal do Rio de Janeiro/Museu Nacional

Silva AG, Guedes-Bruni RR, Lima MPM (1997) Sistemas sexuais e recursos florais do componente arbustivo-arbóreo em mata preservada na reserva ecológica de Macaé de Cima. In: Lima HC, Guedes-Bruni RR (eds) Serra de Macaé de Cima: Diversidade Florística e Conservação em Mata Atlântica. Jardim Botânico, Rio de Janeiro, pp 187–211

Silveira M (1998) Preparo de amostras biológicas para microscopia eletrônica de varredura. In: de Souza W (ed) Técnicas Básicas de Microscopia Eletrônica Aplicada às Ciências Biológicas. Sociedade Brasileira de Microscopia, Rio de Janeiro, pp 33–34

Sinclair JPS, Korte JL, Freeman DC (2013) The pattern of Dioecy in terrestrial, temperate plant succession. Evol Ecol Res 15:545–556

Souza MA, Mendonça CBF, Esteves VG (2010) Palinologia de espécies de Nyctaginaceae Juss. ocorrentes nas restingas do estado do Rio de Janeiro, Brasil. Acta Bot Bras 24:104–110

Spigler RB, Ashman TL (2012) Gynodioecy to dioecy: are we there yet? Ann Bot 109:531–543

StatSoft, Inc. (2005) STATISTICA Data Analysis Software System, Version 7.1. http://www.statsoft.com

Sutherland S, Delph LF (1984) On the importance of male fitness in plants: patterns of fruit-set. Ecology 65:1093–1104

Vamosi SM (2006) A reconsideration of the reproductive biology of the Atlantic Forest in the Volta Velha Reserve. Biodivers Conserv 15:1417–1424

Vamosi JC, Otto SP (2002) When looks can kill: the evolution of sexually dimorphic floral display and the extinction of dioecious plants. Proc Biol Sci 269:1187–1194

Vamosi JC, Otto SP, Barrett SCH (2003) Phylogenetic analysis of the ecological correlates of dioecy. J Evol Biol 16:1006–1018

Vary LB, Gillen DL, Randrianjanahary M et al (2011) Dioecy, monoecy, and their ecological correlates in the Littoral Forest of Madagascar. Biotropica 43:1–9

Wang Y, Lyu T, Shrestha LN et al (2019) Drivers of large-scale geographical variation in sexual systems of woody plants. Glob Ecol Biogeogr 29:546–557

Willson MF (1983) Sexual selection in plants. Am Nat 113:777–790

Chapter 3
Male and Female Sterility in Flowering Plants

Isabella Veríssimo Nader Haddad, Bárbara de Sá-Haiad, and Lygia Dolores Ribeiro de Santiago-Fernandes

Abstract Most flowering plants have perfect flowers with an androecium and gynoecium that produce microgametophytes and megagametophytes, respectively. Failure in the development of the reproductive organ, functional gametophytes, or viable zygotes is known as sterility and is the main mechanism used by plants to produce unisexual flowers. When sterility is due to the failure to produce functional anthers or microgametophytes, it is termed male sterility, whereas a failure to produce a functional gynoecium or megagametophytes is termed female sterility. The phenotypic manifestations of male and female sterility are diverse. Programmed cell death is essential in several processes related to the development of fertile gametophytes, and this program is commonly recruited to interrupt the development of the reproductive organ or gametophyte, causing sterility and sex determination in unisexual flowers. In this chapter, we provide literature data associating the structure and function during gametophyte development, as well as the cellular processes that cause male and female sterility and their significance in reproductive development, focusing on natural populations.

Keywords Autophagy · Flower · Necrosis · Programmed cell death

3.1 Floral Organogenesis

The organs of flowering plants develop from groups of meristematic cells termed meristems. The identity of a meristem is inferred from the structures that it produces: the root apical meristem (RAM) originates roots, while shoot apical meristems (SAM) produce aboveground organs. The pluripotent activity of stem cells allows meristems to generate new meristems with a different identity (Coen and

I. V. N. Haddad (✉) · B. de Sá-Haiad · L. D. R. de Santiago-Fernandes
Departamento de Botânica, Laboratório de Anatomia Vegetal, Museu Nacional, Universidade Federal do Rio de Janeiro (MN/UFRJ), Rio de Janeiro, RJ, Brazil
e-mail: isabellahaddad@gmail.com

M. F. T. Medeiros, B. de Sá Haiad (eds.), *Aspects of Brazilian Floristic Diversity*, https://doi.org/10.1007/978-3-031-07453-0_3

Meyerowitz 1991). For example, during the flowering period, SAM is transformed into inflorescence meristems (IM), which go on to produce the floral meristems (FM), which produce the sepals, petals, stamens, and carpels.

Commonly flowers are organized into four concentric whorls of organs: sepals, petals, stamens, and carpels (Coen and Meyerowitz 1991). However, in some species, one or more floral organs can be absent (Diggle et al. 2011). Floral organogenesis requires floral organ identity specification, growth, and differentiation (Wellmer et al. 2014).

Molecular studies in *Arabidopsis thaliana* L. and *Antirrhinum majus* L. provided a model that explains the molecular basis for the establishment of floral organ identities and their pattern of expression in the FM. In the ABCE model, the expression of E-class genes [*SEPALLATA1* (*SEP1*), *SEPALLATA2* (*SEP2*), *SEPALLATA3* (*SEP3*), and *SEPALLATA4* (*SEP4*)] is required for the development of sepals, petals, stamens, and carpels. The expressions of E-class genes combined with A-class genes [*APETALA1* (*AP1*) and *APETALA2* (*AP2*)] control sepal identity; E-class genes combined with A-class and B-class genes [*APETALA3* (*AP3*) and *PISTILLATA* (*PI*)] control petal identity; E-class genes combined with B-class and C-class genes [*AGAMOUS* (*AG*)] control stamen identity; and E-class genes combined with C-class genes control carpel identity (Wellmer et al. 2014).

3.1.1 Unisexual Flowers

Floral unisexuality evolved independently multiple times during the evolution of different lineages, through different mechanisms, from a hermaphroditic ancestral state (Chakrabarty 2017). Two categories of unisexual flowers are recognized. Type I flowers are morphologically bisexual at initiation and become unisexual by disruption of the development of one reproductive organ, the androecium or gynoecium. Depending on the developmental stage when the abortion occurs, type I flowers can have a vestigial reproductive organ or a nonfunctional reproductive organ, with size and morphology indistinguishable from a functional organ. Unisexuality in type II flowers occurs before the developmental initiation of one of the reproductive organs and results in the absence of this organ or vestiges of it in the flowers (Diggle et al. 2011).

From an evolutionary point of view, the complete suppression of a reproductive organ may involve a two-step process: loss of functionality or abortion and loss of organ initiation (Mitchell and Diggle 2005). Type II flowers can be derived from type I lineages, through positive selection for disruption in the earlier developmental stages. This early developmental disruption allows reallocation of resources that were not used to develop the nonfunctional reproductive organ, increasing the fertility of the functional reproductive organ. However, most unisexual flowers are type I, and the maintenance of the nonfunctional reproductive organ can be attributed to pollinator attraction and/or reward. In animal-pollinated species, staminate and pistillate flowers must retain a sufficient resemblance to each other to attract the same pollinators, and if pollen is the only reward for pollinators, pistillate flowers might retain nonfunctional pollen (Diggle et al. 2011).

Programmed cell death is essential in reproductive development. This programming appears to be commonly recruited to cause sterility in unisexual flowers, through interruption in the development of the nonfunctional organs or gametophytes (Diggle et al. 2011).

3.2 Programmed Cell Death

The cells of multicellular organisms such as plants and animals self-destruct when they are unwanted or unnecessary or have exhausted their functions, through activation of the genetically controlled cellular suicide machinery termed programmed cell death (PCD) (Papini et al. 1999). PCD triggers a sequence of cellular events that causes controlled and organized cell degradation (Lockshin and Zakeri 2004).

In contrast to animals, little is known about the biochemistry underlying plant PCD, and therefore the types of plant PCD are recognized from the cellular morphology (van Doorn 2011). Based on morphology, van Doorn et al. (2011) recognized two types of PCD in plants: "vacuolar cell death" or autophagy and necrosis. Formerly, "vacuolar cell death" was referred to as a "like-autophagy" process. After the identification of the autophagy genes (*atg*), the occurrence of autophagy was recognized in plants, as previously described in yeasts and animals (van Doorn and Papini 2013).

Autophagy is the most common type of PCD in plants and is a highly ordered process that requires a high cellular energy status to prevent a necrotic collapse (Lam 2005). Hydrolases are enzymes that have an essential role in autophagy. Lytic vacuoles, unlike storage vacuoles that store secondary metabolites and proteins, serve as a reservoir for hydrolases that act on the cellular contents, recycling them (Paris et al. 1996).

The autophagic pathways are microautophagy, macroautophagy, massive autophagy, and partial degradation of cytoplasm organelles (van Doorn and Woltering 2005). Microautophagy is the sequestration of small portions of the cytoplasm, excluding large organelles, by invagination or evagination of the tonoplast followed by their degradation by hydrolases inside the vacuole (van Doorn and Papini 2013). Macroautophagy begins with the formation of tubules that originate from the endoplasmic reticulum (ER) or dictyosomes and are termed provacuoles. Provacuoles can merge, thereby sequestering a portion of the cytoplasm, and form a double-membrane structure termed an autophagosome. Both structures, provacuoles and autophagosomes, may occur simultaneously and be arranged concentrically as a macroautophagic structure. Hydrolases can be present inside the provacuoles or between the two membranes of the autophagosomes (Bassham et al. 2006; van Doorn and Papini 2013). The hydrolases degrade the cytoplasm and the inner membrane of autophagosomes, originating vacuoles. The sequestered cytoplasm can be degraded before the inner membrane, which suggests that the hydrolases are actively transported through the inner membrane (van Doorn et al. 2015). Finally, the vacuole can merge with other vacuoles (van Doorn and Papini 2013). Massive autophagy involves the permeabilization or rupture of the tonoplast, which causes cellular degradation through the release of hydrolases from the vacuoles (van Doorn and Woltering 2005). Other cytological markers related to autophagy are the formation of actin cables, nuclear envelope breakup, and, in some cases, nuclear segmentation (van Doorn et al. 2011).

On the other hand, necrosis has been described as a type of uncontrolled PCD that is generally a response to cellular stress (Reape et al. 2008). Necrosis is characterized by early rupture of the plasma membrane, leading to protoplast shrinkage, mitochondrial swelling, and absence of autophagic characteristics (van Doorn et al. 2011).

Programmed cell death has an essential role in several reproductive developmental processes and in different stages of these processes.

3.3 Androecium and Microgametophyte Development

The development of the androecium, particularly anther tissue differentiation and anther dehiscence, was thoroughly described by Goldberg et al. (1993). The differentiated anthers are composed of several specialized nonreproductive tissues, which include the epidermis, endothecium, middle-wall layers, tapetum, and vascular bundle (Goldberg et al. 1993). Considering the importance of the tapetum during microsporogenesis and its role in microgametophyte development and male sterility determination, this will be addressed in detail.

3.3.1 Tapetum

Two tapetum types are recognized: secretory, also termed parietal or glandular, and amoeboid, also termed periplasmodial, invasive, or intrusive. In the amoeboid type, degeneration of the cell walls at an early developmental stage leads to loss of cellular individuality: the cells become arranged in the anther locule, completely surrounding the developing microgametophytes. These cells can remain separate or become fused, originating a syncytium. In the secretory type, which is more common, the cells form a tissue that delimits the anther locule, where the microgametophytes develop. Although tapetum types are structurally different, both have the same functions (Pacini 2010).

Tapetum is an ephemeral tissue with several essential roles in the microgametophyte development, including (1) nourishment, providing proteins, enzymes, hormones, growth substances, water, and other compounds, synthesized from substances "produced by photosynthesis of epidermis, endothecium, and middle layer or conveyed by the vascular bundle of the anther" and released by exocytosis or membrane permeability (Pacini 2010); (2) production and timely release of callase (β-1,3-glucanase), which degrades the callose wall of the microspore tetrads and releases the free microspores (Lu et al. 2014); (3) exine deposition through the release of sporopollenin precursors and compatibility substances that are deposited on the exine (the orbicules, also termed Ubisch bodies, occur only in the secretory tapetum; they are sporopollenin bodies that emerge as pro-orbicules from vesicles derived from the ER. Orbicules are released through exocytosis and polymerized onto the primexine, a microfibrillar polysaccharide matrix, formed by the microspores, that serves as a pattern for the final sculptured pollen wall, the exine (Echlin and Godwin 1968; Gómez et al. 2015)); and (4) release of pollenkitt and tryphine, debris originated by tapetum cell PCD, which becomes embedded on or beneath the exine surface prior to anther dehiscence (Vizcay-Barrena and Wilson 2006). Pollenkitt, a heterogeneous material composed of neutral lipids, flavonoids, carotenoids, proteins, and polysaccharides, is usually formed by elaioplasts differentiated from proplastids and spherosomes derived from the ER (Pacini and Hesse 2005). Tryphine is a mixture of hydrophobic and hydrophilic substances and often contains degenerated cytoplasmic organelles (Pacini et al. 1985). Several functions are associated with pollenkitt, especially protection of the microgametophytes from water loss and UV radiation after anther dehiscence, and enabling adhesion to the bodies of pollinators (Pacini and Hesse 2005). Also, pollenkitt and tryphine facilitate pollen grains adhesion to the stigma, conspecific pollen grains recognition, and rehydration (Gómez et al. 2015). Pollenkitt is reported in almost all flowering plants pollinated by animals, whereas tryphine is restricted to only certain families (Pacini and Hesse 2005).

The tapetum PCD at the microspore tetrad stage is required for proper microgametophyte development (Kawanabe et al. 2006), since premature or late PCD of the tapetum leads to interruption of the microgametophyte development (Kaul 1988). The cytological changes during the tapetum degradation were described in different plant species as a progressive disintegration of the cellular organization, in which there is thinning of the cell wall, decrease in the cellular volume, cytoplasmic polarization, vacuolization (Wu and Cheung 2000), expansion of ER cisterns leading to cytoplasm confinement, chromatin condensation at the nucleus periphery (Papini et al. 1999), and DNA fragmentation (Shi et al. 2009). These processes are usually associated with autophagy, as described in perfect flowers of *Monteverdia obtusifolia* (Mart.) Biral, in which tapetum PCD occurs at the free microspore stage through macroautophagy and massive autophagy (referred as *Maytenus obtusifolia* Mart. *in* Haddad et al. 2019a).

3.3.2 Microsporogenesis

As anther wall tissues differentiate, the archesporial tissue originates the tapetum and the sporogenic tissue, composed of cells that are generally interconnected with each other and connected with tapetum cells by plasmodesmata, which enable the transfer of nutrients from the tapetum to sporogenic cells and among the sporogenic cells. Subsequently, the sporogenic cells deposit the callose wall and differentiate the microsporocytes. The microsporocytes undergo meiosis, producing microspore tetrads in a process termed microsporogenesis (Owen and Makaroff 1995).

The callose wall separates the microsporocytes from the anther tissues, providing mechanical and chemical isolation during meiosis, which is necessary for the transition from the sporophytic to the gametophytic phase, proper gametophytic genome expression, establishment of a selective barrier between genetically different haploid microspores, and prevention of cohesion and fusion of the microspores (Ünal et al. 2013).

The microspore wall begins to form at the tetrad stage, while the callose wall is still present, with deposition of primexine by the microspores. Subsequently, the callase activity promotes callose wall degradation and releases the free microspores (Owen and Makaroff 1995).

3.3.3 Microgametogenesis

The free microspores expand and undergo asymmetrical mitosis, originating bicellular microgametophytes or pollen grains in a process termed microgametogenesis. Bicellular microgametophytes are composed of a large vegetative cell and a small peripheral generative cell (Owen and Makaroff 1995). This asymmetric mitosis is essential for development of the male gametes since symmetrical mitosis would result in failure to generate male gametes (Eady et al. 1995). The vegetative cell generally has an active metabolism, with abundant oil bodies and starch grains, while the generative cell is poor in organelles (Haddad et al. 2019a).

The generative cell undergoes mitosis, originating the two male gametes in tricellular microgametophytes. In most plant species, mature pollen grains are released in the two-cell stage, with the final mitotic division occurring in the pollen tube (Gómez et al. 2015). Each gamete is connected to the other and delimited by an electron-translucent vesiculated region, without a plasma membrane; the cytoplasm lacks organelles, and the nucleus is positioned centrally, with heterochromatin but without a nuclear envelope or nucleoli. The absence of the plasma membrane and cell wall in male gametes may facilitate nutrient transfer between the vegetative cell and the gametes, and the absence of the nuclear envelope may facilitate the fusion of male and female gametes during fertilization (Haddad et al. 2019a). The pollen grain wall has an inner intine, secreted by the microspore and composed of pectin, cellulose, and hemicellulose, and an outer exine, secreted by the tapetum and composed of sporopollenin (Gómez et al. 2015).

3.3.4 PCD Causing Anther Dehiscence

The anther dehiscence program requires activation of many genes that promote anther rupture and pollen grains release through a coordinated PCD of several tissues. Generally, PCD of the tapetum and middle-wall layers occurs prior to anthesis. The endothecium cells are dead at maturity and have cell wall thickenings that provide mechanical movement in the anther dehiscence process. Anther dehiscence is also linked to PCD occurrence in the interlocular septum and in epidermis at the stomium (a specialized site that acts as a fault line) (Goldberg et al. 1993; Wu and Cheung 2000).

3.3.5 PCD of Incompatible Microgametophytes

The anther dehiscence releases pollen grains containing microgametophytes, which arrive on the stigma epidermis, hydrate, and germinate, producing pollen tubes. The incompatibility system can act in the stigma, in the transmitting tissue, in the ovule micropyle, or even postfertilization (Chapman and Goring 2010). Likewise, the pollen tube guidance involves chemical signals from the stigma, transmitting tissue, or ovule micropyle (Mizuta and Higashiyama 2018). The transmitting tissue connects the stigma to the ovary and leads to growth of the appropriate pollen tubes. The transmitting tissue also acts in the recognition of inappropriate pollen tubes, preventing the growth of interspecific and, in some cases, intraspecific pollen tubes, preventing self-pollination (Beers 1997). Prevention of self-pollination, a condition known as self-incompatibility, presumably evolved to prevent the deleterious effects of inbreeding (Dodds et al. 1996).

3.3.6 Staminodes

Nonfunctional stamens are referred to as sterile stamens or staminodes. In one first evolutionary step in stamen reduction, staminodes can become vestigial, with few morphological alterations in comparison to fertile stamens. Vestigial staminodes initiate as regular primordia that abort at a certain developmental stage or have a delayed initiation. In a second evolutionary step, staminodes can assume other functions and became functional staminodes. Functional staminodes, according to the different pollination syndromes, can assume novel biological functions in the flowers, such as nutritional, structural, or attractive (Ronse Decreane and Smets 2001). In flowers of *Clusia lanceolata* Cambess., the sterility of the staminodes is due to the undifferentiated nature of the anthers. Staminodes function to produce resin for pollinators. The resin is released through rupture points on the distal surface of staminodes, associated with disrupted subepidermal resin cavities and canals (Sá-Haiad et al. 2015).

3.4 Gynoecium and Megagametophyte Development

The gynoecium is composed of a single or multiple carpels, free or fused. The fusion processes can be congenital or post-genital. Congenital fusion occurs by the confluence of primordium meristems and does not involve epidermis participation, whereas post-genital fusion occurs by the fusion of two originally separate carpels via the epidermis (Endress 2019). The molecular control of gynoecium development in *A. thaliana* and the role of hormones in gynoecium morphogenesis were treated by Sundberg and Ferrándiz (2018).

3.4.1 Ovule Structure

The ovules are comprised of a megasporangium, termed a nucellus, with vegetative and sporogenous cells; one or, most commonly, two integuments covering the nucellus; and a funiculus that attaches the ovule to the placenta (Esau 1965). The funiculus provides nutrients to the megagametophyte and to the developing embryo and partly determines the position of the micropyle, a small opening at the free pole of the ovule (micropylar pole). The ovule pole opposite to the micropyle is termed the chalazal pole. The nucellus differentiates the megasporocyte that, through megasporogenesis and megagametogenesis, produces the megagametophyte or embryo sac. Subsequently, the integuments will develop into the protective seed coat (Skinner et al. 2004). The ovule development in *A. thaliana* was described in detail by Robinson-Beers et al. (1992).

3.4.2 Megasporogenesis

The megasporocyte undergoes meiosis, producing a dyad of megaspores – a product of meiosis I – and, after meiosis II, a tetrad of megaspores, in a process termed megasporogenesis. The megasporocyte and the megaspores of the dyad show cytoplasmic polarity, as revealed by the asymmetric distribution of organelles and plasmodesmata. Plasmodesmata, plastids, mitochondria, and other organelles accumulate preferentially at the chalazal pole of the microsporocyte. Consequently, after meiosis I, most organelles are inherited by the chalazal megaspore from the dyad. After meiosis II, in most studied species, plasmodesmata are usually observed only between the most chalazal megaspore from the tetrad, the functional megaspore, and the nucellus cells. Thus, the functional megaspore inherits a richer cytoplasm and can receive more nutrients from the nucellus, which suggests that the selection of the functional megaspore starts before meiosis (Reiser and Fischer 1993). In *Eleocharis sellowiana* Kunth, the megasporocyte shows inverse polarity, with plasmodesmata and organelles mainly in the micropylar pole. Nevertheless, the chalazal megaspore is the functional one, suggesting that, in this species, polarity is not decisive in selecting the functional megaspore (Rocha et al. 2015).

3.4.3 Megagametogenesis

More than 15 different types of megagametophyte development were described and summarized by Yadegari and Drews (2004). According to Maheshwari (1950), more than 70% of flowering plants have the *Polygonum* type, first described in *Polygonum divaricatum* L. (Maheshwari 1950).

The functional megaspore develops the mature megagametophyte in a process termed megagametogenesis, in which one or more rounds of mitosis without cytokinesis result in a multinucleate coenocyte. In the *Polygonum* type, the nucleus from the functional megaspore first undergoes two rounds of mitosis, producing a four-nucleate cell with two nuclei positioned at each pole. The third mitosis produces an eight-nucleate cell, with four nuclei positioned at each pole. Two nuclei, one from each pole, migrate toward the center of the megagametophyte and are called polar nuclei. They fuse before or upon the fertilization of the central cell. After cellularization, the final result is a seven-celled megagametophyte, with three antipodal cells at the chalazal pole, one central cell, and, at the micropylar pole, two synergid cells and one egg cell. In some species, the antipodal cells are ephemeral and undergo PCD before fertilization (Yadegari and Drews 2004).

The megagametophyte development is tightly correlated with the development of the surrounding sporophytic ovular tissues. These tissues supply nutrients to the megagametophyte. In species with nucellar degradation during megagametophyte development, the nucellar remnants can provide materials and energy for the megagametophyte maturation (Grossniklaus and Schneitz 1998).

3.4.4 Mature Megagametophyte

3.4.4.1 Antipodal Cells

Several functions have been attributed to the antipodal cells, including (1) transfer of nutrients from the nucellus to the central cell, (2) storage of reserves for the embryo and endosperm, and (3) secretory function. To perform these functions, the antipodal cells have high metabolic activity, as revealed by the rich cytoplasm with mitochondria, dictyosomes, ribosomes, extensive ER, plastids, starch grains, lipids, and proteins. Nutrient transfer from the nucellus to the central cell is facilitated by ingrowths of the cell wall in some areas of the chalazal pole of the antipodal cells and by the plasmodesmata between the central cell and antipodals and between antipodals and the surrounding nucellus cells. The storage function of the antipodals is provided through the accumulation of reserve substances to be used after fertilization. The secretory function is provided by the extensive ER and abundant ribosomes (Willemse and van Went 1984). Persistent antipodals are associated with rich cytoplasm, whereas ephemeral antipodals are associated with cytoplasm with only sparse organelles (Raghavan 2000).

3.4.4.2 Synergid Cells

The synergids have an essential role in mediating fertilization, being responsible for the attraction and reception of the pollen tube (Higashiyama 2002) and for the forwarding of the male gametes to the female gametes, i.e., the egg and central cells. To perform these functions, synergids have an extensive ingrowth of the cell wall at the micropylar pole, termed the filiform apparatus, which acts as a diffusion pathway for the substances involved in attracting the pollen tube. In most species, the filiform apparatus projects to the cytoplasm, enlarging the surface of the plasma membrane to facilitate transport between the apoplast and the symplast (Huang and Russell 1992). The synergids are metabolically active cells with a rich and polarized cytoplasm at the micropylar pole, where mitochondria, extensive ER, ribosomes, dictyosomes, and vesicles are present in high densities (Willemse and van Went 1984). After pollen tube penetration and reception of male gametes, the synergid cytoskeleton reorganizes and forms two "coronas" of actin, which forwards the male to the female gametes (Russell 1993). The synergid PCD can occur before the arrival of the pollen tube, as a response to injury caused by pollen tube penetration or in case of pollination absence (Willemse and van Went 1984).

3.4.4.3 Central Cell

In flowering plants, fertilization of the central cell is essential for development of the triploid secondary endosperm, the tissue that provides nutrients to the developing embryo. In most species, the central cell contains a vacuole that occupies nearly the entire cell volume, and a metabolically active peripheral cytoplasm with numerous mitochondria, dictyosomes, and vesicles, extensive ER, ribosomes and polysomes, and plastids (Willemse and van Went 1984; Thijssen 2003).

3.4.4.4 Egg Cell

The egg cell has a lower metabolic activity than the central cell and synergids, which is revealed by the cytoplasm poor in organelles, with the nucleus at the chalazal pole and a vacuole at the micropylar pole (Willemse and van Went 1984).

Commonly, the cellular boundaries between the central cell, egg cell, and synergids are determined by a plasma membrane and peripheral vesicles that lack cell walls (Thijssen 2003; Haddad et al. 2019b). Haddad et al. (2019a) observed that the male gametes of *Monteverdia obtusifolia* also are not delimited by a plasma membrane or cell wall and the nuclear envelope is absent. Haddad et al. (2019b) interpreted the absence of a cell wall in synergids, female (egg cell and central cell) and male gametes, and the absence of a nuclear envelope in the male gametes as a reproductive strategy to facilitate gametic fusion.

3.5 Male Sterility

Male sterility is a hereditary phenomenon characterized by the failure to produce functional anthers or fertile microgametophytes (Kaul 1988). The genome in plant cells is located in the nucleus and in the cytoplasm, specifically in the mitochondria and chloroplasts. In most species, the DNA of both the mitochondria and chloroplast DNA is inherited uniparentally and usually through the same parent, independently of the nuclear DNA (Hanson and Betolila 2004). The expression between the mitochondrial and nuclear genomes is tightly coordinated, since most genes required for mitochondrial metabolism and for proper mitochondrial functionality are encoded by the nuclear genome, with the nucleus sending signals to regulate the mitochondria, which is known as anterograde signaling (Chen et al. 2017). On the other hand, mitochondria send signals to regulate the expression of some nuclear genes, which is known as retrograde regulation (Chase 2007).

Chloroplast DNA (cp-DNA) of flowering plants is thought to be highly conserved. On the other hand, mitochondrial DNA (mt-DNA) exhibits considerable genetic diversity and characteristically incorporates DNA of nuclear and chloroplastidial origin. The transfer of DNA sequences between nuclear and mitochondrial genomes is two-way, while the transfer from chloroplast to mitochondrial genomes is typically one-way (Chen et al. 2017).

The position of the novel sequences in the mt-DNA can lead to expression of these sequences. Furthermore, the mt-DNA undergoes recombination events that generate different configurations of the mitochondrial genome, even in a single species. Through recombination, the novel sequence can enter a region more favorable for its expression (Hanson and Betolila 2004), and if the mitochondrial population of a cell is connected (i.e., fused), the mitochondrial genome rearrangements spread to the whole mitochondria population (Horn et al. 2014). As a consequence of these changes in the mt-DNA organization, cytoplasmic male sterility arises (Chen et al. 2017).

Male sterility in natural populations occurs only in a limited number of species. However, male sterility occurs more widely than female sterility, probably because microgametophytes are more vulnerable to the environment than megagametophytes, since megagametophytes are protected inside the ovules. Besides that, because a large number of microgametophytes are available in the anther, their fertility can be tested rapidly, and male sterility is easier to detect than megagametophyte fertility (Kaul 1988). Usually, microgametophyte viability is tested by staining with an acetocarmine solution (Dafni et al. 2005).

Based on its origin and inheritance pattern, male sterility can be divided into "nuclear male sterility" and "cytoplasmic male sterility." "Chemical male sterility" is nongenetic male sterility that is induced by applying specific chemicals termed gametocides or chemical hybridizing agents, which cause male sterility without affecting female fertility. Generally, "chemical male sterility" is induced in economically important crop species (Lasa and Bosemark 1993). "Chemical male sterility" is not treated here because it is absent in natural populations.

3.5.1 Nuclear Male Sterility (NMS)

Nuclear male sterility (NMS), also termed "genic male sterility" or "genetic male sterility," is controlled by nuclear genes, usually dependent of a single recessive gene or a dominant gene that is typically expressed in specific sporophytic tissues at different developmental stages (Priyadarshan 2019). In flowering plants, NMS occurs widely. It can originate through spontaneous mutation in natural populations or through artificial induction, such as in several crop species (Lasa and Bosemark 1993). Gene mutations can alter the expression of genes or prevent them from being expressed, which can result in abnormal androecium or microgametophyte development or failure to develop (Kaul 1988). Genetic control of the NMS was extensively treated by Kaul (1988) and the methods and application in plant breeding by Priyadarshan (2019).

3.5.2 Cytoplasmic Male Sterility (CMS)

Cytoplasmic male sterility (CMS), also termed "cytoplasmic-genetic male sterility," is the maternally controlled inability to produce fertile microgametophytes (Priyadarshan 2019). Two types of genes, nuclear and cytoplasmic, control this sterility (Chase 2007). The maternal control is provided by the mitochondrial heritage acquired from the egg cell, since the mitochondria from the microgametophyte are usually excluded in fertilization (Budar and Pelletier 2001). The expression of the mitochondrial genes that cause male sterility can be suppressed by nuclear genes, designated as "fertility-restoring genes" (*Fr*), that lead to restoration of fertility and development of fertile microgametophytes. However, ineffective Fr genes do not restore fertility and the microgametophytes remain sterile (Hu et al. 2014).

In flowering plants, CMS has been recorded in approximately 200 species and is extensively exploited in plant breeding (Hu et al. 2014). The occurrence of CMS in natural populations is revealed by gynodioecy, i.e., the coexistence of hermaphrodite (male fertile) and female individuals (male sterile). The maintenance of CMS in natural populations depends on two parameters: female individuals need to produce at least twice the number of viable seeds produced by hermaphrodite individuals, which is known as female compensation (Budar et al. 2003), and the restorer alleles need to cause a negative pleiotropic effect, termed the "cost of restoration," in the absence of the corresponding sterile cytoplasm (Gouyon et al. 1991).

Female compensation in female individuals is achieved by reducing the investment in energy expended on development of the androecium or microgametophyte. Indeed, in most CMS species, the interruption in the microgametophyte development occurs in the initial developmental stages, saving considerable energy. However, in other CMS species, the interruption occurs in late developmental stages, after considerable energy is expended (Hanson and Betolila 2004).

Hu et al. (2014) suggested that it is advantageous for the plant to interrupt the development of male gametes instead of the female gamete, because the female gamete cannot move. Also, interruption of male gamete development is the easier and more economical way to promote cross-fertilization, increasing the probability that female gametes will receive foreign male gametes, generating offspring with a different nuclear heritage. The mitochondrial CMS genes can only be transmitted maternally, and the *Rfs* genes, which are in the nuclear genome, can be transmitted by male gametes. Hu et al. (2014) also suggested that CMS and *Rfs* genes coevolved in nature.

The products of CMS locate on the inner mitochondrial membrane, which precludes the functioning of the mitochondrial electron transport chain complex, interrupting energy production (Hu et al. 2014). In CMS species, the energy deficit activates the PCD processes in which ATP levels are insufficient to support autophagy, activating the necrosis pathway (Chase 2007).

3.5.2.1 Gametophytic and Sporophytic CMS

The CMS systems are sporophytic or gametophytic. Sporophytic CMS (Sp-CMS), which is dependent on the sporophyte genotype, acts on the sporophytic tissues until the free microspore stage, resulting in sterile microgametophytes. Gametophytic CMS (Ga-CMS) is dependent on the gametophyte genotype and therefore acts after meiosis. In the gametophytic system, each microgametophyte has a different genotype, and only those that carry the restorer genotype will have their fertility restored. In this case, the anthers can have both fertile and sterile microgametophytes (Hu et al. 2014).

3.5.2.2 Ultrastructure Features and Cellular Functionality of the CMS

Microgametophyte development requires a large amount of energy, provided by the large mitochondrial population of the tapetum (Tadege and Kuhlemeier 1997). Structural alterations cause disturbances in the mitochondrial functions and affect male fertility. In some CMS species, structural changes have been reported in the mitochondria of the tapetum and developing microgametophytes, such as hypertrophy and irregular shapes (Ku et al. 2003), enlargement (Polowick and Sawhney 1993), a plate-shaped outline (Polowick and Sawhney 1990), or indistinguishable cristae, and a less dense matrix (Laveau et al. 1989; Ku et al. 2003). In other CMS species, vacuolization occurs in the tapetum cells before the mitochondrial changes, and then the mitochondria elongate and acquire plate-shaped cristae (Bino 1985). The absence of cristae reduces the area available for the enzymatic reactions necessary for the respiration process. A low-density mitochondrial matrix indicates a deficiency in molecules involved in respiration (Haddad et al. 2019a). The lower accumulation of ATP synthase (Sabar et al. 2003) and the ATP/ADP proportions in floral tissues also indicate low respiratory rates (Bergman et al. 2000).

Other changes such as a concentric arrangement of the ER, as well as dilation or fragmentation of the ER cisternae, and a decrease in the ribosome population have been described in the tapetal cells of CMS species (Horner 1977; Grant et al. 1986; Majewska-Sawka et al. 1993; Haddad et al. 2019a). These features were also associated with a lower energy status, since the concentric arrangement is a more stable configuration of the ER and therefore requires less energy than the linear arrangement (Shi et al. 2010).

3.6 Abnormal Development Promoting Male Sterility

The phenotypic manifestations of male sterility are very diverse. Based on floral development, male sterility is divided into (1) "structural or staminal male sterility" in which the androecium is completely absent; (2) "pollen sterility" in which fertile microgametophytes are not produced because of failure in microsporogenesis or microgametogenesis, or abortion of microgametophytes; and (3) "functional male sterility" in which fertile microgametophytes develop but the anthers are indehiscent, or the microgametophytes lack the ability to germinate on compatible stigma (Priyadarshan 2019).

3.6.1 Structural or Staminal Male Sterility

In flowers with any vestiges of the nonfunctional reproductive organ, i.e., absence of androecium in the pistillate flowers, stamen differentiation is controlled by the co-expression of B- and C-class genes. These expressions are regulated directly or indirectly by mitochondria retrograde signaling (Carlsson et al. 2008). Mutations in B- and C-class genes or changes in the expression pattern of these genes through retrograde signaling can result in the absence of stamen primordia, as in the pistillate flowers of *Carica papaya* L. (Ronse Decraene and Smets 1999), *Batis maritima* L. (Ronse Decraene 2005), and *Triuris brevistylis* Donn. Sm. (Espinosa-Matías et al. 2012). In some cases, this also can lead to conversion of the third whorl into a different floral organ type (Goldberg et al. 1993; Hanson and Betolila 2004), such as carpeloid or petaloid structures in CMS species (Carlsson et al. 2008). The carpeloid structures result from the loss expression of B-class genes, and petaloid structures result from expansion of the expression region of A-class genes and contraction of the expression region of C-class genes (Chase 2007).

3.6.2 *Pollen Sterility*

3.6.2.1 Failure in Microsporogenesis

Failure in microsporogenesis can be associated with several types of abnormal structural development, including (1) abnormal microsporangium, (2) formation of abnormal microsporocytes, (3) failure in callose deposition or degradation, and (4) formation of abnormal free microspores.

As previously mentioned, tapetum PCD at the microspore stage is necessary to provide essential materials for microgametophyte development, and a premature or late PCD causes an interruption in the microgametophyte development, resulting in male sterility. The manifestation of male sterility in an early developmental stage can result from abnormal microsporangium development that prevents microsporocyte differentiation. In the pistillate flowers of *Monteverdia obtusifolia*, premature PCD of the tapetum at the sporogenic cell stage leads to PCD of the sporogenic cells through macroautophagy and massive autophagy, preventing microsporocyte differentiation, which has been associated with Sp-CMS (Haddad et al. 2018).

Abnormal microsporocyte differentiation, with normal or abnormal tapetum development, has been observed in several species. In the pistillate flowers of *Clusia fluminensis* Planch. & Triana, the microsporocytes degenerate, impairing microgametophyte development (Sá-Haiad et al. 2015). In the pistillate flowers of *Consolea moniliformis* (L.) A. Berger, *C. millspaughii* (Britton) A. Berger, *C. nashii* (Britton) A. Berger, *C. picardae* (Urb.) Areces, *C. rubescens* (Salm-Dyck ex DC.) Lem., and *C. spinosissima* (Mill.) Lem., the tapetum develops abnormally, and the microsporocytes differentiate, but microsporogenesis does not proceed beyond prophase I, resulting in degeneration of the microsporocytes (Strittmatter et al. 2006). In the pistillate flowers of *Opuntia stenopetala* Engelm., the tapetum shows an abnormal increase in the number and size of the vacuoles at the microsporocyte developmental stage. Besides the increase in number and size of vacuoles, the microsporocytes also have a plasma membrane with blisters and a wrinkled nuclear membrane. Most microsporocytes collapse and do not undergo meiosis. Occasionally some microsporocytes initiate meiosis and reach prophase I but then also collapse (Flores-Rentería et al. 2013).

Deposition of the callose wall by the microsporocytes is a precondition to meiosis and essential for microspore individualization. Callose wall degeneration by callase timely provided by the tapetum is required for release of the microspores from the tetrad. In some species, premature tapetum PCD leads to failure of the callose wall to disintegrate, and the microspores remain in the tetrad, and microgametophyte development does not proceed, as observed in flowers from high-fruiting individuals of *Ocotea tenera* Mez & J. D. Smith ex Mez (Gibson and Diggle 1998).

Most manifestations of male sterility occur during microsporogenesis, or immediately after free microspores are released from the tetrads, which results in abnormal formation of free microspores. Some free microspores of staminate flowers and all free microspores of pistillate flowers of *Garcinia brasiliensis* Mart. are hypertrophied, with a huge vacuole and accumulation of starch grains prior to degeneration (Leal et al. 2012). In *Miconia albicans* (Sw.) Triana, all free microspores are uni- or binucleate and have degenerated cytoplasm, or are completely empty, with abnormal microspore wall development (Cortez et al. 2012). In perfect flowers of *Monteverdia obtusifolia*, the male sterility was associated with Ga-CMS, in which some microspores start a necrosis process, with an accompanying decrease in organelle populations, plasma membrane degradation, protoplasmic shrinkage, and subsequent cytoplasm degradation (Haddad et al. 2019a).

3.6.2.2 Failure in Microgametogenesis

Manifestation of male sterility during microgametogenesis prevents proper bicellular or tricellular microgametophyte formation, as observed by Cortez et al. (2012) in some bicellular microgametophytes with an anomalous wall of *Miconia albicans* and in some bicellular microgametophytes of *M. stenostachya* DC. In both species, symmetric mitosis leads to the formation of microgametophytes with vegetative and generative cells of the same size, or in cases of asymmetric mitosis, the vegetative and/or generative cell contains more than one nucleus and shows ultrastructural abnormalities such as disturbances in the organelles, endomembrane system, and plasma membrane.

3.6.3 Functional Male Sterility

3.6.3.1 Non-dehiscence of Anther

Failures in the dehiscence program lead to confinement of microgametophytes in the anthers and subsequent abortion through a PCD process. This manifestation of male sterility is more prevalent in NMS than in CMS species. Mature anthers of pistillate flowers of *Melicoccus lepidopetalus* Radlk. show an endothecium without fibrous thickenings (Zini et al. 2012), while in *Cardiospermum grandiflorum* Sw. and *Urvillea chacoensis* Hunz., the endothecium wall thickenings are less developed and absent along the stomium margins (Solís et al. 2010). In these anthers, there is persistence of the middle layers, tapetal cells, and interlocular septum, resulting in anthers that are indehiscent at anthesis, although some bicellular microgametophytes were observed (Solís et al. 2010, Zini et al. 2012). In the pistillate flowers of *Clusia fluminensis* (Sá-Haiad et al. 2015) and of *Opuntia stenopetala* (Flores-Rentería et al. 2013), in addition to microsporocyte degeneration, the anthers are indehiscent at anthesis.

3.6.3.2 The Inability of Mature Pollen to Germinate on a Compatible Stigma

This manifestation of sterility represents the most delayed action of male sterility genes. After the arrival of mature pollen grains on the stigma, compatible pollen grains hydrate and germinate, producing pollen tubes. For this to occur, appropriate cell-cell communication is essential to allow the pistil to recognize the microgametophytes (Higashiyama 2010). Self-incompatibility mechanisms are used to avoid self-fertilization, in which the pollen-pistil interactions involve chemical signaling between substances that are produced by the tapetum and accumulate in the pollen coat and substances that cover the stigma (Mizuta and Higashiyama 2018). This signaling mechanism discriminates the inappropriate pollen tubes and prevents the growth of interspecific and, in some cases, intraspecific pollen tubes (Beers 1997). Some microgametophytes of *Miconia stenostachya*, although they have an apparently normal cytoplasm, are not able to develop a pollen tube (Cortez et al. 2012) when reaching the conspecific stigma.

3.7 Female Sterility

Female sterility in natural populations is less common than male sterility (Chakrabarty 2017), which is reflected in the sparse information on female sterility. In flowering plants, female sterility can occur naturally or by induced mutations. The female sterility is determined by sporophytic or gametophytic mutations. Sporophytic mutations can occur during the sporophytic phase of the megagametophyte development (diploid), which includes the megasporangium and megasporocyte differentiation, and megasporogenesis, or occur in the sporophytic tissues that control the megagametophyte development. Gametophytic mutations can occur during the gametophytic phase of megagametophyte development (haploid), which includes the developmental stages after megasporogenesis, i.e., megagametogenesis, or act in the functionality of the mature megagametophyte (Yadegari and Drews 2004).

3.8 Abnormal Development Promoting Female Sterility

In dioecious species, which have coexisting male (female sterile) and female individuals, and in monoecious species, which have staminate and pistillate flowers in the same individual, the staminate flowers show total female sterility. However, in androdioecious species, with male and hermaphrodite individuals, or in gynodioecious species, with female and hermaphrodite individuals, female sterility in the perfect flowers can be partial, and abortion can occur in only some ovules.

The phenotypic manifestations of female sterility cover a wide range, from defects in the stigma to the absence of ovules (Strittmatter et al. 2008). Studies on female sterility in natural populations have treated only a few species and are described below.

In some sterile ovules of *Nicotiana glauca* Graham (Villari and Messina 1996), as well as in *Consolea moniliformis*, *C. millspaughii*, *C. nashii*, *C. picardae*, *C. rubescens*, and *C. spinosissima*, the nucellus development is arrested, which results in the absence of a megagametophyte (Strittmatter et al. 2008). In staminate flowers of *Rauwolfia sellowii* L., female sterility can occur from the functional megaspore stage until the developing megagametophyte stage. Depending on the stage when it occurs, this sterility can result in sterile ovules that do not develop megagametophytes or in sterile ovules with incomplete megagametogenesis (Souto and Koch 2018).

In some sterile ovules of perfect and pistillate flowers of *Monteverdia obtusifolia*, the megagametophyte does not develop, and the integuments can be collapsed or developed. In other sterile ovules, the megagametophyte development is delayed, with nucelli showing dyads, tetrads, or megagametophytes in different developmental stages, or degenerated megagametophytes at anthesis (Haddad et al. 2019b). In sterile ovules of *Nicotiana glauca*, the megagametophyte cells start a vacuolization process, and the synergids do not possess a filiform apparatus (Villari and Messina 1996).

Some staminate flowers of *Consolea moniliformis*, *C. millspaughii*, *C. nashii*, *C. picardae*, *C. rubescens*, and *C. spinosissima* show maturation of acropetal ovules with consequent acropetal abortion, with extensive signs of abortion in ovules from the base of the ovary. Most ovules from the top of the ovary, although nonfunctional, still retain an apparently normal appearance. In other staminate flowers, all ovules present extensive signs of abortion. In some staminate flowers from "leaky" male individuals of *C. moniliformis*, most ovules show extensive signs of abortion, while some remain fertile and capable of being fertilized (Strittmatter et al. 2008).

In the majority of sterile ovules of *Monteverdia obtusifolia*, the mature megagametophyte starts a PCD process with vacuolization and tonoplast rupture, causing release of hydrolase from the vacuoles, which degrades the cytoplasm, plasma membrane, and cell walls through massive autophagy (Haddad et al. 2019b).

Recent studies have made significant progress in understanding the molecular basis of floral development and the molecular control of male sterility, besides providing background information about the cellular changes during PCD processes related to floral development. However, the majority of studies on male sterility are restricted to model or cultivated species, and studies on female sterility are still few. Therefore, studies in natural populations are required for a better comprehension of the cellular processes that promote male and female sterility. In this chapter, we provided literature data associating structure and functionality during gametophyte development, as well as the cellular processes responsible for male and female sterility in natural populations.

References

Bassham DC, Laporte M, Marty F et al (2006) Autophagy in development and stress responses of plants. Autophagy 2(1):2–11

Beers EP (1997) Programmed cell death during plant growth and development. Cell Death Differ 4:649–661

Bergman P, Edqvist J, Farbos I et al (2000) Male-sterile tobacco displays abnormal mitochondrial *atp1* transcript accumulation and reduced floral ATP/ADP ratio. Plant Mol Biol 42:531–544

Bino RJ (1985) Ultrastructural aspects of cytoplasmic male sterility in *Petunia hybrida*. Protoplasma 127:230–240

Budar F, Pelletier G (2001) Male sterility in plants: occurrence, determinism, significance and use. CR Acad Sci Paris 324:543–550

Budar F, Touzet P, De Paepe R (2003) The nucleo-mitochondrial conflict in cytoplasmic male sterilities revisited. Genetica 117:3–16

Carlsson J, Leino M, Sohlberg J et al (2008) Mitochondrial regulation of flower development. Mitochondrion 8:74–86

Chakrabarty S (2017) Female sterility and carpel development in plants. Dissertation, University of Illinois

Chapman LA, Goring DR (2010) Pollen–pistil interactions regulating successful fertilization in the Brassicaceae. J Exp Bot 61(7):1987–1999

Chase CD (2007) Cytoplasmic male sterility: a window to the world of plant mitochondrial-nuclear interactions. Trends Genet 23(2):81–90

Chen Z, Zhao N, Li S et al (2017) Plant mitochondrial genome evolution and cytoplasmic male sterility. Crit Rev Plant Sci 36(1):55–69

Coen ES, Meyerowitz EM (1991) The war of the whorls: genetic interactions controlling flower development. Nature 353:31–36

Cortez PA, Carmello-Guerreiro SM, Teixeira SP (2012) Understanding male sterility in *Miconia* species (Melastomataceae): a morphological approach. Aust J Bot 60:506–516

Dafni A, Keavan PG, Husband BC (2005) Practical pollination biology. Enviroquest, Cambridge

Diggle PK, Di Stilio VS, Gschwend AR et al (2011) Multiple developmental processes underlie sex differentiation in angiosperms. Trends Genet 27(9):368–376

Dodds PN, Clarke AE, Newbigin E (1996) A molecular perspective on pollination in flowering plants. Cell 85:141–144

Eady C, Lindsey K, Twell D (1995) The significance of microspore division and division symmetry for vegetative cell-specific transcription and generative cell differentiation. Plant Cell 7:65–74

Echlin P, Godwin H (1968) The ultrastructure and ontogeny of pollen in *Helleborus foetidus* L. I. The development of the tapetum and Ubisch bodies. J Cell Sci 3(2):161–174

Endress P (2019) The morphological relationship between carpels and ovules in angiosperms: pitfalls of morphological interpretation. Bot J Linn Soc 189:201–227

Esau K (1965) Plant Anatomy, 2nd edn. John Wiley, New York

Espinosa-Matías S, Vergara-Silva F, Vázquez-Santana S et al (2012) Complex patterns of morphogenesis, embryology, and reproduction in *Triuris brevistylis*, a species of Triuridaceae (Pandanales) closely related to *Lacandonia schismatica*. Botany 90:133–1151

Flores-Rentería L, Orozco-Arroyo G, Cruz-García F et al (2013) Programmed cell death promotes male sterility in the functional dioecious *Opuntia stenopetala* (Cactaceae). Ann Bot 112:789–800

Gibson JP, Diggle PK (1998) Flower development and male sterility in *Ocotea tenera* (Lauraceae): a gynodioecious tropical tree. Int J Plant Sci 159(3):405–417

Goldberg RB, Beals T, Sanders PM (1993) Anther development: basic principles and practical applications. Plant Cell 5:1217–1229

Gómez JF, Talle B, Wilson ZA (2015) Anther and pollen development: a conserved developmental pathway. J Integr Plant Biol 57:876–891

Gouyon P-H, Vichot F, Van Damme JMM (1991) Nuclear-cytoplasmic male sterility: single-point equilibria versus limit cycles. Am Nat 137(4):498–514

Grant I, Beversdorf WD, Peterson RL (1986) A comparative light and electron microscopic study of microspore and tapetal development in male fertile and cytoplasmic male sterile oilseed rape (*Brassica napus*). Can J Bot 64:1055–1068

Grossniklaus U, Schneitz K (1998) The molecular and genetic basis of ovule and megagametophyte development. Semin Cell Dev Biol 9:227–238

Haddad IVN, Santiago-Fernandes LDR, Machado SR (2018) Autophagy is associated with male sterility in pistillate flowers of *Maytenus obtusifolia* (Celastraceae). Aust J Bot 66:108–115

Haddad IVN, Sá-Haiad B, Santiago-Fernandes LDR et al (2019a) Pollen grain development and male sterility in the perfect flowers of *Maytenus obtusifolia* Mart. (Celastraceae). Protoplasma 256(3):745–761

Haddad IVN, Sá-Haiad B, Santiago-Fernandes LDR et al (2019b) Megagametophyte development and female sterility in *Maytenus obtusifolia* Mart. (Celastraceae). Protoplasma 256:1667–1680

Hanson MR, Betolila S (2004) Interactions of mitochondrial and nuclear genes that affect male gametophyte development. Plant Cell 16:S154–S169

Higashiyama T (2002) The synergid cell: attractor and acceptor of the pollen tube for double fertilization. J Plant Res 115:149–160

Higashiyama T (2010) Peptide signaling in pollen–pistil interactions. Plant Cell Physiol 51(2):177–189

Horn R, Gupta KJ, Colombo N (2014) Mitochondrion role in molecular basis of cytoplasmic male sterility. Mitochondrion 19:198–205

Horner HT Jr (1977) A comparative light- and electron-microscopic study of microsporogenesis in male-fertile and cytoplasmic male-sterile sunflower (*Helianthus annuus*). Am J Bot 64(6):745–759

Hu J, Huang W, Huang Q et al (2014) Mitochondria and cytoplasmic male sterility in plants. Mitochondrion 19(B):282–288

Huang BQ, Russell SD (1992) Female germ unit: organization, isolation, and function. Int Rev Cytol 140:233–292

Kaul MLH (1988) Male sterility in higher plants, Monographs on theoretical and applied genetics 10. Springer-Verlag

Kawanabe T, Ariizumi T, Kawai-Yamada M et al (2006) Abolition of the tapetum suicide program ruins microsporogenesis. Plant Cell Physiol 47:784–787

Ku S, Yoon H, Suh HS et al (2003) Male-sterility of thermosensitive genic male-sterile rice is associated with premature programmed cell death of the tapetum. Planta 217:559–565

Lam E (2005) Vacuolar proteases livening up programmed cell death. Trends Cell Biol 15(3):124–127

Lasa JM, Bosemark NO (1993) Male sterility. In: Hayward MD, Bosemark NO, Romagosa I, Cerezo M (eds) Plant breeding, Plant Breeding Series. Springer, Dordrecht, pp 213–228

Laveau JH, Schneider C, Berville A (1989) Microsporogenesis abortion in cytoplasmic male sterile plants from *H. petiolaris* or *H. petiolaris fallax* crossed by sunflower (*Helianthus annuus*). Ann Bot 64:137–148

Leal DO, Malafaia C, Cesar R et al (2012) Floral structure of *Garcinia brasiliensis* in relation to flower biology and evolution. Int J Plant Sci 173(2):172–183

Lockshin RA, Zakeri Z (2004) Apoptosis, autophagy, and more. Int J Biochem Cell Biol 36:2405–2419

Lu P, Chai M, Yang J et al (2014) The *Arabidopsis CALLOSE DEFECTIVE MICROSPORE1* gene is required for male fertility through regulating callose metabolism during microsporogenesis. Plant Physiol 164:1893–1904

Maheshwari P (1950) An introduction to the embryology of angiosperms. McGraw-Hill Book Company, Inc, New York

Majewska-Sawka A, Rodriguez-Garcia MI, Nakashima H et al (1993) Ultrastructural expression of cytoplasmic male sterility in sugar beet (*Beta vulgaris* L.). Sex Plant Reprod 6:22–32

Mitchell CH, Diggle PK (2005) The evolution of unisexual flowers: morphological and functional convergence results from diverse developmental transitions. Am J Bot 92(7):1068–1076

Mizuta Y, Higashiyama T (2018) Chemical signaling for pollen tube guidance at a glance. J Cell Sci 131(2):jcs208447
Owen HA, Makaroff CA (1995) Ultrastructure of microsporogenesis and microgametogenesis in *Arabidopsis thaliana* (L.) Heynh. ecotype Wassilewskija (Brassicaceae). Protoplasma 185:7–21
Pacini E (2010) Relationships between tapetum, loculus, and pollen during development. Int J Plant Sci 171(1):1–11
Pacini E, Hesse M (2005) Pollenkitt – its composition, forms and functions. Flora 200:399–415
Pacini E, Franchi GG, Hesse M (1985) The Tapetum: its form, function, and possible phylogeny in Embryophyta. Plant Syst Evol 149:155–185
Papini A, Mosti S, Brighigna L (1999) Programmed-cell-death events during tapetum development of angiosperms. Protoplasma 207:213–221
Paris N, Stanley CM, Jones RL et al (1996) Plant cells contain two functionally distinct vacuolar compartments. Cell 85(4):563–572
Polowick PL, Sawhney VK (1990) Microsporogenesis in a normal line and in the *ogu* cytoplasmic male-sterile line of *Brassica napus*. Sex Plant Reprod 3:263–276
Polowick PL, Sawhney VK (1993) An ultrastructural study of pollen development in tomato (*Lycopersicon esculentum*). II. Pollen maturation. Can J Bot 7(8):1048–1055
Priyadarshan PM (2019) Plant breeding: classical to modern. Springer, Singapore
Raghavan VR (2000) Megasporogenesis and formation of the embryo sac. In: Raghavan VR (ed) Developmental biology of flowering plants. Springer, New York, pp 216–227
Reape T, Molony E, McCabe P (2008) Programmed cell death in plants: distinguishing between different modes. J Exp Bot 59:435–444
Reiser L, Fischer RL (1993) The ovule and the embryo sac. Plant Cell 5:1291–1301
Robinson-Beers K, Pruitt RE, Gasser CS (1992) Ovule development in wild-type *Arabidopsis* two female-sterile mutants. Plant Cell 4:1237–1249
Rocha D, Andrade CGTJ, Vanzela ALL (2015) Female gametophyte development in *Eleocharis sellowiana* (CYPERACEAE): a species with atypical nuclear staining and polarity of the megaspore mother cell. Plant Syst Evol 301:1287–1292
Ronse Decraene LP (2005) Floral developmental evidence for the systematic position of *Batis* (Bataceae). Am J Bot 92(4):752–760
Ronse Decraene LP, Smets EF (1999) The floral development and anatomy of *Carica papaya* (Caricaceae). Can J Bot 77:582–598
Ronse Decraene LP, Smets EF (2001) Staminodes: their morphological and evolutionary significance. Bot Rev 67(3):351–402
Russell SD (1993) The egg cell: development and role in fertilization and early embryogenesis. Plant Cell 5:1349–1359
Sabar M, Gagliardi D, Balk J et al (2003) ORFB is a subunit of F1FO–ATP synthase: insight into the basis of cytoplasmic male sterility in sunflower. EMBO Rep 4:381–386
Sá-Haiad B, Silva CP, Paula RCV et al (2015) Androecia in two *Clusia* species: development, structure and resin secretion. Plant Biol 17(4):816–824
Shi Y, Zhao S, Yao J (2009) Premature tapetum degeneration: a major cause of abortive pollen development in photoperiod sensitive genic male sterility in rice. J Integr Plant Biol 51:774–781
Shi S, Ding D, Mei S et al (2010) A comparative light and electron microscopic analysis of microspore and tapetum development in fertile and cytoplasmic male sterile radish. Protoplasma 241:37–49
Skinner DJ, Hill TA, Gasser CS (2004) Regulation of ovule development. Plant Cell 16:S32–S45
Solís SM, Galati B, Ferrucci MS (2010) Microsporogenesis and microgametogenesis of *Cardiospermum grandiflorum* and *Urvillea chacoensis* (Sapindaceae, Paullinieae). Aust J Bot 58:597–604
Souto LS, Koch I (2018) Explaining ovule sterility in *Rauvolfia* (Apocynaceae). Int J Plant Sci 179(5):368–376

Strittmatter LI, Negrón-Ortiz V, Hickey R (2006) Comparative microsporangium development in male-fertile and male-sterile flowers of *Consolea* (Cactaceae): when and how does pollen abortion occur. Grana 45:81–100
Strittmatter LI, Hickey RJ, Negrón-Ortiz V (2008) Heterochrony and its role in sex determination of cryptically dioecious *Consolea* (Cactaceae) staminate flowers. Bot J Linn Soc 156:305–326
Sundberg E, Ferrándiz C (2018) Gynoecium patterning in Arabidopsis: a basic plan behind a complex structure. In: Østergaard L (ed) Fruit Development and Seed Dispersal, vol 38. Wiley, New York, pp 35–69
Tadege M, Kuhlemeier C (1997) Aerobic fermentation during tobacco pollen development. Plant Mol Biol 35:343–354
Thijssen MH (2003) Ovules, megagametophytes and embryos: Ultrastructural studies after cryofixation. Dissertation, Wageningen University
Ünal M, Vardar F, Aytürk Ö (2013) Callose in plant sexual reproduction. In: Silva-Opps M (ed) Current progress in biological research. IntechOpen, London, pp 319–343
van Doorn WG (2011) Classes of programmed cell death in plants, compared to those in animals. J Exp Bot 62(14):4749–4761
van Doorn WG, Papini A (2013) Ultrastructure of autophagy in plant cells. Autophagy 9:1922–1936
van Doorn WG, Woltering EJ (2005) Many ways to exit? Cell death categories in plants. Trends Plant Sci 10(3):117–122
van Doorn WG, Beers E, Dangl J et al (2011) Morphological classification of plant cell deaths. Cell Death Differ 18:1241–1246
van Doorn WG, Kirasak K, Ketsa S (2015) Macroautophagy and microautophagy in relation to vacuole formation in mesophyll cells of *Dendrobium* tepals. J Plant Physiol 177:67–73
Villari R, Messina R (1996) Ovule and gametophyte development in *Nicotiana glauca* Graham (Solanaceae). G Bot Ital 130:801–809
Vizcay-Barrena G, Wilson ZA (2006) Altered tapetal PCD and pollen wall development in the *Arabidopsis ms1* mutant. J Exp Bot 57(11):2709–2717
Wellmer F, Graciet E, Riechmann JL (2014) Specification of floral organs in *Arabidopsis*. J Exp Bot 65(1):1–9
Willemse MTM, van Went JL (1984) The female gametophyte. In: Johri BM (ed) Embryology of angiosperms. Springer, Berlin, pp 159–196
Wu H, Cheung AY (2000) Programmed cell death in plant reproduction. Plant Mol Biol 44:267–281
Yadegari R, Drews GN (2004) Female gametophyte development. Plant Cell 16:S133–S141
Zini LM, Galati GB, Solís SM et al (2012) Anther structure and pollen development in *Melicoccus lepidopetalus* (Sapindaceae): an evolutionary approach to dioecy in the family. Flora 207:712–720

Part II
Taxonomic, Ecological, and Conservation Aspects

This part will address some taxonomic, ecological, and conservation aspects through the reflection about phytoplankton richness and by viewing the importance of palynology to these kinds of studies considering some *Aspects of Brazilian Floristic Diversity: From Botany to Traditional Communities*.

So, starting the discussion centered on the ecological functioning of freshwater systems, Part II will present a comprehensive analysis to assess how phytoplancton richness varies concerning common major factors (temperature, light, and productivity) over an unprecedentedly large spatial scale. The analysis contributes to a better understanding of how the environment affects phytoplankton species richness, with implications for their ecological functioning in freshwater ecosystems.

Also, in this part, the importance of palynology to taxonomic studies will be seen, as it has a great value as a taxonomic marker. It will be shown that some diagnostic features of pollen grains can actuate in the taxonomic distinguishing of many groups.

Closing the discussions, this Part will present ecological palynology as an ecosystem service that is essential for the maintenance of biodiversity and also for the reproduction of plant species. So, in this regard, ideas about entomopalynology will be developed as a way to understand the relationship established between insects and pollen of different immersive aspects and through different associations. The discussions presented will lead us to the understanding of how valuable entomopalynological research is to the selection of plant species for biological conservation, as in the situation of organic agriculture where decision-makers about the suitability of insecticides are key figures in guiding their use.

Chapter 4
Broadscale Variation of Phytoplankton Richness in Brazilian Inland Waters

Lúcia Helena Sampaio da Silva, Juliana Barreto Oliveira dos Santos, Letícia Barbosa Quesado, Davi Almeida Barreto, Indhira Viana Freire, Mariângela Menezes, Maria da Graça Sophia, Donato Seiji Abe, and Vera Lúcia de Moraes Huszar

Abstract Species richness is a primary biodiversity measure in ecosystem functioning studies. The main factors shaping broadscale patterns of phytoplankton species richness (PhyRich) are temperature, system area, productivity, and less assessed light, flushing, and grazing. Our understanding of how the environment affects PhyRich is limited on broad spatial scales because of the lack of data using the same sampling and analysis methods. We selected three essential factors and hypothesized that PhyRich increases with water temperature, light availability (as turbidity), and intermediate productivity (as total phosphorus) levels. We assessed PhyRich from samples taken by a specially equipped hydroplane at 1045 sites across subcontinental (entire dataset) and regional (hydrographic regions) scales in Brazil (4°N–33°S and 34°–73°W) by quantifying the species number per site (settling technique). We selected models using the Akaike Information Criterion to understand which combination of variables better described PhyRich patterns. On our subcontinental scale, PhyRich was positively related to water temperature and negatively to light availability and productivity, although PhyRich was similar in oligo- and mesotrophic conditions, approaching the expected unimodal relationship. The trend for a slight reduction at the end of all gradients (univariate models) may have implications for the ecological functioning of freshwater systems.

Keywords Light · Productivity · Regional and subcontinental scales · Water temperature

L. H. S. da Silva (✉) · J. B. O. dos Santos · L. B. Quesado · D. A. Barreto · I. V. Freire · M. Menezes · M. da Graça Sophia · V. L. de Moraes Huszar
Departamento de Botânica, Laboratório de Ficologia, Museu Nacional, Universidade Federal do Rio de Janeiro (MN/UFRJ), Rio de Janeiro, RJ, Brazil
e-mail: luciahssilva@mn.ufrj.br

D. S. Abe
Associação Instituto Internacional de Ecologia e Gerenciamento Ambiental, São Carlos, SP, Brazil

M. F. T. Medeiros, B. de Sá Haiad (eds.), *Aspects of Brazilian Floristic Diversity*,
https://doi.org/10.1007/978-3-031-07453-0_4

4.1 Introduction

Species richness has been used as a primary measure of biodiversity and to assess its relationship to ecosystem functioning (Díaz and Cabido 2001; Magurran 2005; Vogt et al. 2010; Magurran and McGill 2011). Except for the neutral theory (Hubbel 2001), the ability to predict patterns of species richness is based on the assumption that different species respond differently to environmental gradients (Tilman 2004; Cadotte et al. 2011). As for macroorganisms, microbial communities also affect ecosystem processes, including respiration and decomposition, autotrophic and heterotrophic production, and nutrient cycling.

Phytoplankton is a highly diverse community composed of 10,000 to 25,000 marine and freshwater species (Sournia et al. 1991; Falkowski et al. 2004; Vargas et al. 2015). The rapid response of phytoplankton to environmental changes may reflect water quality (Padisák et al. 2006; Chen et al. 2021; Zhang et al. 2018). Although quantifying species richness changes (or lack thereof) has been considered only one measure of a much more complex phenomenon of biodiversity change through time and space (Chase et al. 2019), phytoplankton species richness can help to understand ecosystem processes.

The main ecological drivers of phytoplankton species richness are temperature, system area, productivity, and light and, to a lesser extent, hydraulic flow and grazing pressure. Water temperature directly affects individual and population metabolic processes through its direct relationship with the growth of phytoplankton communities (Roland et al. 2010; Weyhenmeyer et al. 2013; Kruk et al. 2017) and its indirect effect on the thermal structure of the water column (Winder and Hunter 2008). Several phytoplankton studies have shown positive relationships between species richness and temperature (Ptacnick et al. 2010; Thomas et al. 2012; Weyhenmeyer et al. 2013; Kruk et al. 2017). This is in line with the metabolic theory of ecology, where a more intense flow of energy in the food chain increases the productivity of systems and consequently the availability of resources, enabling the coexistence of a larger number of species (Hawkins et al. 2003; Segura et al. 2017).

In addition to water temperature, a positive relationship between phytoplankton species richness and ecosystem surface area has been observed in both natural and experimental aquatic ecosystems. In an experimental study, Smith et al. (2005) confirmed that patterns in microbial diversity are highly consistent with those that have been repeatedly reported for macroorganisms. However, in a study of 30 subarctic rock pools, phytoplankton richness was not positively related to the pool volume, i.e., it did not support the species-area relationship (Soininen and Meyer 2014).

A unimodal relationship (higher values at intermediate levels) between phytoplankton species richness and productivity has been reported along the entire nutrient enrichment gradient (Dodson et al. 2000; Smith 2007). This relationship depends on the productivity interval analyzed: it is direct in systems with increasing productivity, inverse when the productivity decreases, and absent at intermediate productivity levels (Smith 2007). However, the mechanisms that lead to this pattern are still widely debated. At low productivity levels, competition for resources and essential variables (e.g., temperature) directly affects individual phytoplankters and population metabolic processes and can be determinant for phytoplankton richness and composition.

Light availability is another critical factor, but the relationship patterns remain unclear. The duration of the open-water season (as a proxy of light) was positively related to the PhyRich along a latitudinal gradient in Sweden, favoring growth and increasing the number of available niches (Weyhenmeyer et al. 2013). In highly productive systems, high biomass can lead to lower PhyRich due to greater competition for light (Zhang et al. 2018). In experimental studies, species richness decreased in treatments with high and low light availability, but the decrease was larger in conditions of high light availability due to photoinhibition (Litchman 1998).

Although hydrology is not as crucial for most large lakes, in small lakes, the runoff during large storm events can potentially increase the rate of loss from communities to the drainage basin. Hydraulic flow tends to be important in reservoirs, especially run-of-river reservoirs (inflow = outflow) and, more markedly, in lotic ecosystems where unidirectional flow is one of the main factors controlling pelagic communities (Reynolds et al. 1994; Schmidt 1994). In floodplain systems, the hydrological pulse is the main regulatory factor for phytoplankton (Zalocar de Demitrovic 2002; Devercelli 2006; Salmaso and Zignin 2010; Talling and Prowse 2010), with phytoplankton showing negative (Huszar et al. 1998; Nabout et al. 2007; Borges and Train 2009) or positive relationships (Train and Rodrigues 1998) to the hydraulic flow. Grazing pressure is an essential factor for controlling phytoplankton, and selective predation may cause low diversity and lead to a negative relationship (Muylaert et al. 2010; Vallina et al. 2014; Kruk et al. 2017).

These responses of phytoplankton species richness to different environmental gradients result in a nonlinear relationship between biodiversity and ecosystem functioning and may differ between local and regional scales, as illustrated by data obtained from several river basins (Chase and Ryberg 2004). The importance of understanding these responses is due mainly to the impact of global changes on biodiversity and ecosystem functioning (Weyhenmeyer et al. 2013). Despite the importance of species richness, our understanding of how the environment affects PhyRich at large spatial scales is limited. This metric can only be used with confidence when phytoplankton sampling and quantification are correctly performed, to provide an accurate survey of the taxa in a system (Várbiró et al. 2017; Borics et al. 2021).

Here, we offer a comprehensive analysis based on 1045 sites (Moss and Moss 2005; Abe et al. 2006) to assess how PhyRich varies concerning common major factors (temperature, light, and productivity). We followed the space-for-time-substitution approach by retrieving samples through a specially equipped hydroplane in motion, over an unprecedentedly large spatial scale in Brazil (*Brasil das Águas* Project). We hypothesized that PhyRich increases with water temperature and light availability (decreases with turbidity as a proxy) and increases at intermediate productivity (total phosphorus as a proxy) levels. We explicitly addressed species richness as estimated from single phytoplankton samples, most directly related to ecosystem processes carried out by phytoplankton.

4.2 Material and Methods

4.2.1 Study Area

The sampling points were based on broad latitudinal and longitudinal scales (4°N–33°S and 34–73°W) in 12 hydrographic regions (HRs; Fig. 4.1) defined by the National Water Resources Council (CNRH) and ANA (National Water Agency). We included as wide a variety of water bodies as possible, such as rivers, coastal lagoons, floodplain lakes, other natural lakes, reservoirs, rivers, and estuaries. Brazil is about 8.5 million km^2 in area, with between 12% and 16% of all freshwater production on the planet. The climate ranges from humid temperate to humid equatorial (Köppen-Geiger classification, updated by Alvares et al. 2013). It varies from areas with sparse (<1 inhabitant per km^2) to high human population densities (>400

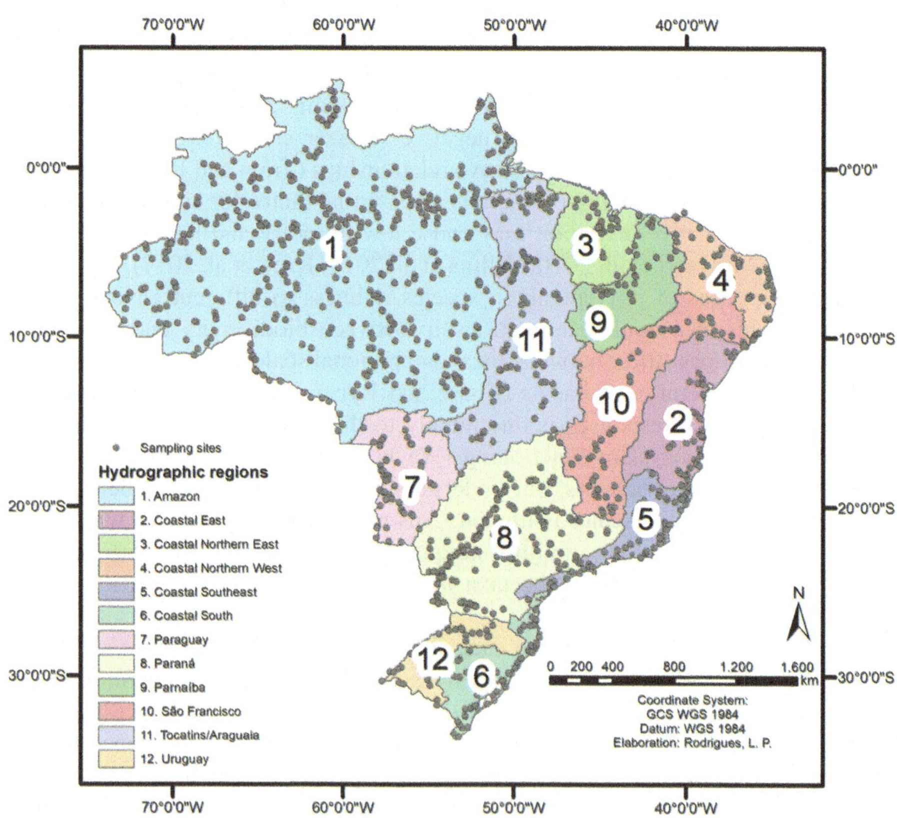

Fig. 4.1 Map of Brazil, showing the sampling sites in the 12 hydrographic regions. Information obtained in the "ANA Metadata Catalog" (https://metadados.snirh.gov.br/geonetwork/srv/por/catalog.search;jsessionid=01A66BC206AAB137A73D5C4F7D761A89#/metadata/0574947a-2c5b-48d2-96a4-b07c4702bbab, accessed in September 21)

Table 4.1 Main features of the 12 Brazilian hydrographic regions, with abbreviations used in the text. Data for population and treated sewage refer to the study period and were extracted from Moss and Moss (2005)

Hydrographic regions/ abbreviation	Area (km^2)	Biome	Human population (million inhabitants and %)	Treated sewage (%)
Amazon (Ama)	3,805,850	Amazon rainforest	7.6 (4.5)	2.3
Coastal east (CoastE)	374,677	Atlantic Forest, Cerrado	13.6 (8)	30
Coastal northeast (CoastNE)	287,348	Atlantic Forest, caatinga	21.6 (12.7)	18.2
Coastal northwest (CoastNW)	254,100	Amazon rainforest, Cerrado	4.7 (2.8)	0.1
Coastal southeast (CoastSE)	229,972	Atlantic forest	25.6 (15.1)	26.4
Coastal south (CoastS)	185,856	Atlantic Forest, pampa	11.6 (6.8)	8.2
Paraguay (Parag)	363,445	Cerrado, Pantanal	1.9 (1.1)	17.2
Paraná (Paran)	879,860	Atlantic Forest, Cerrado	54.6 (32.2)	24.9
Parnaíba (Parnb)	344,112	Cerrado, caatinga	3.6 (2.1)	4.8
São Francisco (SFranc)	638,324	Atlantic Forest, Cerrado, caatinga	12.8 (7.6)	3.2
Tocantins/Araguaia (TocArag)	967,059	Amazon Rainforest, Cerrado	7.9 (4.7)	2.4
Uruguay (Urug)	174,612	Atlantic Forest, pampa	3.8 (2.3)	6.0

inhabitants per km^2; Moss and Moss 2005; Abe et al. 2006; Table 4.1). HRs include areas with different land uses, from large natural forest areas, savanna (Cerrado), and caatinga, to anthropic landscapes such as urban regions, mining, and agriculture (Table 4.1).

4.2.2 Sampling

We followed the space-for-time-substitution approach (Jeppessen et al. 2014; Meehrhof et al. 2012) by retrieving samples through a specially equipped hydroplane in motion, over an unprecedentedly broad spatial scale in Brazil (*Brasil das Águas* project. www.brasildasaguas.com.br). Water temperature, turbidity, and total phosphorus (TP) were analyzed at 1164 and PhyRich at 1045 sampling sites. Sampling was conducted from November 2003 to December 2004, encompassing 177 lakes and reservoirs and 366 watercourses, including the broadest possible variety of protected, agricultural, and urban areas. Samples were taken through a tube 20 cm below the water surface, connected to an autosampler on board a hydroplane

(Lake LA-250 Renegade, Lake Aircraft, USA) in motion. The system was flushed with lake and river water four times before samples were taken.

We describe richness patterns and three of the most reported drivers of PhyRich: (i) temperature, as water temperature (WTemp); (ii) total phosphorus (TP), as a proxy of productivity; and (iii) turbidity (Turb) as a proxy of light availability. Because many sites were sampled in lotic systems and the sampling approach did not allow us to assess zooplankton, we did not evaluate the crucial relationships between PhyRich vs. area and grazing, respectively. We focused on the subcontinental (whole country) and regional (hydrographic regions) scales. WTemp and Turb were measured directly on board the hydroplane with a YSI 660 multiparameter water quality probe (Yellow Springs, USA). Phytoplankton samples were fixed with Lugol's solution. Samples for TP were frozen immediately after sampling, using liquid nitrogen. Upon arrival in the laboratory, they were stored at −20 °C and analyzed within 20 days after sampling.

Maps were constructed using ArcMap 10.8 software. Information obtained in the "ANA Metadata Catalog" (https://metadados.snirh.gov.br/geonetwork/srv/por/catalog.search;jsessionid=01A66BC206AAB137A73D5C4F7D761A89#/metadata/0574947a-2c5b-48d2-96a4-b07c4702bbab, accessed in September/21) was used as a reference to delimit the hydrographic regions. The points indicated on the maps (1045 for PhyRich and 1164 for WTemp, Turb, TP) correspond to the sampling sites. For PhyRich, Turb, and TP, we used the manual classification to define the different categories of values, and for WTemp, we used the Jenks Natural Breaks algorithm (Smith et al. 2018).

4.2.3 Analytical Methods

TP concentrations in water samples were analyzed by digestion with potassium persulfate and subsequent colorimetric reaction by ascorbic acid (Valderrama 1981), using a Varian Cary 50 spectrophotometer. PhyRich was assessed by quantifying the species number from single samples per site, using the settling method (Utermöhl 1958) under an inverted microscope (Zeiss Axiovert 10, Oberkochen, Germany). Only two counters analyzed the phytoplankton samples.

4.2.4 Data Analyses

All data obtained (1164 samples) were included for environmental variables and, for regression and GLM analysis, only samples containing phytoplankton data (1045 samples). Trophic states were assessed according to Nürnberg (1996) based on TP concentrations: oligotrophic <10 μg/L; mesotrophic 10–30 μg/L; eutrophic 30–100 μg/L; and hypereutrophic >100 μg/L.

To perform the cumulative species curve at the subcontinental scale (all sample points), we used the estimator Chao 2 (Chao 1987), which includes rare species, as the species number found in one sample (unique) and as the species number found in two samples (duplicates). To compare PhyRich at the regional scale (for each HR), we used the rarefaction method. This allowed us to compare the number of species between communities when the abundance of individuals was not the same (Gotelli and Chao 2013). Rarefaction calculates the expected number of species in each HR, based on a value where all samples reach a standard size, or comparisons based on the community with fewer individuals as a comparative basis. The analyses were carried out in R 4.1.1 (R Core Team 2021). The cumulative species curve was constructed using the vegan package (Oksanen et al. 2020). Sample-based rarefaction curves were performed using the iNEXT package (Hsieh et al. 2016).

We tested the statistical differences of the variables among HRs using the non-parametric Kruskal-Wallis test (Kruskal and Wallis 1952), followed by Dunn's post hoc pairwise comparisons. Statistical analyses were performed in SigmaPlot 12.5 software at the significance level of $p < 0.05$. Turb and TP were $\log_{10}$-transformed before the analysis. To avoid zeros in Turb values, we added 0.2 (lowest observed value) before the $\log_{10}$ transformation.

To describe the pattern of phytoplankton richness along the gradients of WTemp, Turb, and TP, we regressed with a generalized linear model (GLM). We plotted 95% confidence intervals on the fitted function, with the function *geom_smooth* and the generalized additive model (GAM) in the R package *ggplot2*. The significance of the relationship was calculated with GAM function in the R package *mgcv*. We used TP as a proxy for productivity (Ptacnick et al. 2010) and Turb for light availability.

To understand the PhyRich patterns at subcontinental and regional scales, we performed multiple combinations of environmental variables. First, we $\log_{10}$ transformed the data to reduce the dispersion of Turb (plus its minimum value) and TP. Then, from each environmental gradient tested, we removed the variation of the other variables using a generalized linear model (GLM) with a Gaussian error distribution, with the function *glm* in the R package *stats*. We selected models using the Akaike Information Criterion (AIC), with the regression residuals from the first step. For each dataset, we first constructed a full model with all variables (i.e., WTemp, TP, and Turb). Then, the best models, which retained the most information with the fewest variables, were selected using a model selection table based on ΔAICc (difference between the AICc of a given model and the AICc of the best model – Akaike Information Criterion) (Burnham and Anderson 2004), with the function *aictab* in the R package *AICcmodavg*. We also used the Dsquared (D2) model selection, which is the explained deviance of the model expressed as a percentage and is considered a pseudo r^2 of the best-selected model and a measure of model fit (Guisan and Zimmermann 2000).

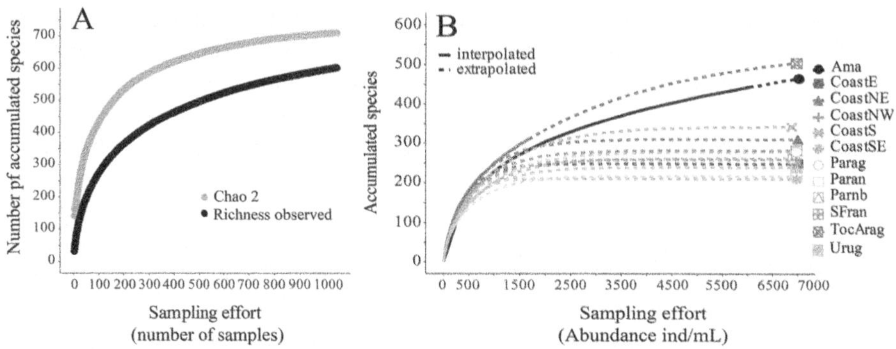

Fig. 4.2 (**a**) Measured and estimated (Chao 2) cumulative species curves for the subcontinental scale (n = 1045) in Brazil; (**b**) sample-based rarefaction curves for phytoplankton samples in the 12 hydrographic regions in Brazil. The curves show the observed (interpolated; solid lines) and extrapolated (dashed lines) cumulative phytoplankton species numbers. *Ama* Amazon, *CoastE* coastal east, *CoastNE* coastal northeast, *CoastNW* coastal northwest, *CoastSE* coastal southeast, *CoastS* coastal south, *Parag* Paraguay, *Paran* Paraná, *Parnb* Parnaíba, *SFranc* São Francisco, *TocArag* Tocantins/Araguaia, *Urug* Uruguay

4.3 Results

The combined phytoplankton dataset from the 12 hydrographic regions (HRs) contained 605 species (morphospecies). They belonged to 10 algal taxonomic classes: 102 cyanobacteria, 18 cryptomonads, 22 dinoflagellates, 85 diatoms, 43 chrysophyceans, 10 xanthophyceans, 30 euglenoids, 3 raphidophyceans, 204 chlorophyceans, and 68 zygnematophyceans. Of the total morphospecies, only 20 were observed in at least 1 sample in the 12 HRs. At the continental scale, we observed many rare species; 134 occurred only at 1 site (singletons) and 86 twice (doubletons). Accumulation (Fig. 4.2a) and rarefaction (Fig. 4.2b) curves, both based on Chao 2 estimator, showed no tendency to stabilize at an asymptote. It also estimated the possibility of finding 100 more species with a greater sampling effort. The rarefaction curves did not show differences between the cumulative number of phytoplankton species in the TocArag and Ama HRs. Nonetheless, these two basins showed almost twice as many cumulative numbers of species as the other HRs (Fig. 4.2b).

4.3.1 Phytoplankton Richness

Phytoplankton richness (PhyRich) at the subcontinental scale ranged between 0 and 62 spp./sample, with a median of 10 spp./sample. CoastE and CoastSE showed the significantly lowest PhyRich (median = 3 spp./sample), and TocArag and Urug showed the highest (median = 16 spp./sample; $p < 0.001$; Figs. 4.3a and 4.4a).

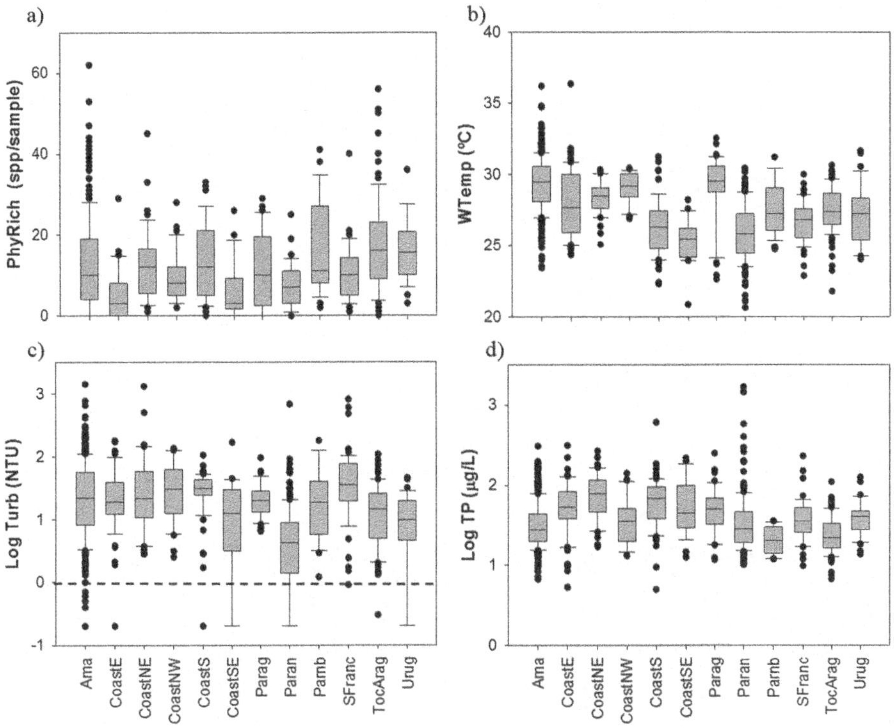

Fig. 4.3 (**a**) Phytoplankton richness, PhyRich; (**b**) water temperature, WTemp; (**c**) turbidity, Turb; and (**d**) total phosphorus, TP, expressed by box-whisker plots, where the horizontal lines inside the box plots indicate the median and the boundaries of the box plots indicate the 25th and 75th percentiles. Whiskers above and below indicate the 90th and 10th percentiles; dots are outliers in the 12 Brazilian hydrographic regions. *Ama* Amazon, *CoastE* coastal east, *CoastNE* coastal northeast, *CoastNW* coastal northwest, *CoastSE* coastal southeast, *CoastS* coastal south, *Parag* Paraguay, *Paran* Paraná, *Parnb* Parnaíba, *SFranc* São Francisco, *TocArag* Tocantins/Araguaia, *Urug* Uruguay

4.3.2 Environmental Variables

Variables described in this section are some of the most often reported to drive PhyRich over a broad spatial scale. WTemp ranged from 20.6 °C (Paran) to 36.4 °C (CoastE). Median values were significantly higher in Parag, Ama, and CoastNW and lower in CoastSE, CoastS, and Paran (Figs. 4.3b and 4.4b).

The range of Turb was between 0.2 (Ama, CoastE, CoastS, Paran, and Urug) and 1429 NTU (Ama). Two hundred and forty-four sites had Turb <5 NTU, 655 from 5 to 50, and 212 > 50 NTU over the subcontinental scale. For the regional scale, median values were highest in SFranc, CoastS, and CoastNW and lowest in Paran and Urug (Figs. 4.3c and 4.4c). Our data also showed that clear waters influenced sampling sites in the Amazon HR (Fig. 4.4c).

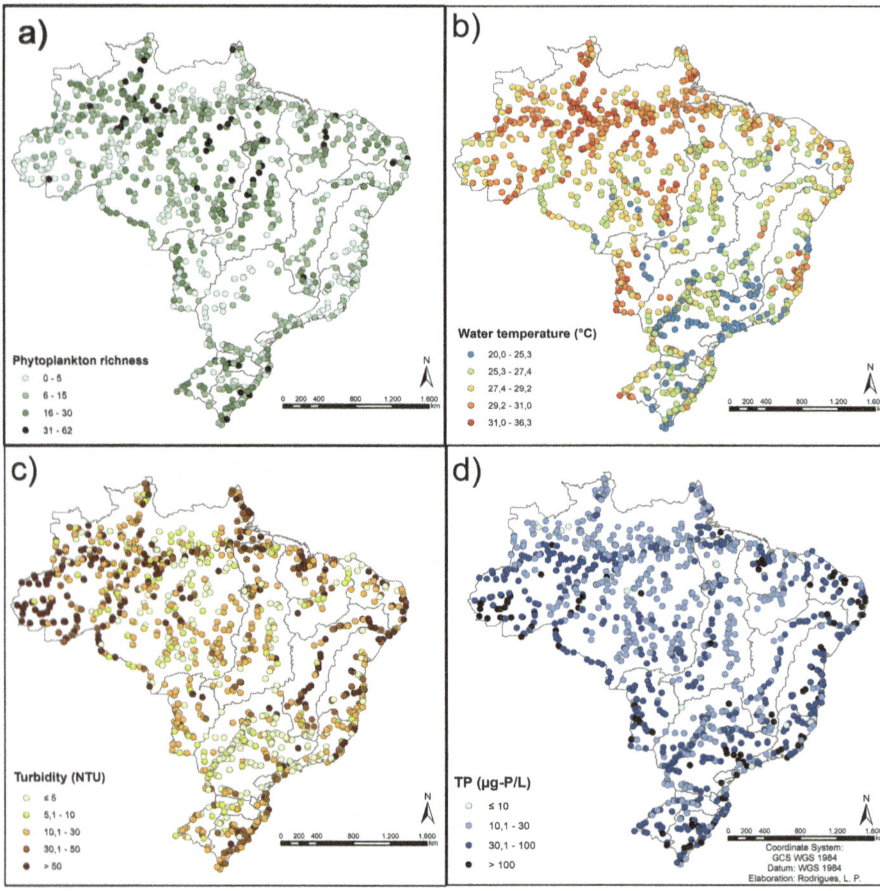

Fig. 4.4 (**a**) Spatial structure of phytoplankton richness, PhyRich; (**b**) water temperature, WTemp; (**c**) turbidity, Turb; (**d**) total phosphorus, TP. Values increase from the lighter to the darker color (n = 1045 in **a** and n = 1164 in **b, c, d**)

Based on TP concentrations, 21 sites were oligotrophic, 528 mesotrophic, 490 eutrophic, and 109 hypereutrophic. TP concentrations ranged from 5 µg/L in Ama, CoastS, CoastE, SFranc, and TocArag to 1671 µg/L (Paran). Median values were at mesotrophic levels (20.4–28.1 µg/L) in Ama, Parnb, and TocArag. The other nine HRs showed median values at eutrophic levels (35.4–79.2 µg/L; Fig. 4.3d). Furthermore, some sampling sites in the coastal RHs, Paran, and Parag were hypereutrophic (Fig. 4.4d).

4.3.3 *Relationships Between Species Richness and Environmental Variables*

The univariate relationships between PhyRich and environmental drivers showed different patterns. PhyRich was similar from 20 to ~28 °C, increased to 34 °C, and then decreased slightly to the end of gradient (Fig. 4.5a). PhyRich increased slightly until turbidity ~40 NTU L and then reduced to the end of the gradient (Fig. 4.5b). PhyRich remained similar until TP concentrations ~30 μg/L and then decreased to ~350 μg/L (Fig. 4.5c). At the beginning and end of the WTemp and TP gradients, we observed greater variability at the 95% confidence range.

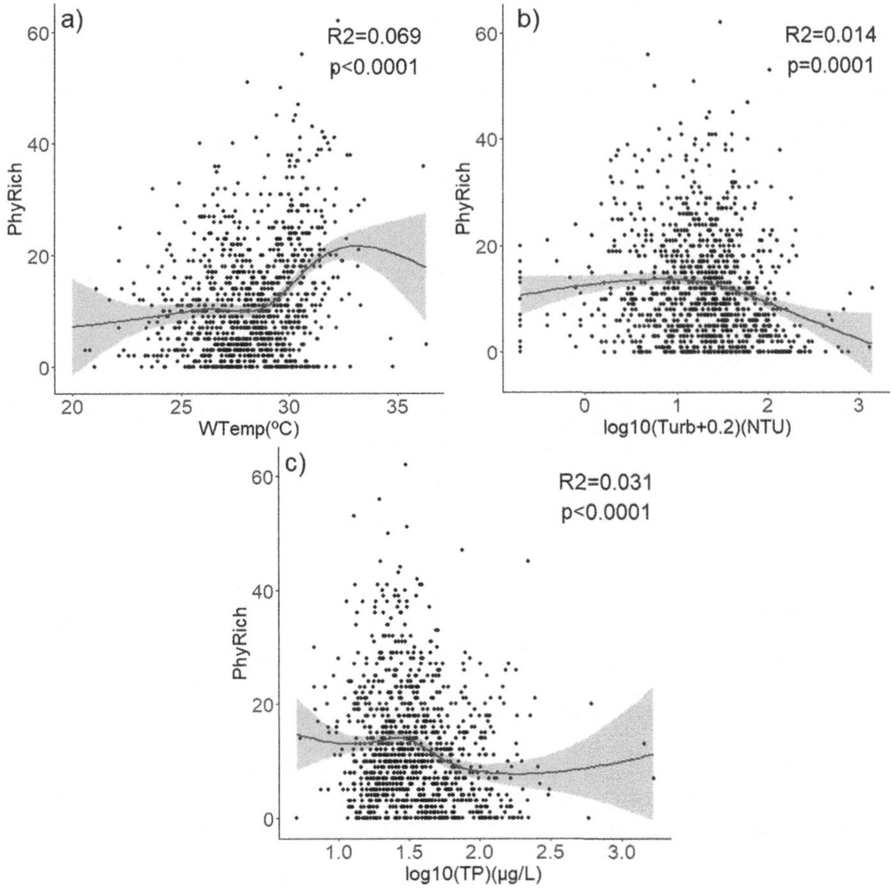

Fig. 4.5 The relationship between phytoplankton richness (PhyRich, taxa/sample) and (**a**) water temperature (Wtemp, °C, $n = 1025$), (**b**) turbidity (Turb, NTU, $n = 1017$), and (**c**) total phosphorus (TP, μg/L, $n = 1045$) obtained by GLMs (Generalized linear model regressions). The gray area is the approximate 95% confidence interval of the fitted function

The full GLM for the subcontinental scale selected WTemp, Turb, and TP (Table 4.2). PhyRich was significant and positively associated with WTemp and negatively with TP and Turb. For the regional scale, the three factors were significant only in the Ama HR, following a similar signal of the entire dataset. WTemp and Turb, both positively, explained the variability in PhyRich in the Parnb HR. In CoastNW, TocArag, and Parag, PhyRich was explained by WTemp as the only significant variable, and Urug and CoastE only by TP.

CoastNW showed the highest PhyRich variance explained by the environmental conditions (D2 = 35.303%), followed in descending order by Parnb (D2 = 31.357%), Ama (D2 = 25.160%), and Urug (D2 = 24.011%). For 5 of the 12 HRs, no model significantly explained PhyRich variability: CoastNE, CoastSE, CoastS, SFranc, and Paran (Table 4.2).

Table 4.2 The selected models for subcontinental and regional scales, where a significant combination of the environmental variables (autocorrelation removed) best explained the variation in phytoplankton richness

Regions	Model	AICcWt	D2	*n*
	Subcontinental scale			
	12.183*** + 1.255 Wtemp*** – 2.606 log(Turb+0.2)*** – 7.586 log(TP) ***	0.999	10.192	1017
	Regional scale			
CoastNW	9.513*** + 2.399 Wtemp** – 6.734 log(Turb+0.2) + 5.890 log(TP)	0.481	35.303	39
Parnb	16.423*** + 3.671 Wtemp* + 10.790 log(Turb+0.2)* + 34.530 log(TP)	0.282	31.357	26
Ama	12.781*** + 2.659 Wtemp*** – 5.806 log(Turb+0.2)*** – 15.518 log(TP)***	1.000	25.160	456
Urug	16.056*** + 0.453 Wtemp +0.662 log(Turb+0.2) – 20.439 log(TP)**	0.560	24.011	36
TocArag	17.302*** + 2.737 Wtemp*** + 3.527 log(Turb+0.2) – 3.768 log(TP)	0.344	14.961	96
Parag	12.318*** + 1.617 Wtemp* – 3.385 log(Turb+0.2) + 0.506 log(TP)	0.541	14.823	44
CoastE	5.217*** + 0.607 Wtemp – 0.199 log(Turb+0.2) – 6.917 log(TP)*	0.395	10.867	69

AICcWt AIC weight for the most parsimonious among the candidate model set, *D2* explained model deviance (%). All ΔAICc were zero. *WTemp* water temperature, *Turb* turbidity, *TP* total phosphorus. PhyRich in the five remaining hydrographic regions was not significantly explained by any variable

Ama Amazon Rainforest, *CoastE* coastal east, *CoastNE* coastal northeast, *CoastNW* coastal northwest, *CoastS* coastal south, *CoastSE* coastal southeast, *Parag* Paraguay, *Paran* Paraná, *Parnb* Parnaíba, *SFranc* São Francisco, *TocArag* Tocantins/Araguaia, *Urug* Uruguay. Signif. codes: $^{***}p = <0.001$; $^{**}p = <0.01$; $^{*}p = <0.05$

4.4 Discussion

Species richness (PhyRich), the number of morphospecies per site, is a primary measure of biodiversity in ecosystem functioning studies. The main factors shaping the broadscale patterns of PhyRich are temperature, system area, productivity, and, less often assessed, light, flushing, and grazing (Litchman 1998; Train and Rodrigues 1998; Smith et al. 2005; Nabout et al. 2007; Smith 2007; Muylaert et al. 2010; Weyhenmeyer et al. 2013; Vallina et al. 2014). Here, we evaluated PhyRich along with three of these gradients: temperature, productivity (total phosphorus as a proxy), and light availability (turbidity as a proxy). Our dataset is unprecedented in its broad spatial scale and inclusion of difficult-to-reach sites in Brazil. We hypothesized that PhyRich increases with water temperature and light availability (i.e., decreases with turbidity) and increases at intermediate productivity levels. Our hypotheses were partially confirmed since our broad spatial scale study showed that PhyRich was positively related to Wtemp and negatively to turbidity (lower light availability) and TP. Regarding the productivity gradient, PhyRich remained similar under oligo- and mesotrophic conditions, approximating the expected unimodal relationship. On a regional scale, we observed similar tendencies: for temperature, PhyRich initially increased to 34 °C and then slightly decreased; for turbidity, it smoothly increased until ~40 NTU and then decreased to the end of the spectrum; and for total phosphorus, it remained similar from 5 to 30 μg/L and then decreased until ~350 μg/L.

Phytoplankton is a highly diverse group composed of 10,000 to 25,000 extant species in marine and freshwater ecosystems (Sournia et al. 1991; Falkowski et al. 2004; Vargas et al. 2015). The 605 morphospecies found here comprise about 17% of the total number of freshwater phytoplankton species (~4000 spp.; Reynolds (2006). On our subcontinental scale, the most important major taxonomic groups were chlorophyceans and cyanobacteria, followed by zygnematophyceans, comprising ~60% of the total morphospecies. Of this total, only 20 (3%) were found in at least one sample in the 12 HRs. On a large scale, common species drive the species richness patterns (Jetz and Rahbek 2002; Lennon et al. 2004; Vázquez and Gaston 2004) because the richness of common species is more closely related to environmental variations than the richness of rare species (Jetz and Rahbek 2002; Kreft et al. 2006; Rahbek et al. 2007). Despite the extensive sampling program, we found no trend for the cumulative species curve to stabilize at an asymptote. With a greater sampling effort, we estimate that 100 more species could be found. The role of rare species in communities has been widely discussed through maintaining ecosystem processes and perhaps as a hidden driver of microbiome function (e.g., Jain et al. 2014; Mouillot et al. 2013; Jusset et al. 2017; Säterberg et al. 2019). However, the role of these species remains controversial because it may vary with local abundance, habitat specificity, and geographical spread (Rabinowitz 1981). The role of rare species can also be interpreted as complementarity or redundancy of the specific function. Some authors have suggested that rare species have low functional redundancy and are likely to support the most vulnerable ecosystem functions, with no other species carrying similar trait combinations. Rarity can result from different mechanisms, such as stochastic processes, inherent trade-offs in life-history strategies, and biotic and abiotic interactions (Jusset et al. 2017). On our subcontinental scale, of the 605 species, 22% were singletons and 14% doubletons.

The lowest PhyRich on our subcontinental scale – algae not detected by the method employed – occurred in ~10% of the samples, mainly in Ama (Amazonas main system, Madeira and Purus rivers) and CoastE HRs (Doce River). The highest PhyRich (62 species/sample) was found in a floodplain lake in the Ama HR (Xingu River watershed). Floodplain lakes are usually rich in phytoplankton species due to their strong interaction with the floodplain, modulated by the regular pulse of the hydrometric level, such as in the TocArag (Nabout et al. 2007) and Ama HRs (Huszar and Reynolds 1997; Cardoso et al. 2017).

The univariate relationships between PhyRich and the selected environmental drivers showed different patterns. On our subcontinental scale, the water temperature range was relatively narrow (20.6 to 36.4 °C). Despite this somewhat limited range, we found an increase of PhyRich with increasing temperature. PhyRich remained steady at the beginning of the gradient and then increased as the temperature rose to 34 °C. This finding contradicts the view that the low amplitude of temperature and irradiance through the entire annual cycle in tropical systems allows a permanently high biological activity, i.e., the alleged "endless summer" (Kilham and Kilham 1990). Even in warmer climates, the temperature matters and influences phytoplankton responses in tropical systems (Lewis 1987, 1996; Sarmento 2012; Silva et al. 2014; Rangel et al. 2016).

The positive influence of temperature as a determinant factor for diversity is well recognized in macroorganisms (Allen et al. 2002; Currie et al. 2004; Mittelbach et al. 2007). Particularly for phytoplankton, the temperature is among the major factors acting directly on growth rates and nutrient stoichiometry and indirectly on the lake mixing regime, shaping the spatial and temporal distribution of populations in freshwater systems (Schabhüttl et al. 2013; Kruk et al. 2017). Several phytoplankton studies have shown positive relationships between species richness and temperature (Ptacnick et al. 2010; Weyhenmeyer et al. 2013; Thomas et al. 2012; Kruk et al. 2017). This finding agrees with the metabolic theory of ecology, where a larger flow of energy in the food chain increases the productivity of systems and consequently the availability of resources, enabling more species to coexist (Hawkins et al. 2003; Segura et al. 2017). Temperature acts as an essential selective agent in species adaptation, influencing the occurrence of certain functional traits and maximizing their performance and maintenance in the system (Thomas et al. 2012; Schabhüttl et al. 2013).

Our models also indicated that temperature was essential to explain PhyRich at the regional scale, particularly for HRs above latitude 20°S (Ama, CoastNW, Parnb, TocArag, and part of Parag). Water temperature was positively related to PhyRich, but the largest increase was observed between 28 and 34 °C, with a tendency to decline in higher temperatures. However, the slight decrease at the end of the spectrum was driven by only three lotic sites with high temperatures and low PhyRich, located in the CoastE (Jequitinhonha River) and Ama (Negro River) HRs.

Besides temperature, light availability is another critical factor, favoring growth and increasing the number of available niches (Weyhenmeyer et al. 2013). The relationship between light and PhyRich remains controversial, but there is some convergence toward a decrease of PhyRich with decreasing light (Weyhenmeyer et al. 2013; Zhang et al. 2018), although Litchman (1998) found a reduction of PhyRich at both low and high light levels.

Turbidity, as a proxy for light availability, was an essential factor for explaining PhyRich, mainly on our subcontinental scale. PhyRich smoothly increased until turbidity was ~40 NTU and then decreased toward the end of the spectrum (~1000 NTU). The highest PhyRich at the intermediate levels may be linked to the productivity gradient based on the photoinhibition process. High light availability in clear water and low nutrient conditions are on one side of the gradient. On the other, turbid water promotes self-shading conditions in enriched waters (Zhang et al. 2018). On both sides, PhyRich was lower, as experimentally shown by Litchman (1998), who found a decrease of PhyRich in treatments with high and low light availability. At low light intensities, species that disappeared had low initial densities and low growth rates, explaining their exclusion. With high light availability, species had high growth rates, and some species may have been competitively excluded (Litchman 1998).

Our models indicated a decrease in PhyRich with increasing turbidity at the subcontinental scale and the Ama HR at the regional scale. The negative relationship in the Ama HR may be related to the broad spectrum of turbidity. One of the factors contributing to the increase in turbidity in this HR is the flux of suspended matter transported by the Amazon River to the ocean and the Andean contributions, transported mainly by the Solimões and Madeira rivers (Abe et al. 2006; Silva et al. 2013). On the other hand, waters draining the old Precambrian terrains have low suspended-matter content (Junk et al. 2011). Random field data may explain the positive relationship between PhyRich and turbidity in the Parnb HR since high PhyRich values were observed in samples from less- or more-enriched sites and with a wide variety of turbidity values. For example, in the eutrophic Parnaíba River, the smallest and largest PhyRich values (0 and 18 spp., respectively) were observed, with Turb between 13 and 603 NTU. Higher PhyRich values were observed in conditions of intermediate light availability and may be related to different niches provided by variations in light availability in the water column. Since light incidence decreases with increasing depth, different species have various light absorption capacities. Besides, light availability was not significantly related to PhyRich in any other HR.

In addition to temperature and light availability, productivity is essential for explaining PhyRich variability. Over our sizeable total phosphorus gradient (5–1671 μg/L), used as a proxy of productivity, 45% of the sites were mesotrophic and were primarily located in HRs where more pristine areas occur (Ama, Parnb, and TocArag); and 42% were eutrophic and located mainly in the other nine HRs. The remaining sites were hypereutrophic (~10%), located in the most populated and agricultural areas (coastal RHs, Paran, and Parag), and oligotrophic (~3%), sparsely distributed in all HRs except CoastNE and Urug.

Over large gradients, unimodal patterns of PhyRich have been observed (Dodson et al. 2000; Smith et al. 2005; Stomp et al. 2011), but some positive (Korhonen et al. 2011; Weyhenmeyer et al. 2013), negative, or nonsignificant relationships (Smith 2007) can occur over smaller gradients. This apparent contradiction has been attributed to the scale of a productivity gradient and synergy with other environmental factors (Smith et al. 2005; Cermeño et al. 2013; Rodríguez-Ramos et al. 2015; Zhang et al. 2018). Over our wide productivity gradient at the subcontinental scale, PhyRich was negatively related to productivity, remaining similar in oligo- and mesotrophic conditions (TP concentrations <30 μg/L) and decreasing toward eutrophic and hypereutrophic waters (~350 μg/L), approximating the expected unimodal relationship. The competition for resources explains the lower species richness at the extremes of the productivity gradient, with low productivity on the one hand and limitation by light in highly productive systems on the other (Zhang et al. 2018).

At the regional scale, PhyRich tended to decrease with increasing trophic states in Ama, Parnb, and TocArag, the HRs with the highest proportions of relatively pristine areas and with lower human population densities (see Table 4.1). In Ama, sites with high Turb and TP concentrations are primarily of natural origin, resulting from the weathering of rocks in the Andes where their sources are located (Abe et al. 2006; McClain and Naiman 2008; Almeida et al. 2015). These rivers exert a significant influence on the concentration of fine P-rich suspended sediments in the Amazon basin. The good water conditions in TocArag and Parnb, with low population densities (Moss and Moss 2005), may explain the high PhyRich.

On the other hand, the Paran and coastal regions, except CoastS, showed a trend toward lower PhyRich and high Turb and TP concentrations. These densely populated regions have poor sanitation and extensive agricultural and industrial areas (Moss and Moss 2005; Abe et al. 2006). The CostS and Parag HRs showed relatively high PhyRich. Notably, CoastS was also marked by high concentrations of TP due to the extensive agricultural areas with heavy use of fertilizers, as well as areas where the high TP concentration may be related to natural enrichment processes due to the influence of a large community of migratory birds that feed in these systems (Moss and Moss 2005). In the Parag HR, the high TP concentrations may be due to the low percentage of treated sewage and the influence of extensive cultivation areas, mainly soybean and cotton, as well as intensive livestock ranching (Abe et al. 2006). At some sites in Parag, high TP concentrations may also be of natural origin; in that region, the many bird roosts on the riverbanks can significantly contribute to phosphorus input (Moss and Moss 2005; Abe et al. 2006).

Over our wide TP gradient, the high data variability within the same hydrographic region may explain the D2 values of our results. Furthermore, the causality of field data, bias from potential unmeasured variables, and the inherent chaotic behavior of phytoplankton communities must be considered to explain the PhyRich results (Beninca et al. 2008; Kosten et al. 2012). Despite this, we still observed higher PhyRich at sites with intermediate concentrations. In low TP concentrations, competition for resources may explain the lower PhyRich. In contrast, at sites with high TP concentrations, light limitation due to either high phytoplankton biomass or high concentrations of other solids in suspension may have determined the lower

PhyRich. Massive phytoplankton blooms, especially cyanobacteria, can accumulate in surface waters, reducing light availability to other phytoplankton species and thus biodiversity (Chorus et al. 2000; Smith 2003; Huisman et al. 2018; Zhang et al. 2018). The negative relationship between PhyRich and Turb in Ama may be related to the high correlation of TP with the fine suspended sediments transported from the Andes to Amazonian rivers (Devol et al. 1995; Abe et al. 2006). For 5 of the 12 HRs, no model significantly explained PhyRich variability (CoastNE, CoastSE, CoastS, SFranc, and Paran).

In summary, by combining the variables that best described the PhyRich patterns on subcontinental and regional scales, WTemp most frequently explained PhyRich. On our subcontinental scale, PhyRich was positively related to water temperature and negatively to turbidity (lower light availability) and productivity, although PhyRich was similar in oligo- and mesotrophic conditions, approaching the expected unimodal relationship. Furthermore, we observed a tendency toward a slight reduction in the univariate models at the end of all gradients, with implications for the ecological functioning of freshwater systems.

Across a subcontinental scale, these findings contribute to a better understanding of how the environment affects phytoplankton species richness, with implications for their ecological functioning in freshwater ecosystems. In addition, our data include information from locations that have never been reached previously, many of them in a pristine state of conservation, serving as a reference for future studies that enable the identification of impacts in these areas.

Acknowledgments We thank Gerard Moss (*in memoriam*) and Margi Moss for sampling, Dr. Janet W. Reid (JWR Associates) for revising the English text, and Leonardo Preza Rodrigues for the map charting. VLMH is partially supported by the National Council for Scientific and Technological Development (CNPq), Brazil.

References

Abe D, Tundisi JG, Matsumura-Tundisi T, JEM T, Sidagis-Galli C, Teixeira-Silva V, Afonso GF, Albarici FL, PHA VH, Moss G, Moss M (2006) Monitoramento da qualidade ecológica das águas interiores superficiais e do potencial trófico em escala continental no Brasil com o uso de hidroavião. In: Tundisi JG, Matsumura-Tundisi T, JEM T, Sidagis-Galli C (eds) Eutrofização na América do Sul: causas consequências e tecnologias para gerenciamento e controle. Instituto Internacional de Ecologia, São Carlos, pp 225–239

Allen AP, Brown JH, Gillooly JF (2002) Global biodiversity, biochemical kinetics, and the energetic-equivalence rule. Science 297:1545–1548. https://doi.org/10.1126/science.1072380

Almeida R, Tranvik L, Huszar VLM, Sobek S, Mendonça R, Barros NO, Boemer G, Arantes JRJD, Roland F (2015) Phosphorus transport by the largest Amazon tributary (Madeira River, Brazil) and its sensitivity to precipitation and damming. Inland Waters 5:275–282. https://doi.org/10.5268/IW-5.3.815

Alvares CA, Stape JL, Sentelhas PC, Gonçalves JLM, Sparovek G (2013) Köppen's climate classification map for Brazil. Meteorol Z 22:711–728. https://doi.org/10.1127/0941-2948/2013/0507

Beninca E, Huisman J, Heerkloss R, Jöhnk KD, Branco P, Van Nes E, Scheffer M, Ellner SP (2008) Chaos in a long-term experiment with a plankton community. Nature 451:822–825. https://doi.org/10.1038/nature06512

Borges PAF, Train S (2009) Phytoplankton diversity in the upper Paraná River floodplain during two years of drought (2000 and 2001). Brazilian Journal of Biology = Revista Brasileira de Biologia 69(2 Suppl):637–647. https://doi.org/10.1590/S1519-69842009000300018

Borics G, Abonyi A, Salmaso N, Ptacnik R (2021) Freshwater phytoplankton diversity: models, drivers and implications for ecosystem properties. Hydrobiologia 848:53–75. https://doi.org/10.1007/s10750-020-04332-9

Burnham KP, Anderson DR (2004) Model selection and multimodel inference: a practical information-theoretic approach. Springer-Verlag, New York. https://doi.org/10.1007/b97636

Cadotte MW, Carscadden K, Mirotchnick N (2011) Beyond species: functional diversity and the maintenance of ecological processes and services. J Appl Ecol 48:1079–1087. https://doi.org/10.1111/j.1365-2664.2011.02048.x

Cardoso SJ, Nabout JC, Farjalla VF, Lopes PM, Bozelli RL, Huszar VLM, Roland F (2017) Environmental factors driving phytoplankton taxonomic and functional diversity in Amazonian floodplain lakes. Hydrobiologia 802:115–130. https://doi.org/10.1007/s10750-017-3244-x

Cermeño P, Rodríguez-Ramos T, Dornelas M, Figueiras F, Marañón E, Teixeira IG, Vallina SM (2013) Species richness in marine phytoplankton communities is not correlated to ecosystem productivity. Mar Ecol Prog Ser 488:1–9. https://doi.org/10.3354/meps10443

Chao A (1987) Estimating the population size for capture-recapture data with unequal catchability. Biometrics 43:783–791. https://doi.org/10.2307/2531532

Chase JM, Ryberg WA (2004) Connectivity, scale-dependence, and the productivity-diversity relationship. Ecol Lett 7:676–683. https://doi.org/10.1111/j.1461-0248.2004.00622.x

Chase JM, McGill BJ, Thompson PL, Antão LH, Bates AE, Blowes SA, Dornelas M, Gonzalez A, Magurran AE, Supp SR, Winter M, Bjorkman AD, Bruelheide H, Byrnes JEK, Cabral JS, Elahi R, Gomez C, Guzman HM, Isbell F, Myers-Smith IH, Jones HP, Hines J, Vellend M, Waldock C, O'Connor M (2019) Species richness change across spatial scales. Oikos 128:1079–1091. https://doi.org/10.1111/oik.05968

Chen M, Si Y, Han L, Liu X, Huang B, Kang CK (2021) Effect of prey selectivity and trophic cascades induced by mesozooplankton on the dynamics of phytoplankton. Mar Ecol Prog Ser 662:35–51. https://doi.org/10.3354/meps13627

Chorus I, Falconer IR, Salas HJ, Bartram J (2000) Health risks caused by freshwater cyanobacteria in recreational waters. J Toxicol Environ Health Part B 3(4):323–347. https://doi.org/10.1080/109374000436364

Currie DJ, Mittelbach GG, Cornell HV, Field R, Guégan JF, Hawkins BA, Kaufman DM, Kerr JT, Oberdorff T, O'Brien E, Turner JRG (2004) Predictions and tests of climate-based hypotheses of broad-scale variation in taxonomic richness. Ecol Lett 7:1121–1134. https://doi.org/10.1111/j.1461-0248.2004.00671.x

Devercelli M (2006) Phytoplankton of the middle Paraná River during an anomalous hydrological period: a morphological and functional approach. Hydrobiologia 563:465–478. https://doi.org/10.1007/s10750-006-0036-0

Devol AH, Forsberg BR, Rickey JE, Pimentel TP (1995) Seasonal variation in chemical distribution in the Amazon (Solimões) river: a multi-year time series. Global Biogeochem Cycles 9:307–328. https://doi.org/10.1029/95GB01145

Diaz S, Cabido M (2001) Vive la difference: plant functional diversity matters to ecosystem processes. Trends Ecol Evol 16:646–655. https://doi.org/10.1016/S0169-5347(01)02283-2

Dodson SI, Arnott SE, Cottingham KL (2000) The relationship in lake communities between primary productivity and species richness. Ecology 81(10):2662–2679. https://doi.org/10.1890/0012-9658(2000)081[2662:TRILCB]2.0.CO;2

Falkowski PG, Katz ME, Knoll AH, Quigg A, Raven J, Schofield O, Taylor FJR (2004) The evolution of modern eukaryotic phytoplankton. Science 305(5682):354–360. https://doi.org/10.1126/science.1095964

Gotelli NJ, Chao A (2013) Measuring and estimating species richness, species diversity, and biotic similarity from sampling data. Encyclopedia of Biodiversity 5:95–211. https://doi.org/10.1016/B978-0-12-384719-5.00424-X

Guisan A, Zimmermann NE (2000) Predictive habitat distribution models in ecology. Ecol Model 135(2–3):1–186. https://doi.org/10.1016/s0304-3800(00)00354-9

Hawkins BA, Field R, Cornell HV, Currie DJ, Guegan JF, Kaufmann DM (2003) Energy, water, and broad-scale geographic patterns of species richness. Ecology 84:3105–3117. https://doi.org/10.1890/03-8006

Hsieh TC, Ma KH, Chao A (2016) iNEXT: an R package for rarefaction and extrapolation of species diversity (Hill numbers). Method Ecol Evol 7:1451–1456. https://doi.org/10.1111/2041-210X.12613

Hubbell SP (2001) The unified neutral theory of biodiversity and biogeography. Princeton University Press, Princeton

Huisman J, Geoffrey AC, Paerl HW, Ibelings BW, Verspagen JMH, Visser PM (2018) Cyanobacterial blooms. Nat Rev Microbiol 16:471–483. https://doi.org/10.1038/s41579-018-0040-1

Huszar VLM, Reynolds CS (1997) Phytoplankton periodicity and sequences of dominance in an Amazonian floodplain lake (Lago Batata, Pará, Brasil): responses to gradual environmental change. Hydrobiologia 346:169–181. https://doi.org/10.1023/A:1002926318409

Huszar VLM, Silva LHS, Domingos P, Marinho M, Melo S (1998) Phytoplankton species composition is more sensitive than OECD criteria to the trophic status of three Brazilian tropical lakes. Hydrobiologia 369(370):59–71. https://doi.org/10.1007/978-94-017-2668-9

Jain M, DFB F, Prager CM, Hart GM, CM DV, Ahrestani FS, Palmer MI, Bunker DE, JMH K, Jouseau CF, Naeem S (2014) The importance of rare species: a trait-based assessment of rare species contributions to functional diversity and possible ecosystem function in tall-grass prairies. Ecol Evol 4(1):104–112. https://doi.org/10.1002/ece3.915

Jeppesen E, Meerhoff M, Davidson TA, Trolle D, Søndergaard M, Lauridsen TL, Beklioğlu M, Brucet S, Volta P, González-Bergonzoni I, Nielsen A (2014) Climate change impacts on lakes: an integrated ecological perspective based on a multi-faceted approach, with special focus on shallow lakes. J Limnol 73(s1):84–107. https://doi.org/10.4081/jlimnol.2014.844

Jetz W, Rahbek C (2002) Geographic range size and determinants of avian species richness. Science 297:1548–1551. https://doi.org/10.1126/science.1072779

Junk WJ, Piedade MTF, Schöngart J, Cohn-Haft M, Adeney JM, Wittmann F (2011) A classification of major naturally-occurring Amazonian lowland wetlands. Wetlands 31:623–640. https://doi.org/10.1007/s13157-011-0190-7

Jusset A, Bienhold C, Chatzinotas A, Gallien L, Gobet A, Kurm V, Küsel K, Rilling MC, Rivett DW, Salles JF, van der Heijden MGA, Youssef NH, Zhang X, Wei Z, Hol WHG (2017) Where less may be more: how the rare biosphere pulls ecosystems strings. ISME J 11:853–862. https://doi.org/10.1038/ismej.2016.174

Kilham SS, Kilham P (1990) Endless summer: internal loading processes dominate nutrient cycling in tropical lakes. Freshw Biol 23:379–389. https://doi.org/10.1111/j.1365-2427.1990.tb00280.x

Korhonen JJ, Wang J, Soininen J (2011) Productivity-diversity relationship in lake plankton communities. PLoS One 6:e22041. https://doi.org/10.1371/journal.pone.0022041

Kosten S, Huszar VLM, Bécares E, Costa LS, van Donk E, Hansson L-A, Jeppesen E, Kruk C, Lacerot G, Mazzeo N, De Meester L, Moss B, Lürling M, Nõges T, Romo S, Scheffer M (2012) Warmer climates boost cyanobacterial dominance in shallow lakes. Glob Chang Biol 18:118–126. https://doi.org/10.1111/j.1365-2486.2011.02488.x

Kreft H, Sommer JH, Barthlott W (2006) The significance of geographic range size for spatial diversity in Neotropical palms. Ecography 29:21–30. https://doi.org/10.1111/j.2005.0906-7590.04203.x

Kruk C, Devercelli M, Huszar VLM, Hernández E, Beamud G, Diaz M, Silva LHS, Segura AM (2017) Classification of Reynolds phytoplankton functional groups using individual traits and machine learning techniques. Freshw Biol 62(10):1681–1692. https://doi.org/10.1111/fwb.12968

Kruskal WH, Wallis WA (1952) Use of ranks in one-criterion variance analysis. J Am Stat Assoc 47(260):583–621. https://doi.org/10.2307/2280779

Lennon JJ, Koleff K, Greenwood JJD, Gaston KJ (2004) Contribution of rarity and commonness to patterns of species richness. Ecol Lett 7:81–87. https://doi.org/10.1046/j.1461-0248.2004.00548.x

Lewis WMJ (1987) Tropical limnology. Annu Rev Ecol Syst 18:159–184. https://doi.org/10.1146/annurev.es.18.110187.001111

Lewis WMJ (1996) Tropical lakes: how latitude makes a difference. In: Schiemer F, Boland KT (eds) Perspectives in tropical limnology. SPB, Amsterdam, pp 43–64

Litchman E (1998) Population and community responses of phytoplankton to fluctuating light. Oecologia 117:247–257. https://doi.org/10.1007/s004420050655

Magurran AE (2005) Species abundance distributions: pattern or process? Funct Ecol 19:177–181. https://doi.org/10.1111/j.0269-8463.2005.00930.x

Magurran AE, McGill BJ (2011) Biological diversity: Frontiers in measurement and assessment. Oxford University Press, Oxford

McClain ME, Naiman RJ (2008) Andean influences on the biogeochemistry and ecology of the Amazon River. Bioscience 58:325–338. https://doi.org/10.1641/B580408

Meerhoff M, Teixeira-de Mello F, Kruk C, Alonso C, González-Bergonzoni I, Pacheco JP, Arim M, Beklioğlu M, Brucet S, Goyenola G, Iglesias C, Lacerot G, Mazzeo N, Kosten S, Jeppesen E (2012) Environmental warming in shallow lakes: a review of potential changes in community structure as evidenced from space-for-time substitution approaches. Adv Ecol Res 46:259–349. https://doi.org/10.1016/B978-0-12-396992-7.00004-6

Mittelbach GG, Schemske DW, Cornell HV, Allen AP, Brown JM, Bush MB, Harrison SP, Hurlbert AH, Knowlton N, Lessios HA, McCain CM, McCune AR, McDade LA, McPeek MA, Near TJ, Price TD, Ricklefs RE, Roy K, Sax DF, Schluter D, Sobel JM, Turelli M (2007) Evolution and the latitudinal diversity gradient: speciation, extinction and biogeography. Ecol Lett 10:315–331. https://doi.org/10.1111/j.1461-0248.2007.01020.x

Moss G, Moss M (2005) Brasil das águas – Revelando o azul do verde e amarelo. Supernova Editora, São Paulo

Mouillot D, Bellwood DR, Baraloto C, Chave J, Galzin R, Harmelin-Vivien M, Kulbicki M, Lavergne S, Lavorel S, Mouquet N, Paine CET, Renaud J, Thuiller W (2013) Rare species support vulnerable functions in high-diversity ecosystems. PLoS Biol 11(5):e1001569. https://doi.org/10.1371/journal.pbio.1001569

Muylaert K, Pérez-Martínez C, Sánchez-Castillo P, Lauridsen TL, Vanderstukken M, Declerck SAJ, Van der Gucht K, Conde-Porcuna J-M, Jeppesen E, De Meester L, Vyverman W (2010) Influence of nutrients, submerged macrophytes and zooplankton grazing on phytoplankton biomass and diversity along a latitudinal gradient in Europe. Hydrobiologia 653:79–90. https://doi.org/10.1007/s10750-010-0345-1

Nabout JC, Nogueira IS, Oliveira LG, Morais RR (2007) Phytoplankton diversity (alpha, beta, and gamma) from the Araguaia River tropical floodplain lakes (Central Brazil). Hydrobiologia 557:455–461. https://doi.org/10.1007/s10750-006-0393-8

Nürnberg GK (1996) Trophic state of clear and colored, soft and hardwater lakes with special consideration of nutrients, anoxia, phytoplankton and fish. Lake Reserv Manag 12:432–447. https://doi.org/10.1080/07438149609354283

Oksanen J, Blanchet FG, Friendly M, Kindt R, Legendre P, McGlinn D, Minchin PR, O'Hara RB, Simpson GL, Solymos P, Stevens MHH, Szoecs E, Wagner H (2020) Vegan: community ecology package. R Package Version 2:5–7

Padisák J, Borics G, Grigorszky I, Soróczki-Pintér E (2006) Use of phytoplankton assemblages for monitoring ecological status of lakes within the water framework directive: the assemblage index. Hydrobiologia 553:1–14. https://doi.org/10.1007/s10750-005-1393-9

Ptacnik R, Andersen T, Brettum P, Lepisto L, Willen E (2010) Regional species pools control community saturation in lake phytoplankton. Proc R Soc B: Biol Sci 277:3755–3764. https://doi.org/10.1098/rspb.2010.1158

Rabinowitz D (1981) Seven forms of rarity. In: Synge H (ed) The biological aspects of rare plant conservation. John Wiley & Sons Ltd., Chichester, pp 205–217

Rahbek C, Gotelli NJ, Colwell RK, Entsminger GL, Rangel TF, Graves GR (2007) Predicting continental-scale patterns of bird species richness with spatially explicit models. Proc R Soc B: Biol Sci 274:165–174. https://doi.org/10.1098/rspb.2006.3700

Rangel LM, Soares MC, Paiva R, Silva LHS (2016) Morphology-based functional groups as effective indicators of phytoplankton dynamics in a tropical cyanobacteria-dominated transitional river–reservoir system. Ecol Indic 64:217–227. https://doi.org/10.1016/j.ecolind.2015.12.041

Reynolds CS (2006) The ecology of freshwater phytoplankton (ecology; biodiversity and conservation). Cambridge University Press, Cambridge

Reynolds CS, Descy JP, Padisák J (1994) Are phytoplankton dynamics in rivers so different from those in shallow lakes? Hydrobiologia 289:1–7. https://doi.org/10.1007/BF00007404

Rodríguez-Ramos M, Marañón E, Cermeño P (2015) Marine nano- and microphytoplankton diversity: redrawing global patterns from sampling-standardized data. Glob Ecol Biogeogr 24:527–538. https://doi.org/10.1111/geb.12274

Roland F, Lobão L, Vidal L, Jeppesen P, Paranhos R, Huszar VL (2010) Relationships between pelagic bacteria and phytoplankton abundances in contrasting tropical freshwaters. Aquat Microb Ecol 60:261–272. https://doi.org/10.3354/ame01429

Salmaso N, Zignin A (2010) At the extreme of physical gradients: phytoplankton in highly flushed, large rivers. Hydrobiologia 639:21–36. https://doi.org/10.1007/s10750-009-0018-0

Sarmento H (2012) New paradigms in tropical limnology: the importance of the microbial food web. Hydrobiologia 686:1–14. https://doi.org/10.1007/s10750-012-1011-6

Säterberg T, Jonsson T, Yearsley J, Berg S, Ebenman B (2019) A potential role for rare species in ecosystem dynamics. Sci Rep Nat 9:11107. https://doi.org/10.1038/s41598-019-47541-6

Schabhüttl S, Hingsamer P, Weigelhofer G, Hein T, Weigert A, Striebel M (2013) Temperature and species richness effects in phytoplankton communities. Oecologia 171:527–526. https://doi.org/10.1007/s00442-012-2419-4

Schmidt A (1994) Main characteristics of phytoplankton of the southern Hungarian section of the river Danube. Hydrobiologia 289:97–108. https://doi.org/10.1007/978-94-017-2670-2_9

Segura AM, Calliari D, Lan BL, Fort H, Widdicombe CE, Harmer R, Arim M (2017) Community fluctuations and local extinction in a planktonic food web. Ecol Lett 20:471–476. https://doi.org/10.1111/ele.12749

Silva MSR, Miranda SÁF, Domingos RND, Silva SLR, Santana GP (2013) Classificação dos rios da Amazônia: uma estratégia para preservação desses recursos. Holos Environ 13(2):163–174. https://doi.org/10.14295/holos.v13i2.7344

Silva LHS, Huszar VLM, Marinho MM, Rangel LM, Brasil J, Domingues CD, Branco CC, Roland F (2014) Drivers of phytoplankton, bacterioplankton, and zooplankton carbon biomass in tropical hydroelectric reservoirs. Limnologica 48:1–10. https://doi.org/10.1016/j.limno.2014.04.004

Smith VH (2003) Eutrophication of freshwater and coastal marine ecosystems a global problem. Environ Sci Pollut Res 10(2):126–139. https://doi.org/10.1065/espr2002.12.142

Smith VH (2007) Microbial diversity-productivity relationships in aquatic ecosystems. FEMS Microbiol Ecol 62:181–186. https://doi.org/10.1111/j.1574-6941.2007.00381.x

Smith VH, Foster BL, Grover JP, Holt RD, Leibold MA, de Noyelles F Jr (2005) Phytoplankton species richness scales consistently from laboratory microcosms to the world's oceans. Proc Natl Acad Sci U S A 102:4393–4396. https://doi.org/10.1073/pnas.0500094102

Smith MJ, Goodchild MF, Longley PA, Associates (2018) Geospatial analysis. A comprehensive guide to principles, techniques and software tools. 6th edn. http://www.spatialanalysisonline.com

Soininen J, Meyer S (2014) Phytoplankton richness is related to nutrient availability, not to pool size, in a subarctic rock pool system. Hydrobiologia 740:137–145. https://doi.org/10.1007/s10750-014-1949-7

Sournia A, Chrétiennot-Dinet M-J, Ricard M (1991) Marine plankton: how many species in the world oceans? J Plankton Res 13:1093–1099. https://doi.org/10.1093/plankt/13.5.1093

Stomp M, Huisman J, Mittelbach GG, Litchman E, Klausmeier CA (2011) Large-scale biodiversity patterns in freshwater phytoplankton. Ecology 92(11):2096–2107. https://doi.org/10.1890/10-1023.1

Talling JF, Prowse GA (2010) Selective recruitment and resurgence of tropical river phytoplankton: evidence from the Nile system of lakes, rivers, reservoirs and ponds. Hydrobiologia 637(1):187–195. https://doi.org/10.1007/s10750-009-9995-2

Thomas MK, Kremer CT, Klausmeier CA, Litchman EA (2012) A global pattern of thermal adaptation in marine phytoplankton. Science 338:1085–1088. https://doi.org/10.1126/science.1224836

Tilman D (2004) Niche trade-offs, neutrality, and community structure: a stochastic theory of resource competition, invasion, and community assembly. Proc Natl Acad Sci U S A 101(30):10854–10861. https://doi.org/10.1073/pnas.0403458101

Train S, Rodrigues LC (1998) Temporal fluctuations of the phytoplankton community of the Baía River, in the upper Paraná River floodplain, Mato Grosso do Sul, Brazil. Hydrobiologia 361:125–134. https://doi.org/10.1023/A:1003118200157

Utermöhl H (1958) Zur Vervollkomnung der quantitativen Phytoplankton-Methodik. Verhandlungen der Internationalen Vereinigung für Theoretische und Angewandte Limnologie 9:1–38. https://doi.org/10.1080/05384680.1958.11904091

Valderrama JC (1981) The simultaneous analysis of total nitrogen and phosphorus in natural waters. Mar Chem 10:109–122. https://doi.org/10.1016/0304-4203(81)90027-X

Vallina SM, Follows M, Dutkiewicz S, Montoya JM, Cermeno P, Loreau M (2014) Global relationship between phytoplankton diversity and productivity in the ocean. Nat Commun 5:4229. https://doi.org/10.1038/ncomms5299

Várbíró G, Görgényi J, Tóthmérész B, Padisák J, Hajnal É, Borics G (2017) Functional redundancy modifies species–area relationship for freshwater phytoplankton. Ecol Evol 7:9905–9913. https://doi.org/10.1002/ece3.3512

Vargas C, Audic S, Henry N, Decelle J, Mahé F, Logares R, Lara E, Berney C, Le Bescot N, Probert I, Carmichael M, Poulain J, Romac S, Colin S, Aury J-M, Bittner L, Chaffron S, Dunthorn M, Engelen S, Flegontova O, Guidi L, Horák A, Jaillon O, Lima-Mendez G, Lukes J, Malviya S, Morard R, Mulot M, Scalco E, Siano R, Vincent F, Zingone A, Dimier C, Picheral M, Searson S, Kandels-Lewis S, Acinas SG, Bork P, Bowler C, Gorsky G, Grimsley N, Hingamp P, Iudicone D, Not F, Ogata H, Pesant S, Raes J, Sieracki ME, Speich S, Stemmann L, Sunagawa S, Weissenbach J, Wincker P, Karsenti E (2015) Eukaryotic plankton diversity in the sunlit ocean. Science 348(6237):1261605. https://doi.org/10.1126/science.1261605

Vázquez L-B, Gaston KJ (2004) Rarity, commonness, and patterns of species richness: the mammals of Mexico. Glob Ecol Biogeogr 13:535–542. https://doi.org/10.1111/j.1466-822X.2004.00126.x

Vogt DR, Murrell DJ, Stoll P (2010) Testing spatial theories of plant coexistence: no consistent differences in intra- and interspecific interaction distances. Am Nat 175(1):73–84. https://doi.org/10.1086/648556

Weyhenmeyer GA, Peter H, Willen E (2013) Shifts in phytoplankton species richness and biomass along a latitudinal gradient–consequences for relationships between biodiversity and ecosystem functioning. Freshw Biol 58:612–623. https://doi.org/10.1111/j.1365-2427.2012.02779.x

Winder M, Hunter DA (2008) Temporal organization of phytoplankton communities linked to chemical and physical forcing. Oecologia 156:179–192. https://doi.org/10.1007/s00442-008-0964-7

Zalocar de Domitrovic Y (2002) Structure and variation of the Paraguay River phytoplankton in two periods of its hydrological cycle. Hydrobiologia 472:177–196. https://doi.org/10.1023/A:1016304803431

Zhang M, Straile D, Chen F, Shi X, Yanga Z, Cai Y, Yu J, Konga F (2018) Dynamics and rivers of phytoplankton richness and composition along productivity gradient. Sci Total Environ 625:275–284. https://doi.org/10.1016/j.scitotenv.2017.12.288

Chapter 5
The Importance of Palynology to Taxonomy

Vania Gonçalves-Esteves, Ana Carolina Mezzonato-Pires, Elysiane Barros Marinho, Raquel Maria Batista Souza de Souza, Roberto Lourenço Esteves, Simone Cartaxo-Pinto, and Claudia Barbieri Ferreira Mendonça

Abstract Given the wide morphological variability of pollen grains and spores, palynology can provide important contributions to several branches of science. Palynological information, alone or in conjunction with other data, is particularly useful for the taxonomic delimitation of species, genera, families, and higher-rank taxa. The pollen character with the highest relevance as a taxonomic marker is the type of structure resulting from sporogenesis: if a spore, it characterizes the large groups of vascular (Pteridophyta *s.l.*) and avascular (Bryophyta *s.l.*) cryptogams; if a pollen grain, it characterizes gymno- and angiosperms. Pollen unit, polarity, aperture, and sexine ornamentation are other important pollen characters; these traits are genetically determined and do not respond to variations in environmental conditions. Palynology applied to taxonomy has been the major field of research of the Álvaro Xavier Moreira Laboratory of Palynology, Brazil. Palynotaxonomy has proven useful in the study of the families Asteraceae, Passifloraceae, Podostemaceae, Vitaceae, and Leguminosae, among others. It is noteworthy the growing use of palynology to support cladistic and multidisciplinary studies seeking to establish relationships and degrees of kinship between different groups of plants to trace the evolutionary history of taxa.

V. Gonçalves-Esteves (✉) · E. B. Marinho · R. M. B. S. de Souza · S. Cartaxo-Pinto · C. B. F. Mendonça
Departamento de Botânica, Laboratório de Palinologia, Museu Nacional, Universidade Federal do Rio de Janeiro (MN/UFRJ), Rio de Janeiro, RJ, Brazil
e-mail: esteves.vr@gmail.com

A. C. Mezzonato-Pires
Departamento de Botânica, Universidade Federal de Juiz de Fora (UFJF), Juiz de Fora, MG, Brazil

R. L. Esteves
Departamento de Biologia Vegetal, Universidade do Estado do Rio de Janeiro (UERJ), Rio de Janeiro, RJ, Brazil

M. F. T. Medeiros, B. de Sá Haiad (eds.), *Aspects of Brazilian Floristic Diversity*, https://doi.org/10.1007/978-3-031-07453-0_5

Keywords Laboratory of Palynology · National Museum · Palynology · Palynotaxonomy

The interest in the knowledge of pollen grains in the Department of Botany of the National Museum, Rio de Janeiro, Brazil, began in 1949 from the initiative of Professor Álvaro Xavier Moreira, who recognized the taxonomic value of these structures. Professor Moreira began to study the pollen diversity of Asteraceae and, in 1958, published his first work, entitled "*Novo índice a ser introduzido na terminologia palinológica*" (A new index to be introduced to palynological terminology), which was followed by several other studies on the palynology of Asteraceae genera. After Professor Moreira's death in 1988, the laboratory was placed under the direction of Professor Vania Gonçalves L. Esteves, selected through civil service examination in 1989. In this new scenario, research efforts moved beyond the Asteraceae to include other families occurring in the *restingas* (coastal vegetation) of Rio de Janeiro State, as part of the departmental project "Botanical studies in the Restinga of Carapebus, Carapebus, Rio de Janeiro." In 2011, the now called Álvaro Xavier Moreira Laboratory of Palynology was joined by Professor Cláudia Barbieri Ferreira Mendonça, also hired by civil service examination. Currently, the team includes laboratory technician Pedro Cesar Teixeira de Souza, postdoctoral fellows, as well as several graduate and undergraduate students at different academic stages (Junior Scientific Initiation, Scientific Initiation, master's, and doctoral levels) (Fig. 5.1). Throughout most of its history, the laboratory had pollen morphology as its main line of research. Nowadays, palynotaxonomy has become the major focus

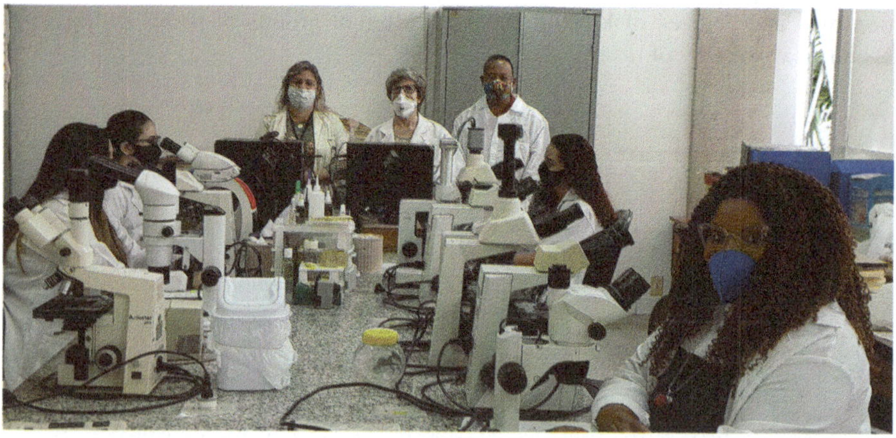

Fig. 5.1 Laboratório de Palinologia, Departamento de Botânica, Museu Nacional: students in front of the microscopes and in the background, from right to left, Professor Claudia BF Mendonça, Professor Vania G.L. Esteves, and the Laboratory Technician, Pedro Cesar T. de Souza

of investigations, followed by different fields of applied palynology, such as aeropalynology, melissopalynology, paleopalynology, and ecological palynology (animal-plant interactions).

Palynology is of great value to several branches of science. For instance, given the great morphological variability in pollen grains and spores, palynological information can be used, alone or in combination with other types of data, for taxonomic delimitation of higher-order taxa, families, genera, and species (Melhem et al. 2003). The pollen character that holds the greatest relevance as a taxonomic marker is the type of structure resulting from sporogenesis: spores characterize the large groups of vascular (Pteridophyta *s.l.*) and avascular (Bryophyta *s.l.*) cryptogams, whereas pollen grains characterize gymnosperms and angiosperms (Fig. 5.2). Other highly relevant pollen characters are (in descending order of importance) pollen unit, polarity, aperture (shape, number, and distribution), and sexine ornamentation. These traits are genetically controlled and not affected by environmental conditions. Information such as pollen shape and size also find application as diagnostic characters in the taxonomic treatment of many groups. However, unlike the previously mentioned traits, pollen shape and size can be influenced by the environment and/or affected by sample preparation (Erdtman 1952; Raynal and Raynal 1973; Salgado-Labouriau 1973; Salgado-Labouriau et al. 1965; Wortley et al. 2015).

In the last decade, growing attention has been paid to aperture type and number. As discussed by Furness and Rudall (2004), angiosperm pollen has undergone a gradual increase in aperture number during the evolutionary process, suggesting that this region of pollen grains is under strong selective pressure. Judd et al. (1999) and APG II (2003) noted that tricolpate pollen grain is a derived character in a large number of angiosperm families, being a synapomorphy and indicating monophyly of the eudicot clade. On the other hand, inaperturate pollen occurs mainly in basal

Fig. 5.2 Examples of cryptogam spores and phanerogam pollen grains

angiosperms, including ANA grade (APG IV 2016), Magnoliidae, and monocots, and is probably related to the thin exine of this group (Lu et al. 2015).

Inaperturate pollen also occurs, albeit far less frequently, in eudicots, both in basal and core lineages. In the latter, a lack of aperture is usually associated with infertility and functional dioecy, representing a reward offered by functionally female flowers to pollinators. Such an association is not observed in basal groups of angiosperms, eudicots, or monocots. Inaperturate but fertile pollen grains occur in different core eudicots, having evolved independently in different lineages. This condition may be functionally related to aquatic habitat, parasitism, heterostyly, anemophily, or arrangement in pollinia (Furness et al. 2006). Research has shed some light on the development of fertile inaperturate pollen. These pollen grains lack the characteristic colpal shield of endoplasmic reticulum and other organelles during the tetrad stage and, consequently, exhibit vesicles and uninterrupted deposition of primexine between the plasma membrane and the callose wall around micropores (Furness 2007).

According to Furness et al. (2006), endoapertures (inner part of compound apertures) are common in eudicots but rare in basal eudicot lineages, in which they are restricted to the families Menispermaceae, Sabiaceae, and Buxaceae. Sampson (2007), in analyzing pollen grains of the basal angiosperms Trimeniaceae, Winteraceae, and Monimiaceae, found examples of pollen differences and similarities within species, between species, and between genera. The author observed that different individuals of a Trimeniaceae species showed dimorphic pollen (inaperturate or polyporate), whereas other species had disulcate or polyporate pollen, indicating that apertures evolved from disulcate to polyporate, passing through an inaperturate stage. Zhang et al. (2017) analyzed several studies on the relationship between pollen characters and evolution in basal angiosperms. The results suggested that the herbaceous habit is associated with spheroidal pollen and the arborescent habit with oblate pollen, but no correlations were found between anemophily and aperture number. By contrast, Yang et al. (2020) reported a significant association between ornamentation and pollination type. Doyle (2005), contradicting some of his assertions in previous work with Walker (Walker and Doyle 1975), accepted that monosulcate, boat-shaped pollen grains with granular or atectate exine are more derived than those with columellate exine and continuous tectum (more basal), followed by pollen with foveolate-reticulate exine. These are but a few examples of the importance of palynology to taxonomy.

In the past, many scholars were interested in the palynology of Asteraceae but analyzed the family without drawing associations with taxonomy. We highlight the findings of Wodehouse (1928), who observed that the exine of Asteraceae pollen was marked by large depressions separated by walls. Wodehouse's studies contributed to increasing the knowledge of the wide variability of pollen exomorphology in Asteraceae by recognizing the three main surface patterns of the family: psilate, echinate, and lophate. Erdtman (1952), in examining the pollen characters of about 400 species belonging to 155 genera, found a great diversity of aperture, size, shape, and exine ornamentation, concluding that the family Asteraceae has a eurypalynological character. Stix (1960) was the first to describe, in detail, the pollen morphology of the family, proposing about 40 pollen types based on exine layers. Skvarla and Turner (1966) identified four types of pollen structure by transmission electron microscopy. These authors laid the foundations for our palynological knowledge of the family.

On the basis of molecular studies, Asteraceae was divided into 10 subfamilies and 35 tribes (Baldwin et al. 2002; Panero and Funk 2002), although, for practical purposes, the most widely accepted classification is that of Bremer (1994), which includes 3 subfamilies and 17 tribes (Asteroideae, with 10 tribes; Barnadesioideae, with 1 tribe; and Cichorioideae, with 6 tribes). Currently, 44 tribes and 13 subfamilies of Asteraceae have been recognized based on phylogenetic studies (Panero et al. 2014). The number of palynological studies on the family is small, considering its taxonomic, economic, and numerical importance, but, since 2009, more attention has been given to the topic, as explained by Wortley et al. (2012).

Steetz (1864) and Bentham (1873) were pioneers in the use of palynology as a subsidy to taxonomy. Wagenitz (1976) postulated that differences in pollen grains may be derived from modifications of a basal type of pollen found, in part, in Cardueae, Mutisieae, and most members of the tribe Anthemideae. According to the author, within the family, the emergence of different pollen types can be explained by three lines of evolution: (i) reduction of spines, (ii) reduction or loss of internal columellae (resulting in cavity formation), and (iii) formation of a pattern of ridges (lophate pollen). Studies have recognized the importance of these morphological features for understanding phylogenetic and taxonomic relationships in the family (Skvarla et al. 1977; Blackmore et al. 2009). Blackmore et al. (2009) hypothesized that echinate ornamentation is a synapomorphy in Asteraceae, whereas microechinate is a plesiomorphy; prolate and absence of cavea are likely plesiomorphic traits.

Jeffrey (1988, 2007) and Bolick and Keeley (1994) defined or redefined taxonomic categories of Asteraceae using palynology and other sources. Kingham (1976) and Kelley and Jones (1977) described several pollen types of the tribe Vernonieae on the basis of exine ornamentation of echinolophate pollen, chromosome number, and chemical data.

Palynology has shown to be a useful tool in the circumscription of genera. Robinson (1987, 1988a, b, 1990, 1996, 1999a, b), for instance, used palynological information to fragment the genus *Vernonia* Schreb. into smaller and, probably, monophyletic units. Through analysis of general morphology and, mainly, palynological data, the author recognized Liabeae as an independent tribe comprising taxa formerly assigned to Vernonieae and Senecioneae (subfamily Asteroideae). Gamerro (1991), using principally pollen traits, determined that *Moquinia* DC., a genus historically belonging to Mutisieae, was to be placed in the tribe Vernonieae.

The pollen grains of subfamilies Asteroideae and Barnadesioideae (sensu Bremer 1994) are similar in aperture type and sexine ornamentation (tricolporate with spiny sexine, with or without cavea) but differ in shape, size, and spine characteristics (size and distance between spines). The large variation in pollen characteristics within the subfamily Cichorioideae (pollen grains triporate, tricolporate, spiny, psilolophate, subechinolophate, echinolophate, with or without cavea) stimulated the use of pollen data for the taxonomy of this group.

According to Blackmore (1986) and Blackmore et al. (2009), lophate ornamentation, which is quite common in the subfamily Cichorioideae, has evolved many times, although the tendency to develop lophae might have evolved only once. In the subtribe, reversals to the non-lophate state are also common, even within genera. Under light microscopic examination, basal taxa have very similar microechinate ornamentation; the exine, however, shows great taxonomic and phylogenetic value because of its thickness and diversity of layers (Tellería 2008, 2017). In more derived taxa (Asteroideae), exine structure is practically uniform under light microscopy, generally exhibiting only a thin layer of columellae. In Cichorioideae, columellae are aggregated under spines; this apomorphy suggests the occurrence of a single transition to aggregation during the history of the group and a subsequent reversion toward homogeneity of columella distribution in the tribe Moquinieae (Blackmore et al. 2009).

The Álvaro Xavier Moreira Laboratory of Palynology has traditionally investigated the pollen grains of Asteraceae, as the family was a central interest of the founder. Our research group has produced several studies, dissertations, theses, monographs, and scientific communications on the family (Fig. 5.3a–d). We highlight the following studies: Gonçalves-Esteves (1976, 1977a, b) and Gonçalves-Esteves and Esteves (1986, 1988a, b, 1989a, b) on representatives of the tribe Heliantheae; Moreira et al. (1981) and Pastana (1989) on species of the tribe Mutisieae; Mendonça and Gonçalves-Esteves (2000) and Mendonça et al. (2000, 2002) on taxa of the tribes Eupatorieae, Vernonieae, Astereae, Inuleae, Heliantheae, and Tageteae occurring in the *restinga* of Carapebus, Rio de Janeiro State; Peçanha et al. (2001, 2008) on *Piptocarpha* R.Br., *Albertinia brasiliensis* Spreng., *Eremanthus bicolor* (DC.) Baker, and *Vanillosmopsis erythropappa* (DC.) Sch.Bip. individuals occurring in Rio de Janeiro State; Esteves and Gonçalves-Esteves (2003) on redefining the genus *Strophopappus* DC. using pollen characteristics; and Carrijo et al. (2005) on pollen grains of *Stilpnopappus* Mart. ex DC. and *Strophopappus*. Esteves et al. (2005) described a new taxon of *Vernonia* using pollen traits. Mendonça et al. (2007a, b, c, 2009, 2010) developed a series of palynotaxonomic studies of Vernonieae species occurring in southeastern Brazil aiming to analyze the pollen grains of about 94 species belonging to the following 10 genera on the basis of circumscription proposed by Robinson (1999a, b): *Albertinia* Spreng., *Chrysolaena* H.Rob., *Cololobus* H.Rob., *Cyrtocymura* H.Rob., *Dasyanthina* H.Rob., *Echinocoryne* H.Rob., *Lepidaploa* (Cass.) Cass., *Lessingianthus* H.Rob., *Stenocephalum* Sch.Bip., and *Vernonanthura* H.Rob. This study aimed to palynologically characterize species, provide information so that taxonomists could accept or not the circumscription of genera segregated from *Vernonia s.l.*, and underscore the importance of palynology to the taxonomy of the tribe.

Souza et al. (2008) contributed to and published the palynological results of the doctoral thesis of Anabella Deble at the Federal University of Santa Maria, Rio Grande do Sul State. Later that year, Pereira published part of his dissertation on the palynology of *Trixis* P.Browne species occurring in Rio de Janeiro State (Pereira et al. 2008). Magenta et al. (2010), in conducting a taxonomic revision of *Viguiera* Kunth (Heliantheae), used pollen data to support evidence of the homogeneity of the genus. Mendonça et al. (2010) reported part of the data of their thesis on the infrageneric relationship of *Lessingianthus*, analyzed from a palynological perspective. Loeuille et al. (2012) demonstrated the palynological homogeneity of *Eremanthus* Less. and applied the results to the taxonomy of the genus. Other studies of note include Carrijo et al. (2013), Abreu et al. (2015), Souza-Souza et al. (2016), Souza et al. (2016), Siniscalchi et al. (2017), Moreira et al. (2018), and Soares et al. (2020) (Fig. 5.3).

Palynological studies on Podostemaceae have proven useful in solving taxonomic problems and have advanced the knowledge of the morphology of the group by associating morphological traits to phylogenetic relationships that remain to be elucidated. Podostemaceae, the largest family of strictly aquatic angiosperms, comprises three subfamilies (Tristichoideae, Weddellinoideae, and Podostemoideae) with pantropical distribution, totaling about 50 genera (several of which are monotypic) and 300 species. In the Neotropical region, the family is well represented by all subfamilies, with circa 20 genera and 150 species (Bove and Philbrick 2010). Members of the family are known to have a high degree of local endemism. Brazil, Guyana, and Suriname are the most species-rich Neotropical countries, and distribution is not uniform. Of these countries, Brazil has the greatest diversity and endemism of Podostemaceae. Podostemaceae individuals are annual or perennial submerged aquatic herbs that grow in river rapids and waterfalls and may resemble lichens, bryophytes, or algae (Philbrick et al. 2010).

The Álvaro Xavier Moreira Laboratory of Palynology has conducted research aiming to characterize the pollen morphology of Neotropical genera of Podostemaceae using palynological data to support taxonomic and phylogenetic analyses. Such efforts have resulted in a better understanding of the systematics, biology, and evolution of pollen characters (Fig. 5.3e, f). In Podostemaceae, pollen grains are monads, dyads, or tetrads, small or medium, and isopolar or apolar, with polar area small or very small; aperture colporate, colpate, or porate; and ornamentation spinulose with granules/perforations or rugulate. Dyads are rare in angiosperms, having been reported only in Podostemaceae and Scheuchzeriaceae and occasionally in representatives of Ericaceae and *Mimosa* (Leguminosae). In Podostemaceae, dyad pollen is characteristic of the Paleotropical and Neotropical genera *Diamantine*, *Ceratolacis*, *Cipoia*, *Lophogyne*, and *Podostemum.* Pollen grains are united in the equatorial region in *Chives* and in the polar region in *Ceratolacis*, *Diamantina*, *Lophogyne*, and *Podostemum*. A striking trait that is exclusive to Podostemaceae pollen is the presence of a single aperture in the equatorial region of dyad pollen in *Chives*. Diffusion of knowledge about these pollen types is expected to stimulate their recognition in several research areas by inferring the paleoenvironment of their occurrence: rapids and waterfalls. The findings discussed here were reported in a monograph, a dissertation, and articles that adopted a holistic approach to the family Podostemaceae (Sá-Haiad et al. 2010; Abreu et al. 2012; Marinho et al. 2014, 2018).

The laboratory's research group performs studies that seek to contextualize pollen data for visualization of the evolution of pollen characters within a family based on phylogenetic analyses. The family Passifloraceae has been the subject of

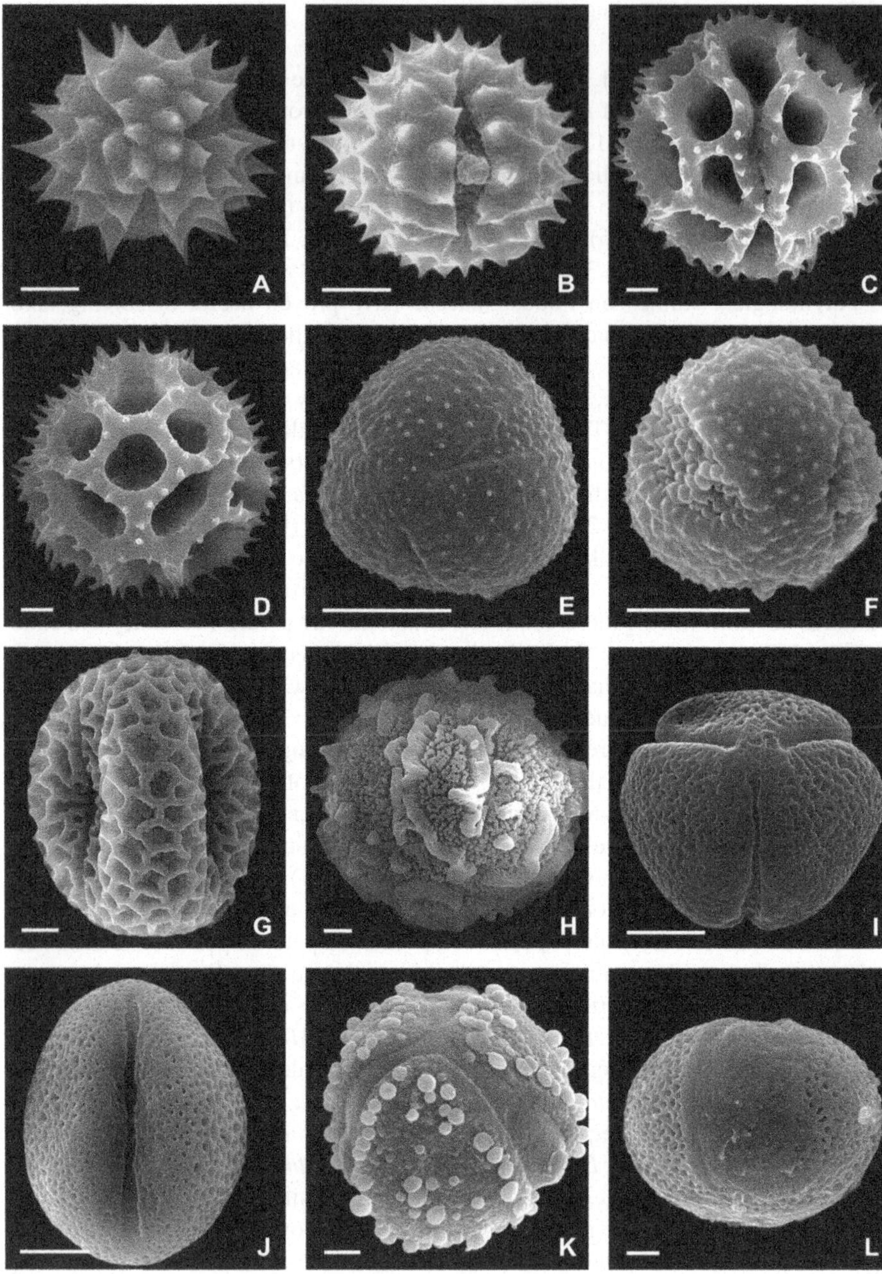

Fig. 5.3 Electromicrographs of pollen grains in scanning electron microscopy. Asteraceae: *Koanophyllon solidaginoides* (Kunth) R.M. King & H. Rob., (**a**) surface polar view, (**b**) equatorial view, aperture. *Echinocoryne pungens* (Gardner) H. Rob., (**c**) equatorial view, aperture, (**d**) equatorial view, showing the mesocolpus. Podostemaceae: *Castelnavia princeps* Tul. & Wedd., (**e**) polar view, surface. *Podostemum flagelliforme* (Tul. & Wedd.) C.T. Philbrick & Novelo, (**f**) equatorial view. Passifloraceae: *Passiflora emarginata* Bonpl., (**g**) equatorial view, showing the mesocolpus. *Passiflora cauliflora* Harms, (**h**) equatorial view, surface. Vitaceae: *Vitis bourgeana* Planch., (**i**) polar view, surface. *Ampelocissus barbata* (Wall.) Planch., (**j**) equatorial view, aperture. Leguminosae (Caesalpinioideae): *Bauhinia acreana* Harms, (**k**) polar view, surface. *Paubrasilia echinata* (Lam.) Gagnon, H.C. Lima & G.P. Lewis, (**l**) equatorial view, aperture

dissertations, theses, and postdoctoral studies aimed at associating its pollen morphology with taxonomy. Such efforts resulted in several publications. Passifloraceae *s.s.* is a botanical family of great economic importance. With a pantropical distribution, Passifloraceae comprises 700–750 species distributed in 16 genera and 2 tribes. Since the publication of the APG II system in 2003, the family has been grouped together with Turneraceae and Malesherbiaceae in Passifloraceae *s.l.* In many taxa of the family Passifloraceae, pollen grains exhibit diagnostic morphology, which underscores the importance of the complex taxonomic classification of the group. The team of the Álvaro Xavier Moreira Laboratory of Palynology has completed almost 15 years of studies on the family (Fig. 5.3g, h).

The first article of the group was published in 2007 as part of the dissertation of Michaele Alvim Milward-de-Azevedo entitled "*Passiflora* L. subgenus *Decaloba* (DC.) Rchb. in the Southeast region of Brazil." In 2010 and 2014, two other studies were published: the first expanded the sampling of *Decaloba*, describing all species occurring in Brazil; the second focused on the palynotaxonomy of the section *Xerogona* (Raf.) Killip, belonging to the subgenus *Decaloba* (Milward-de-Azevedo et al. 2004, 2010, 2014).

Between 2015 and 2020, eight articles were published as part of the master's, doctoral, and postdoctoral studies of Ana Carolina Mezzonato-Pires. In 2015, Mezzonato-Pires described the pollen of representatives of *Passiflora* occurring in the Atlantic Forest and *P.* subg. *Astrophea* (DC.) Mast., with a focus on sexine characters (Mezzonato-Pires et al. 2015a, b). In 2016, the author described pollen morphology and a new species of *Passiflora*, demonstrating the importance of pollen characters (Mezzonato-Pires et al. 2016). A new approach to the subgenus *Astrophea* was presented in 2017, via discussion of the systematics of this group (Mezzonato-Pires et al. 2017). In 2018, Gabriel Dos Santos Almeida published the results of his monograph, aimed at elucidating the proximity of Turneraceae and Passifloraceae *s.s.* by analyzing species of the genus *Piriqueta* Aubl. (Almeida et al. 2018). Later that year, Mezzonato-Pires and collaborators (Mezzonato-Pires et al. 2018a) analyzed pollen grain characters and described seven new records of *Passiflora* in Brazil, which substantiated the morphological description of the studied species. In 2019, data from postdoctoral studies began to be published, including work on the palynological description of non-Brazilian genera (*Barteria* Hook.f. and *Paropsia* Noronha ex Thouars) (Mezzonato-Pires et al. 2018b). Also in 2019, three other genera occurring in the Old World (*Adenia* Forssk., *Crossostemma* Planch. ex Benth., and *Schlechterina* Harms) were analyzed for pollen characters (Mezzonato-Pires et al. 2019). Finally, in 2020, the pollen morphology of the few genera occurring in the Neotropical region, excepting *Passiflora* (*Ancistrothyrsus* Harms, *Dilkea* Mast., and *Mitostemma* Mast.), was published as part of the monograph written by Gabriela Passos (Mezzonato-Pires et al. 2020). The most recent study of the research group was published in 2020 and written by the doctoral student Priscila De Freitas Cruz. It is focused on the section *Granadillastrum* (Triana & Planch.) Folio & J.M. MacDougal, belonging to the subgenus *Passiflora* (Cruz et al. 2020). A manuscript on the heterostyly of representatives of the genus *Piriqueta* is currently in press (as part of the dissertation of Gabriel Henrique Teixeira), and about five more articles on the family are being developed for publication (Teixeira et al. 2021, in press).

Vitaceae has also been widely studied by the research group because of the family's great economic importance. *Vitis vinifera* L., as well as other species and hybrids of *Vitis*, are used for the production of grapes, wines, juices, and raisins. The family is represented by 16 genera and about 950 species and distributed mainly in pantropical regions. Taxa are generally found in forest areas (Lombardi 2000; Wen 2007; Wen et al. 2018). In Brazil, the family is represented by the genera *Cissus* L. and *Clematicissus* Planch. Some species are ecologically important as vines in tropical and temperate forests (Lombardi 2000; Wen 2007; Lu et al. 2017).

The palynology of *Cissus* was the subject of the monograph and dissertation of Simone Cartaxo Pinto, and the results were reported in Cartaxo-Pinto et al. (2017). The author's interest in the palynological knowledge of the family continued to grow, and now, as a PhD student, Cartaxo-Pinto develops palynotaxonomic studies on other species of *Cissus* and genera of the family, with publications in important journals (Fig. 5.3i, j). According to Cartaxo-Pinto et al. (2017), the genus *Cissus* is heterogeneous with regard to pollen aperture; most species are tricolporate, but a few are syncolporate. Polarity (isopolar or heteropolar) and sexine ornamentation are important characters that contribute to taxonomic identification. Other genera of the family also show this heterogeneity, as confirmed by Cartaxo-Pinto et al. (2021a, b, in press).

Leguminosae (Fabaceae) is of great palynological interest because of the large variability of its representatives. According to the Flora do Brasil 2020 project, the family comprises 795 genera and almost 20,000 species, ranking among the 3 largest and most diverse families of plants (Lewis et al. 2005; LPWG 2017). Leguminosae has representatives in practically all biomes in Brazil as well as on almost all continents. Because of their high adaptability, Leguminosae individuals are highly diverse, occur in different phytogeographic domains, and represent a great portion of species of the Brazilian flora. The LPWG Group (2017) showed that the family is divided into six monophyletic subfamilies (Caesalpinioideae, Cercidoideae, Detarioideae, Dialioideae, Duparquetioideae, and Papilionoideae), five of which occur in Brazil. Mimosoideae, traditionally recognized as a subfamily, became a distinct clade within Caesalpinioideae and informally referred to as "Mimosoid clade," awaiting further classification (LPWG 2017).

Palynological studies were developed in the laboratory to contribute to the taxonomy of the family (Fig. 5.3k, l). Studies began with representatives found in the *restingas* of Rio de Janeiro State, including pollen grains of members of the subfamily Faboideae (Gonçalves-Esteves and Crespo 1994) and the tribes Aeschynomeneae and Phaseoleae (Souza et al. 2004). In 2008, Ana Dú Bocage, as part of her thesis project, investigated species of the genus *Acacia* Mill. (Dú Bocage et al. 2008). The doctoral thesis of Fábio França Moreira focused on the palynology of tree and shrub species of *Bauhinia* L., and the results were reported in Moreira et al. (2013). The master's student Fabiana Carvalho de Souza analyzed pollen grains of *Poecilanthe* Benth. and applied the knowledge to the taxonomy of the genus (Souza et al. 2014). Priscila de Freitas Cruz, in a scientific initiation project, developed a monograph on the palynology of Mimosoideae species (currently Mimosoid clade) occurring in the Atlantic Forest of Rio de Janeiro State; the results can be found in Cruz et al. (2018).

5.1 Final Considerations

Overall, research has shown that pollen data increasingly support taxonomic, cladistic, and multidisciplinary studies that seek to determine relationships and degrees of kinship between different groups of plants and trace the evolutionary lines of various taxa. As a consequence, palynotaxonomy has been valorized, supported by the fact that pollen morphology is widely studied from an evolutionary point of view and documented in fossil records of dated material. The team of the Álvaro Xavier Moreira Laboratory of Palynology has been intensively dedicated to the generation of informative palynological data as a means to expand taxonomic knowledge and provide subsidies to other areas of palynology.

Acknowledgments The authors thank the national and international herbaria that provided the material used in our studies and members of the laboratory team who performed studies not contemplated in this work. We also thank the Rio de Janeiro Research Foundation (FAPERJ), the Brazilian National Council for Scientific and Technological Development (CNPq), and the Brazilian Federal Agency for Support and Evaluation of Graduate Education (CAPES) for the grants and financial support provided over the years.

References

Abreu VHR, Bove CP, Philbrick CT et al (2012) Pollen morphology of the aquatic Brazilian endemic genus *Castelnavia* Tul. & Wedd. (Podostemaceae). Plant Syst Evol 298:1455–1461

Abreu VHR, Santos JC, Esteves RL et al (2015) Pollen morphology of *Praxelis* (Asteraceae, Eupatorieae, Praxelinae) in Brazil. Plant Syst Evol 301:599–608

Almeida GS, Mezzonato-Pires AC, Mendonça CBF et al (2018) Pollen morphology of selected species of *Piriqueta* Aubl. (Passifloraceae sensu lato). Palynology 43(1):43–52

APG II (Angiosperm Phylogeny Group) (2003) An update of the Angiosperm Phylogeny Group classification for the orders and families of flowering plants: APG II. Bot J Linn Soc 141:399–436

APG IV (2016) The Angiosperm Phylogeny Group. An update of the Angiosperm Phylogeny Group classification for the orders and families of flowering plants: APG IV. Bot J Linn Soc 181:1–20

Baldwin BG, Wessa BL, Panero JL (2002) Nuclear rDNA evidence for major lineages of Helenioid Heliantheae (Compositae). Syst Bot 27:161–198

Bentham G (1873) Compositae. In: Bentham G, Hooker JD (eds) Genera plantarum, 2(1). Lovell Reeve & Co., London

Blackmore S (1986) The identification and taxonomic significance of lophate pollen in Compositae. Can J Bot 64:3101–3112

Blackmore S, Wortley AH, Skvarla JJ et al (2009) Evolution of pollen in the Compositae. In: Funk VA, Susanna A, Stuessy TF, Bayer RJ (eds) Systematics, evolution and biogeography of Compositae. IAPT, Wien, pp 101–130

Bolick M, Keeley S (1994) Pollen morphology and classification of the *Vernonieae* (Compositae). Acta Bot Gall 141(2):279–284

Bove CP, Philbrick CT (2010) Podostemaceae neotropical. In: Milliken W, Klitg RDB, Baracat A (eds) (2009) Neotropikey – Chave interativa e recursos de informação para plantas de floração dos neotrópicos. http://www.kew.org/science/tropamerica/neotropikey/families/Podostemaceae.htm. Accessed 10 May 2021

Bremer K (1994) Asteraceae: cladistics and classification. Timber Press, Portland

Carrijo TT, Mendonça CBF, Esteves RL et al (2005) Palinotaxonomia de espécies de *Stilpnopappus* Mart. ex DC. e *Strophopappus* DC. (Compositae). Hoehnea 32(2):259–268

Carrijo TT, Garbin ML, Leite WP et al (2013) Pollen morphology of some related genera of Vernonieae (Asteraceae) and its taxonomic significance. Plant Syst Evol 299:1275–1283

Cartaxo-Pinto S, Mendonça CBF, Lopes RC et al (2017) Pollen morphology of species of *Cissus* (Vitaceae): an evaluation of ornamentation. Palynology 41:359–369

Cartaxo-Pinto S, Paulo GHC, Jackes BR et al (2021a) A palynological perspective on the tribe Viteae of the grape family (Vitaceae). Grana (in press)

Cartaxo-Pinto S, Jackes BR, Mendonça CBF et al (2021b) A palynological study of species of *Cissus* L. (Vitaceae) and their taxonomic importance. Botany (in press)

Cruz PF, Mendonça CBF, Gonçalves-Esteves V (2018) Pollen morphology of selected species of Mimosoideae (Leguminosae) occurring in the Atlantic Forest of the state of Rio de Janeiro, Brazil. Braz J Bot 41:197–208

Cruz PF, Mezzonato-Pires AC, Mendonça CBF et al (2020) Pollen morphology of *Passiflora* section *Granadillastrum* (Passifloraceae) with a detailed description of the aperture. Grana 59(5):319–334

Doyle JA (2005) Early evolution of angiosperm pollen as inferred from molecular and morphological phylogenetic analyses. Grana 44:227–251

Dú Bocage AL, Souza MA, Miotto STS et al (2008) Palinotaxonomia de espécies de *Acacia* (Leguminosae-Mimosoideae) no semi-árido brasileiro. Rodriguesia 59:587–596

Erdtman G (1952) Pollen morphology and plant taxonomy. Angiosperms. Almquist & Wiksell, Stockholm

Esteves RL, Gonçalves-Esteves V (2003) Redelimitação de *Stilpnopappus* Mart. ex DC. (Vernoniae-Asteraceaae). Bradea 9(14):77–92

Esteves RL, Mendonça CBF, Gonçalves-Esteves V (2005) Uma nova espécie de *Vernonia* (Asteraceae-Vernonieae) do Mato Grosso do Sul, Brasil. Bradea 11(1):17–28

Flora do Brasil (2020) Fabaceae. In: Jardim Botânico do Rio de Janeiro. http://floradobrasil.jbrj.gov.br/reflora/floradobrasil/FB115. Accessed 10 May 2020

Furness CA (2007) Why does some pollen lack apertures? A review of inaperturate pollen in eudicots. Bot J Linn Soc 155:29–48

Furness CA, Rudall PJ (2004) Pollen aperture evolution – a crucial factor for eudicot success? Trends Plant Sci 9(3):154–158

Furness CA, Magallo S, Rudall PJ (2006) Evolution of endoapertures in early-divergent eudicots, with particular reference to pollen morphology in Sabiaceae. Plant Syst Evol 263:77–92

Gamerro JC (1991) Identidad de *Pseudostifftia* com *Moquinia* (Compositae) y consideraciones sobre la ubicacion tribal del taxon. Darwin 30:123–136

Gonçalves-Esteves V (1976) Contribuição ao estudo palinológico da tribo Heliantheae (Compositae). Rev Bras Biol 36:157–166

Gonçalves-Esteves V (1977a) Contribuição ao estudo palinológico da tribo Heliantheae (Compositae) II. Rev Bras Biol 37:399–403

Gonçalves-Esteves V (1977b) Contribuição ao estudo palinológico da tribo Heliantheae (Compositae) III. Rev Bras Biol 37:837–841

Gonçalves-Esteves V, Crespo SEM (1994) Estudo polínico em plantas de restinga do Estado do Rio de Janeiro – Leguminosae A.L. Juss. – Faboideae Hutch. Bol Museu Nacional Zoologia 96:01–13

Gonçalves-Esteves V, Esteves RL (1986) Contribuição ao estudo polínico da tribo Heliantheae (Compositae) IV. Bol Museu Nacional. Zoologia 74:01–14

Gonçalves-Esteves V, Esteves RL (1988a) Contribuição ao estudo polínico da tribo Heliantheae (Compositae) V. Bol Museu Nacional N Ser Bot 77:01–11

Gonçalves-Esteves V, Esteves RL (1988b) Contribuição ao conhecimento de *Vernonia beyrichii* Less. (Compositae) – Aspectos morfológicos e palinológicos. Bol Museu Nacional Zoologia 27:67–112

Gonçalves-Esteves V, Esteves RL (1989a) Contribuição ao estudo polínico da tribo Heliantheae (Compositae) VI. Bol Museu Nacional N Ser Bot 80:01–11

Gonçalves-Esteves V, Esteves RL (1989b) Contribuição ao estudo polínico da tribo Heliantheae (Compositae) VII. Bol Museu Nacional. Zoologia 82:01–11

Jeffrey C (1988) The Vernonieae in east tropical Africa; notes on Compositae, V. Kew Bull 43:195–277

Jeffrey C (2007) Compositae. Introduction with key to tribes. In: Kadereit JW, Jeffrey C (eds) The families and genera of vascular plants. Flowering plants eudicots: Asterales. v. VIII. Springer-Verlag, Berlin, pp 61–87

Judd WS, Campbell CS, Kellogg EA et al (1999) Plant systematics: a phylogenetic approach. Inc. Publish, Sunderland

Kelley SC, Jones SB (1977) Taxonomic implications of external pollen morphology to *Vernonia* (Compositae) in the West Indies. Am J Bot 64(5):576–584

Kingham DL (1976) A study of the pollen morphology of tropical African and certain other Vernonieae (Compositae). Kew Bull 31:9–26

Lewis GP, Schrire B, Mackinder B et al (2005) Legumes of the world. Royal Botanic Gardens, Kew

Loeuille B, Souza-Souza RMB, Abreu VHR et al (2012) Pollen morphology of the genus *Eremanthus* Less. (Vernonieae, Asteraceae). Acta Bot Bras 26:46–57

Lombardi JA (2000) Vitaceae (*Ampelocissus, Ampelopsis* e *Cissus*) flora neotropica. N Y Bot Gard Monograph 80:1–251

LPWG (2017) New subfamily classification of the Leguminosae based on a taxonomically comprehensive phylogeny. Taxon 66:44–77

Lu L, Wortley AH, Li DZ et al (2015) Evolution of angiosperm pollen: 2. The basal angiosperms. Ann Mo Bot Gard 100:227–269

Lu LM, Cox CJ, Mathews S et al (2017) Optimal data partitioning, multispecies coalescent and Bayesian concordance analyses resolve early divergences of the grape family (Vitaceae). Cladistics 34:57–77. https://doi.org/10.1111/cla.12191

Magenta MAG, Nunes AD, Mendonça CBF et al (2010) Palynotaxonomy of Brazilian *Viguiera* (Asteraceae) species. Bol Soc Argent Bot 45:285–299

Marinho EB, Abreu VHR, Bove CP et al (2014) Pollen morphology of *Podostemum*: the type genus of the Family Podostemaceae. Palynology 38:162–170

Marinho EB, Bove CP, Mendonça CBF et al (2018) Pollen morphology of *Mourera* (Podostemaceae). Palynology 42:1–7

Melhem TS, Cruz-Barros MAV, Corrêa MAS et al (2003) Morfologia polínica em plantas de Campos do Jordão (São Paulo, Brasil). Bol Inst Bot 16:1–104

Mendonça CBF, Gonçalves-Esteves V (2000) Palinologia de espécies da tribo Eupatorieae (Compositae Giseke) ocorrentes na restinga de Carapebus, Carapebus, Rio de Janeiro. Rev Bras Bot 23:195–205

Mendonça CBF, Esteves RL, Gonçalves-Esteves V (2000) Morfologia polínica de algumas espécies da tribo Vernonieae (Compositae Giseke) ocorrentes na restinga de Carapebus, Rio de Janeiro. Hoehnea 27(2):131–142

Mendonça CBF, Gonçalves-Esteves V, Esteves RL (2002) Palinologia de espécies de Asteroideae (Compositae Giseke) ocorrentes na restinga de Carapebus, Carapebus, Rio de Janeiro. Hoehnea 29(3):233–240

Mendonça CBF, Gonçalves-Esteves V, Souza MA et al (2007a) Estudio polínico de *Albertinia brasiliensis* Spreng., de espécies de *Cololobus* H. Rob., *Cyrtocymura* H. Rob. y *Dasyanthina serrata* (Less.) H. Rob. (Vernoniinae-Compositae) que ocurren el sureste del Brasil. Polibotánica 23:77–92

Mendonça CBF, Souza MA, Gonçalves-Esteves V et al (2007b) Palinotaxonomia de espécies de *Chrysolaena* H. Rob., *Echinocoryne* H.Rob. e *Stenocephalum* Sch. Bip. (Vernonieae-Compositae) ocorrentes no sudeste do Brasil. Acta Bot Bras 21:627–639

Mendonça CBF, Gonçalves-Esteves V, Esteves RL (2007c) Palinotaxonomia de espécies de *Lepidaploa* (Cass.) Cass. (Vernoniinae-Compositae) ocorrentes no Sudeste do Brasil. Rev Bras Bot 30:69–86

Mendonça CBF, Gonçalves-Esteves V, Esteves RL et al (2009) Palynotaxonomy of *Vernonanthura* H. Rob. (Vernonieae, Compositae) species from Southeast, Brazil. Rev Bras Bot 32:625–640

Mendonça CBF, Carrijo TT, Esteves RL et al (2010) *Lessingianthus* H. Rob. (Vernonieae-Asteraceae): generic and infrageneric relationships based on pollen morphology. Nord J Bot 28:1–10

Mezzonato-Pires AC, Milward-de-Azevedo MA, Mendonça CBF et al (2015a) Pollen morphology of selected species of *Passiflora* L. (Passifloraceae) from the Atlantic Forest. Acta Bot Bras 29:391–399

Mezzonato-Pires AC, Mendonça CBF, Gonçalves-Esteves V (2015b) Pollen morphology and detailed sexine *Passiflora* subgenus *Astrophea* (Passifloraceae). Plant Syst Evol 301:2189–2202

Mezzonato-Pires AC, Gonçalves-Esteves V, Bernacci LC (2016) A new species of *Passiflora* subgenus *Astrophea* (Passifloraceae) from the Brazilian Amazon. Phytotaxa 288(1):077–084

Mezzonato-Pires AC, Mendonça CBF, Milward-de-Azevedo MA et al (2017) The systematic value of pollen morphology of *Passiflora* subgenus *Astrophea* (Passifloraceae). Phytotaxa 298:1–19

Mezzonato-Pires AC, Milward-de-Azevedo MA, Mendonça CBF et al (2018a) Taxonomy, palynology and distribution notes of seven species of *Passiflora* L. (Passifloraceae s.s.) newly recorded from Brazil. PhytoKeys 95:1–14

Mezzonato-Pires AC, Sousa HCF, Mendonça CBF et al (2018b) Pollen morphology of *Barteria* and *Paropsia*: implications for the systematics of Passifloraceae sensu stricto. Palynology 43(3):373–382

Mezzonato-Pires AC, Teixeira GHGSF, Gonçalves-Esteves V et al (2019) Pollen diversity of *Adenia*, *Crossostemma* and *Schlechterina* (Passifloraceae s.s.): types of apertures and implications for evolutionary studies. Palynology 44(3):529–538

Mezzonato-Pires AC, Nascimento GSPP, Mendonça CBF et al (2020) Palynology of three Neotropical genera of Passifloraceae sensu stricto: *Ancistrothyrsus* Harms, *Dilkea* Mast. and *Mitostemma* Mast. Palynology 45(2):259–268

Milward-de-Azevedo MA, Gonçalves-Esteves V, Baumgratz JFA (2004) Palinotaxonomia das espécies de *Passiflora* L. subg. *Decaloba* (DC.) Rchb. (Passifloraceae) no Sudeste do Brasil. Rev Bras Bot 27:655–665

Milward-de-Azevedo MA, Souza FC, Baumgratz JFA et al (2010) Palinotaxonomia de *Passiflora* L. subg. *Decaloba* (DC.) Rchb. (Passifloraceae) no Brasil. Acta Bot Bras 24:133–145

Milward-de-Azevedo MA, Souza FC, Gonçalves-Esteves V et al (2014) Palinotaxonomy of *Passiflora* section *Xerogona* (Passifloraceae). Phytotaxa 159:1–10

Moreira AX, Gonçalves-Esteves V, Leite NAS et al (1981) Estudos palinológicos de espécies da tribo Mutisieae (Compositae) I. Bol Museu Nacional N Ser Bot 64:01–31

Moreira FF, Vaz AMSF, Mendonça CBF et al (2013) The systematic value of pollen morphology in trees and shrubs species of *Bauhinia* L. (Caesalpinioideae- subg. Bauhinia- sect. Pauletia) occurring in Brazil. Acta Bot Bras 27:400–417

Moreira GL, Cavalcanti TB, Mendonça CBF et al (2018) Pollen morphology of Brazilian species of Verbesina L. (Heliantheae – Asteraceae). Acta Bot Bras 32:128–134

Panero JL, Funk VA (2002) Toward a phylogenetic subfamilial classification for the Compositae (Asteraceae). Proc Biol Soc Wash 115:909–922
Panero JL, Freire SE, Espinar AL et al (2014) Resolution of deep nodes yields an improved backbone phylogeny and a new basal lineage to study early evolution of Asteraceae. Mol Phylogenet Evol 80:43–53
Pastana GII (1989) Contribuição preliminar aos tipos polínicos da tribo Mutisieae (Compositae). Acta Bot Bras 21(1):55–66
Peçanha AF, Mendonça CBF, Gonçalves-Esteves V et al (2001) Palinotaxonomia de espécies de *Piptocarpha* R. Br. (Compositae, Vernonieae) do Estado do Rio de Janeiro. Bol Museu Nacional Nov Sér Bot 112:1–14
Peçanha AF, Esteves RL, Gonçalves-Esteves V (2008) Palinotaxonomia de *Albertinia brasiliensis* Spreng., *Eremanthus bicolor* (DC.) Baker e *Vanillosmopsis erythropappa* (DC.) Sch. Bip. (Compositae -Vernonieae). Acta Bot Bras 22:313–321
Pereira ACM, Esteves RL, Mendonça CBF et al (2008) Estudo polínico de espécies de *Trixis* (Mutisieae- Asteraceae) ocorrentes no Estado do Rio de Janeiro. Arq Mus Nac (Rio de J) 66:477–484
Philbrick CT, Bove CP, Stevens HI (2010) Endemism in neotropical Podostemaceae. Ann Missouri Bot Gard 97:425–456
Raynal A, Raynal J (1973) Une techinique de préparation des graines de pollen fragiles. Adansonia 11(1):77–79
Robinson H (1987) Studies in the *Lepidaploa* Complex (Vernonieae: Asteraceae) III- Two new genera *Cyrtocymura* and *Eirmocephala*. Proc Biol Soc Wash 100(4):844–855
Robinson H (1988a) Studies in the *Lepidaploa* Complex (Vernonieae: Asteraceae) IV- The new genus *Lessingianthus*. Proc Biol Soc Wash 101(4):929–951
Robinson H (1988b) Studies in the *Lepidaploa* Complex (Vernonieae: Asteraceae) V- The new genus *Chrysolaena*. Proc Biol Soc Wash 101(4):952–958
Robinson H (1990) Studies in the *Lepidaploa* Complex (Vernonieae: Asteraceae) VII- The new genus, *Lepidaploa*. Proc Biol Soc Wash 103:464–498
Robinson H (1996) The status of generic and subtribal revisions in the Vernonieae. In: Hind DJN, Beentje HJ (eds) Compositae: systematics. Proceedings of the international compositae conference, Kew, 1994 DJN Hind (Ed Chief), vol 1. Royal Botanic Gardens, Kew, pp 511–529
Robinson H (1999a) Revision of Paleotropical Vernonieae (Asteraceae). Proc Biol Soc Wash 112:216–219
Robinson H (1999b) Generic and subtribal classification of American Vernonieae. Smithson Contrib Bot 89:1–758
Sá-Haiad B, Torres CA, Abreu VHR et al (2010) Floral structure and palynology of *Podostemum weddellianum* (Podostemaceae, Malpighiales). Plant Syst Evol 190:141–149
Salgado-Labouriau ML (1973) Contribuição à Palinologia dos Cerrados. Academia Brasileira de Ciências, Rio de Janeiro
Salgado-Labouriau ML, Vanzolini PE, Melhem TS (1965) Varuation of polar axes and equatorial diameters in pollen grains of two species of *Cássia*. Grana Palynol 6:166–176
Sampson FB (2007) Variation and similarities in pollen features in some basal angiosperms, with some taxonomic implications. Plant Syst Evol 263:59–75
Siniscalchi CM, Souza-Souza RMB, Loeuille B et al (2017) The systematic value of pollen morphology in *Chresta* Vell. ex DC. (Vernonieae, Asteraceae). Rev Palaeobot Palynol 244:182–191
Skvarla JJ, Turner BL (1966) Systematic implications from electron microscopic studies of Compositae pollen – a review. Ann Mo Bot Gard 53(2):220–256
Skvarla JJ, Turner BL, Patel VC et al (1977) Pollen morphology in the Compositae and in morphologically related families. In: Heywood VH, Harborne JB, Turner BL (eds) The biology and chemistry of the Compositae. Academic Press, London, pp 141–248
Soares PN, Gonçalves-Esteves V, Semir J et al (2020) *Chrysolaena glandulosa* (Vernonieae, Asteraceae): a new species from Brazil. Phytotaxa 439:295–230

Souza FC, Souza MA, Mendonça CBF et al (2004) Estudo polínico de espécies de Aeschynomeneae e Phaseoleae (Papilionoideae-Leguminosae Juss.) ocorrentes nas restingas do Estado do Rio de Janeiro. Arq Mus Nac (Rio de J) 62:357–366

Souza PCT, Souza FC, Oliveira-Deble AO et al (2008) Estudo polínico de nove espécies da subtribo Baccharinae Less. (Astereae-Asteraceae) ocorrentes no Sul do Brasil. Balduinia 12:01–11

Souza FC, Meireles JE, Mendonça CBF et al (2014) Pollen diversity and its implications to the systematics of *Poecilanthe* (Fabaceae, Papilionoideae, Brongniartieae). Plant Syst Evol 300:1759–1770

Souza MA, Mendonça CBF, Esteves RL et al (2016) Pollen morphology of *Graphistylis* B. Nord. species (Asteraceae) of Brazil. Acta Bot Bras 30:138–146

Souza-Souza RMB, Loeuille B, Mendonça CBF et al (2016) Pollen morphology of genus *Paralychnophora* (Vernonieae – Asteraceae). Palynology 40:280–288

Steetz J (1864) Crystallopollen and Ambassa. In: WCH Peters (ed.) Naturwissenschaftliche Reise nach Mossambique auf Befehl seiner Majestãt des Königs Friedrich Willhelm IV. Botaik 6(2):363–364

Stix E (1960) Pollenmorpholosgische untersuchungen na Compositen. Grana Palynol 2:41–114

Teixeira GHGSF, Mezzonato-Pires AC, Rocha LNG et al (2021) Pollen morphology and heterostyly in *Piriqueta* (Passifloraceae s.l.): taxonomic implications. Plant Syst Evol (in press)

Tellería MC (2008) Taxonomic significance of pollen types in the Guyana Highland-centred composite genera of Mutisioideae (Asteraceae). Bot J Linn Soc 156:327–340

Tellería MC (2017) Spines vs. microspines: an overview of the sculpture exine in selected basal and derived Asteraceae with focus on Asteroideae. J Plant Res 130:1023–1033

Wagenitz G (1976) Systematic and phylogeny of the Compositae. Plant Syst Evol 125:29–46

Walker JW, Doyle JA (1975) The basis of angiosperm phylogeny: palynology. Ann Mo Bot Gard 62:270–664

Wen J (2007) The families and genera of vascular plants. Vitaceae. Flowering plants. Eudicots IX:467–479

Wen J, Lu LM, Nie ZL et al (2018) A new phylogenetic tribal classification of the grape family (Vitaceae). J Syst Evol 56(4):262–272

Wodehouse RP (1928) The phylogenetic value of pollen grain characters. Ann Bot 42:891–934

Wortley AH, Blackmore S, Chissoe WF et al (2012) Recent advances in Compositae (Asteraceae) palynology, with emphasis on previously unstudied and unplaced taxa. Grana 51(2):158–179

Wortley AH, Wang H, Lu L et al (2015) Evolution of angiosperm pollen. 1. Introduction. Ann Mo Bot Gard 100:177–226

Yang LE, Lu L, Burgess K et al (2020) Evolution of angiosperm pollen: 8 Lamiids. Ann Mo Bot Gard 105:323–376

Zhang MY, Lu L, Wortley AH et al (2017) Evolution of angiosperm pollen: 4. Basal eudicots. Ann Mo Bot Gard 102:141–182

Chapter 6
Ecological Palynology

Claudia Barbieri Ferreira Mendonça, Leila Nunes Morgado, Gabriel Henrique Gomes de Souza Freitas Teixeira, Elen de Lima Aguiar-Menezes, Thiago Sampaio de Souza, Vinicius José Fernandes, Alice Teodorio Lixa, Roberto Lourenço Esteves, and Vania Gonçalves-Esteves

Abstract Ecological palynology is an interdisciplinary science dedicated to the study of pollen grains used by animals, particularly insects, to meet basic needs for nutrition, reproduction, and, ultimately, survival and species perpetuation. Animal–pollen interactions play an important role in nature: that of pollination. An important branch of ecological palynology is entomopalynology, which investigates the relationship between insects and pollen from different perspectives and through different associations. Bees and flowers are notable examples of such interactions. Flowers visited by bees can be traced by examining pollen grains stored in hives or present in bee products, including honey, propolis, and royal jelly. Pollen grains found in honey can provide information on botanical origin and geographical distribution. This type of investigation belongs to a field of science called melissopaly-

C. B. F. Mendonça (✉) · G. H. G. de Souza Freitas Teixeira · V. Gonçalves-Esteves
Departamento de Botânica, Laboratório de Palinologia, Museu Nacional, Universidade Federal do Rio de Janeiro (MN/UFRJ), Rio de Janeiro, RJ, Brazil
e-mail: cb.mendonca@gmail.com

L. N. Morgado · T. S. de Souza
Instituto de Investigação em Tecnologias Agrárias e do Ambiente, Faculdade de Ciências Agrárias e do Ambiente, Universidade dos Açores, Angra do Heroísmo, Ilha Terceira, Açores, Portugal

E. de Lima Aguiar-Menezes
Departamento de Entomologia e Fitopatologia, Instituto de Ciências Biológicas e da Saúde, Universidade Federal Rural do Rio de Janeiro (UFRRJ), Seropédica, RJ, Brazil

Programa de Pós-Graduação em Fitotecnia, Universidade Federal Rural do Rio de Janeiro (UFRRJ), Seropédica, RJ, Brazil

V. J. Fernandes · A. T. Lixa
Programa de Pós-Graduação em Fitotecnia, Universidade Federal Rural do Rio de Janeiro (UFRRJ), Seropédica, RJ, Brazil

R. L. Esteves
Departamento de Biologia Vegetal, Universidade do Estado do Rio de Janeiro (UERJ), Rio de Janeiro, RJ, Brazil

M. F. T. Medeiros, B. de Sá Haiad (eds.), *Aspects of Brazilian Floristic Diversity*,
https://doi.org/10.1007/978-3-031-07453-0_6

nology. A more recent application of entomopalynology has been the examination of pollen ingested by entomophagous insects, such as ladybugs (Coccinellidae) and green lacewings (Chrysopidae), to identify trap and attractive plants. Conservation tactics contribute to the preservation of natural enemies, stimulate the use of more selective or less toxic insecticides, and, when combined with effective sampling of pests and natural enemies, can guide decision-makers on the suitability of insecticides for organic agriculture.

Keywords Conservation · Entomopalynology · Laboratory of Palynology · Melissopalynology · National Museum · Organic agriculture

Palynology is an interdisciplinary science that deals with the study of pollen grains, spores of bryophytes, pteridophytes s.l., and, less commonly, other recent and fossil palynomorphs such as foraminifera, diatoms, dinoflagellates, fungal spores, etc. The prefix *palyno* refers to the Greek word *paluno*, which can be translated as "to sprinkle," whereas the suffix *logy* refers to the word "study." Emerged as a branch of geoscience in Scandinavia in the early twentieth century, palynology became an established field of research at the end of World War II (Edwards and Pardoe 2018). One of the applied areas of this science, known as ecological palynology, can be combined with behavioral data from different organisms to help clarify relationships of interdependence between plant species and pollinators or herbivores (consumers of nectar and/or pollen). Entomopalynology, on the other hand, explores, from different perspectives, the relationship between insects and pollen grains and, augmented by the study of interaction networks, allows not only to taxonomically differentiate plant species foraged by insects but also to identify floral preferences and variations in the resources used by insects throughout the year.

Bees are perhaps the most important agents in insect–plant interactions. These insects can be classified according to their level of organization, ranging from species with solitary habits, which are the most diverse, to social species characterized by high levels of organization among individuals. Able to travel great distances in search of food, bees are exclusively dependent on floral resources, such as pollen and nectar, although some groups also rely on less common resources (e.g., oils and floral resins). Pollen grains contain a wide variety of substances, including carbohydrates (13–55%), proteins (10–40%), fibers (0.3–20%), lipids (1–13%), vitamins, minerals, polyphenols, sterols, and terpenes (Belina-Aldemita et al. 2019). The chemical composition of pollen is mainly determined by its botanical origin but also influenced by factors such as geographical origin, climate, soil type, time of foraging, and forager species (Bauermann 2006; Węglińska et al. 2020).

The higher the fidelity to a plant species, the greater the chances of bees transferring pollen between plants of the same species, resulting in greater pollination efficiency for plants that require cross-pollination (Baker and Hurd 1968; Michener 1979). Cross-pollination increases gene exchange between populations (Pires-O'Brien and O'Brien 1995). According to Kiill and Costa (2003), knowledge of the aspects of floral biology associated with pollinator behaviors that result in cross-pollination allows developing strategies for the conservation of plants (flora) and animals (pollinators).

In melissopalynology, different methods of palynological research are applied to identify the source of floral resources used by bees. The most commonly analyzed samples include pollen grains transported in specialized structures of bees or found in honey sediments (Gutiérrez and García 2007; Silva et al. 2010; Morgado et al. 2011; Mouga and Dec 2012; Morgado and Lorenzon 2014; Morgado and Rocha 2016; Silva et al. 2014; Oliveira and Santos 2014), propolis (Morgado et al. 2018), and royal jelly (Morgado and Barth 2011). Some studies analyzed pollen residues found in trap nests of solitary bees that nest in pre-existing cavities (Dorea et al. 2009, 2010).

The Laboratory of Palynology of the National Museum, Federal University of Rio de Janeiro (UFRJ), Brazil, conducts research on different areas of palynology. In 2016, a scientific initiation student produced a monograph on the diversity of corbicula pollen of the stingless bee *Tetragonisca angustula* Latreille (Apidae, Trigonini) (Silva 2016). Two colonies located in different parts of the Botanical Garden of the National Museum were sampled. Sampling was performed between 8:00 and 12:00 h by capturing worker bees that returned to the nest. Pollen was removed from the corbiculae of bees and stored in bottles identified with the nest, date, and time of collection. Collections were carried out on a monthly basis from August 2015 to August 2016, totaling 12 collections per nest. The material was treated by the acetolysis method (Erdtman 1952). During the study period, *T. angustula* individuals collected 94,052 pollen grains, classified into 21 pollen types belonging to 9 plant families (Fig. 6.1). Worker bees from nest 1 collected pollen from the following families and species: Solanaceae, *Cestrum nocturnum* (21.7%), Solanaceae sp. 1 (7.8%), sp. 7 (7.4%); Leguminosae, *Caesalpinia echinata* (12.7%), *Delonix regia* (16.9%); Euphorbiaceae, *Croton* sp. (11.7%); Myrtaceae, *Eugenia uniflora* (4.7%); Chenopodiaceae (3.7%); Malvaceae (Bombacoideae), *Pachira aquatica* (2.6%), sp. 8 (2.5%), sp. 2 (2.3%); Cactaceae, *Rhipsalis* sp. (1.9%), sp. 3 (1.4%); Melastomataceae, *Tibouchina holosericea* (1.1%), sp. 4 (0.7%), sp. 2; and Asteraceae (0.3%), sp. 5 (0.2%), sp. 1 (0.1%), *Mikania guaco* (0.5%), sp. 1 (0.4%). Bees from nest 2 collected pollen from the following plants: Leguminosae, *D. regia* (31.3%), *C. echinata* (4%), sp. 7 (3.8%), sp. 3 (3.4%); Solanaceae, *C. nocturnum* (28.1%); Cactaceae, *Rhipsalis* sp. (10%); Myrtaceae, *E. uniflora* (5.4%), sp. 1; Asteraceae (4.2%); Malvaceae (Bombacoideae), *P. aquatica* (2.6%); Chenopodiaceae (1.9%), sp. 5 (1.5%), sp. 8 (1.3%); Euphorbiaceae, *Croton* sp. (1%); and Melastomataceae, *T. holosericea* (0.7%). For both nests, the most frequent families were Solanaceae and Fabaceae, regardless of the season. Overall, the results showed that *T. angustula* has a generalist feeding habit.

Another study developed at the Laboratory of Palynology was part of a doctoral thesis of a graduate student enrolled at Unicamp (Fonseca 2012) and resulted in a publication (Fonseca et al. 2016). The study involved monitoring hummingbird visits to plants to investigate their action as pollinators, the role of pollen as a food source, and the location of pollen on the body of birds. Three species of hummingbirds from the Atlantic Forest were analyzed. This research provided relevant

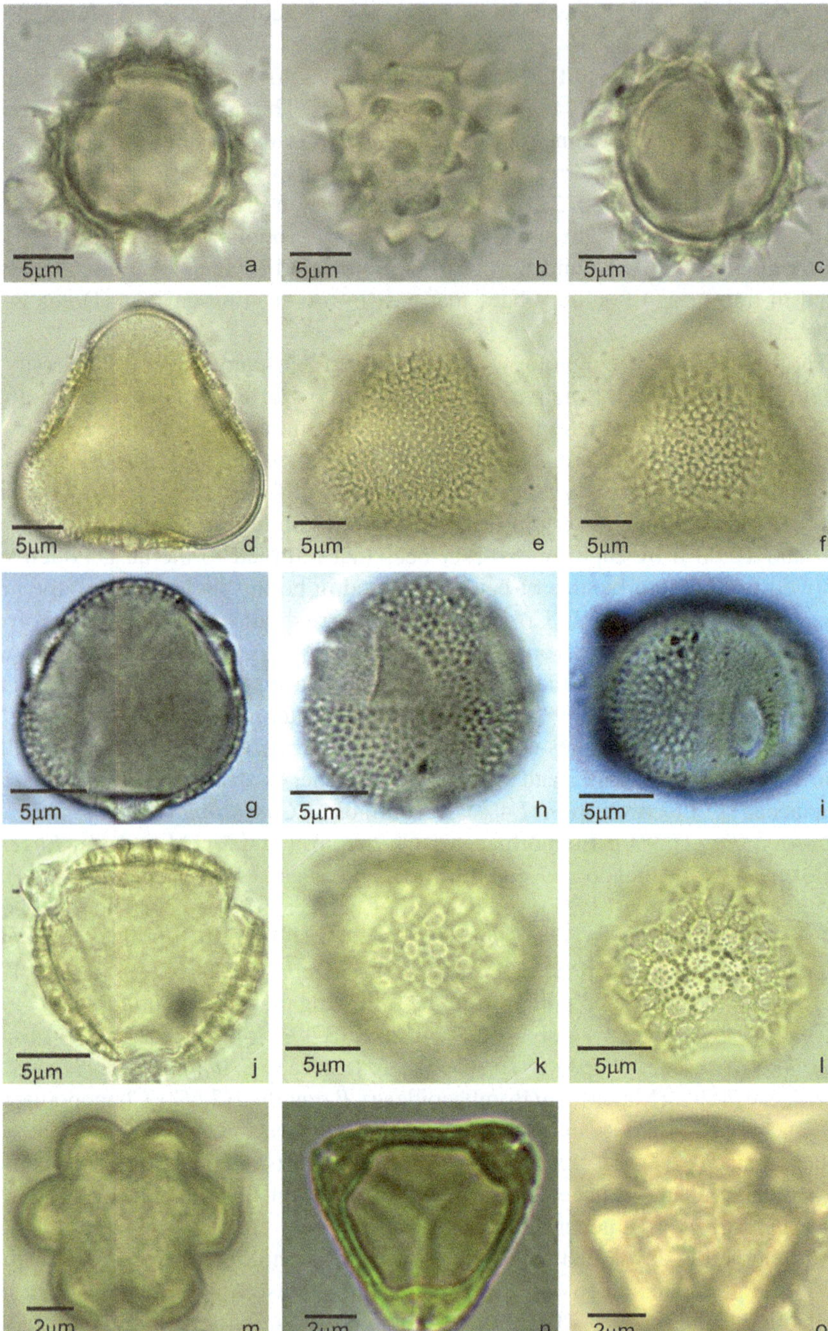

Fig. 6.1 Photomicrographs of pollen grains collected by bees. *Mikania guaco*: (**a**) polar view, optical section; (**b**) surface; (**c**) equatorial view. *Pachira aquatica*: polar view, (**d**) optical section; (**e**, f) LO analysis. *Caesalpinia echinata*: polar view, (**g**) optical section; (**h**) surface; (**i**) equatorial view. *Delonix regia*: (**j**) optical section; (**k**, **l**) LO analysis. *Tibouchina holosericea*: (**m**) polar view, outline. *Eugenia uniflora*: (**n**) polar view, optical section. *Cestrum nocturnum*: (**o**) polar view, outline

information on little-known topics: the pollination potential of hummingbirds and their role in the reproduction of Atlantic Forest plants. Hummingbirds were found to visit several plant species, favoring the occurrence of heterospecific pollen deposition (HPD), as various types of pollen were concomitantly transferred by the birds. The study evaluated whether HPD occurred under natural conditions in *Canistropsis seidelii* (L.B.Sm. & Reitz) Leme (Bromeliaceae) and *Psychotria nuda* (Cham. & Schltdl.) Wawra (Rubiaceae). In another part of the investigation, the reproductive system of *Nidularium innocentii* Lem. (Bromeliaceae) was examined and pollinated with a pollen mixture to simulate natural pollination conditions and assess the effects of HDP. Hummingbirds were found to carry heterospecific pollen mixtures on various parts of the body (forehead, frontal head, neck, lower and upper parts of the beak, among others) and deposit these pollen grains onto stigmas of different plant species. Artificial pollination of *N. innocentii* stigmas with a pollen mixture produced negative effects. In summary, it was concluded that hummingbirds carry different types of pollen on the same region of the body, potentially leading to HDP at the community level. Plant and hummingbird communities in the Atlantic Forest have several similarities in terms of temporal organization and interaction patterns. These findings suggest that HDP may be a widespread phenomenon in these communities (Fonseca et al. 2016).

The laboratory developed a postdoctoral project on pollen residues collected from trap nests. The study was conducted in Atlantic Forest fragments on Ilha Grande, Angra dos Reis, southern Rio de Janeiro State, Brazil. Ilha Grande, the largest island in the state, is a fragment of a 190 km^2 coastal massif and part of a group of islands in Ilha Grande Bay (Oliveira 2002). Geographically, the island is located at coordinates 23°11′S and 44°12′W, about 150 km southwest from the city of Rio de Janeiro. The relief is rugged, reaching 1031 m in elevation at Serra do Retiro and 959 m at Pico do Papagaio (Maciel et al. 1984).

Ilha Grande comprises different habitats, including lagoon and alluvial deposits, beaches, restingas, marsh and mangrove sediments, and forest fragments on slopes, with great variability and richness of tropical species. This mountainous region has a predominance of rocky shores without vegetation and coves with sandy beaches where streams that drain the island flow into the sea. The climate is tropical hot and humid, with an annual rainfall of about 2240 mm, maximum rainfall occurring in January (350 mm), and minimum rainfall occurring in July (75 mm). The average annual temperature is 25.5 °C, with the maximum average in February (25.7 °C) and the minimum average in July (19.6 °C) (CNAAA 2007).

Currently, Ilha Grande is an important region for environmental preservation and research. The island contains four conservation units, Praia do Sul Biological Reserve (located on the ocean side of the island), Ilha Grande State Park, Tamoios Environmental Protection Area, and Aventureiro Marine Park (Bastos et al. 2009). For the study of pollen residues collected by solitary bees, we distributed 300 bamboo trap nests (Fig. 6.2) at different preselected points.

Field observations of nesting activity were made fortnightly for 10 months. Colonized trap nests were stored in microperforated bags until adult emergence for

Fig. 6.2 Trap nests made of bamboo cane. Note the founded nests

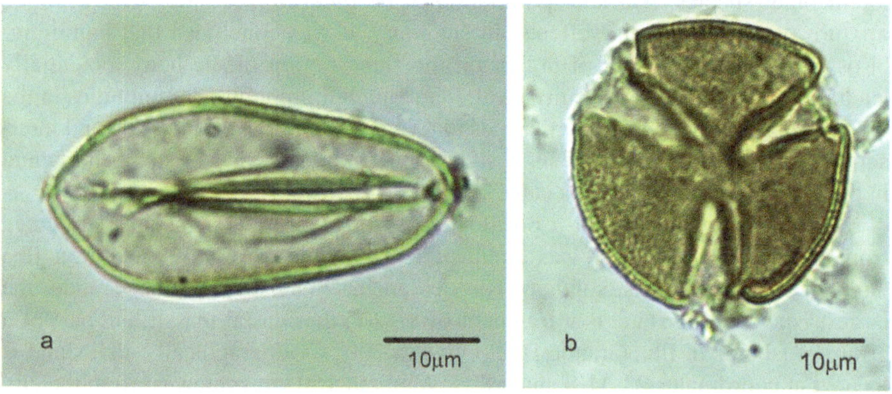

Fig. 6.3 (**a**) Pollen of *Syagrus romanzoffiana*, distal polar view, optical section; (**b**) pollen of *Solanum* sp., polar view, optical section

analysis of the pollen content of residual material. In the field, colonized nests were replaced by empty nests.

Of the 300 trap nests placed in the field, only 3.33% were colonized by bees collecting pollen of *Syagrus romanzoffiana* (Cham.) Glassman (Arecaceae) and the genera *Cestrum* L. and *Solanum* L. (Solanaceae) (Fig. 6.3).

The following bee species occupied the trap nests: *Euglossa cordata* L. ($n = 12$), *Euglossa pleosticta* Dressler ($n = 2$), and *Megachile guaranitica* Schrottky ($n = 22$). Nest structures differed between genera; whereas *Euglossa* spp. built their nests with earth and clay, *M. guaranitica* used leaves (Fig. 6.4).

It should be noted that solitary bees belong to a seasonal group that requires a longer period for nesting and adult emergence. Such behavior can explain the low nest occupation (3.33%) and, consequently, the low variety of pollen types found in trap nests. We observed that *E. cordata*, *E. pleosticta*, and *M. guaranitica* interacted with *S. romanzoffiana*, *Cestrum* sp., and *Solanum* sp., using pollen produced by these plants.

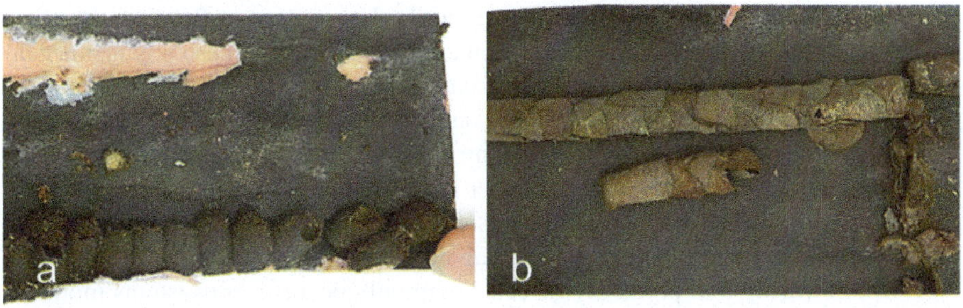

Fig. 6.4 (**a**) Cells constructed by *Euglossa* using wax and resin, (**b**) cells constructed by *Megachile guaranitica* using cut and stacked leaves

One of the strategies to understand the nest-building behavior of the native bee *T. angustula*, commonly known as jataí, was to analyze the pollen content of propolis. Jataí uses propolis to close openings and build the entrance to the colony. Propolis also serves as an antiseptic material and is deposited inside colony cells. The study was developed at a stingless bee unit located in Vila do Abraão, Ilha Grande, Angra dos Reis, Rio de Janeiro State, Brazil. Several medicinal plant species occur in the area, including *Clusia criuva* Cambess. (Clusiaceae), *Norantea brasiliensis* Choisy (Marcgraviaceae), *Anacardium occidentale* L., *Schinus terebinthifolius* Raddi (Anacardiaceae), *E. uniflora* L. (Myrtaceae), and *Cedrela fissilis* Vell. (Meliaceae). We analyzed 60 propolis samples and recorded 128 pollen types belonging to 51 botanical families. The most frequent families were Fabaceae (n = 13), Asteraceae (n = 11), Bignoniaceae (n = 7), Euphorbiaceae (n = 7), Malvaceae (n = 7), Anacardiaceae (n = 5), Malpighiaceae (n = 5), and Solanaceae (n = 5) (Morgado et al. 2018).

The laboratory maintains important partnerships with other institutions through collaborative research. We highlight collaborations with Dr. Maria Cristina Gaglianone from the State University of Northern Rio de Janeiro (UENF) for the development of dissertations and theses and with Dr. Elen Menezes from the Federal Rural University of Rio de Janeiro (UFRRJ). In collaborative research with Dr. Gaglianone, we sought to identify the pollen types used for nesting by *Tetrapedia diversipes* Klug (Apidae) at Atlantic Forest sites under regeneration in Rio de Janeiro State, Brazil (Menezes et al. 2012). Another important study comprised the identification of pollen grains used as alternative resources by two bee species (*Bombus morio* Swederus and *Exomalopsis analis* Spinola) in tomato (*Solanum lycopersicum* L.) fields. The results confirmed that both bee species were generalists; in addition to tomato, bees pollinated and used the floral resources of other plant species, even during the peak of tomato flowering. This finding underscored the importance of maintaining a diversity of flowering plants near agricultural sites (Hautequestt et al. 2020). The partnership with Dr. Elen Menezes has resulted in student co-orientations and important publications on natural enemies of agricultural pests. Some of the studies will be discussed below.

Predatory insects are a subject of study at the Laboratory of Palynology, given their participation in the trophic chains of natural ecosystems. These organisms generally occupy the third trophic level but can be found at higher levels. Predatory insects are therefore carnivorous, feeding on their prey during at least one of the life stages. These biotic mortality agents contribute to the regulation of prey populations. Predatory insects can prey on other insect species, including phytophagous insects, in which case they are defined as entomophagous insects (from the Greek word "entomon" meaning insect and "phagein" meaning eat, devour).

It is noteworthy that predatory insects commonly occur in ecosystems modified by anthropogenic activities, particularly in plant production systems or agroecosystems, where they feed on different phytophagous arthropods, some of which constitute pests of agricultural crops. Such insects are referred to as natural enemies or biological control agents of agricultural pests (Aguiar-Menezes and Menezes 2005; Aguiar-Menezes et al. 2013). Predatory insects comprise different taxonomic orders of the class Insecta. Members of this diverse group may differ in morphological characteristics and postembryonic development (hemimetabolism or holometabolism). In general, after the egg phase (embryonic period), hemimetabolous insects undergo two distinct feeding stages, the nymph and the adult, whereas holometabolous insects undergo larval, pupal, and adult stages. In holometabolous individuals, feeding is restricted to larval and adult stages. Furthermore, the type of food may or may not differ between stages. It is estimated that 25% of the known insect species have at least one predatory life stage (Aguiar-Menezes et al. 2013).

Whether in the carnivorous phase or not, insects may consume other foods as part of their diet (Wäckers 2005; Lundgren 2009a, b). Pollen is an important source of protein for certain species of entomophagous insects, also providing dietary carbohydrates, lipids, minerals, vitamins, and water (Nielsen et al. 1955; Roulston et al. 2000; Lundgren 2009a, b; Nicolson and Human 2013). In the noncarnivorous life stage of some predatory insects, such as adult chrysopids (Neuroptera: Chrysopidae) and syrphids (Diptera: Syrphidae), pollen is essential for survival and reproduction (Freitas 2002; Ambrosino et al. 2006; Albuquerque 2009; Lundgren 2009b). Floral pollen may also serve as a dietary supplement for insects that feed on prey with low nutritional quality, allowing predatory insects to survive, as occurs in some species of predatory ladybugs (Almeida and Ribeiro-Costa 2019; Resende et al. 2015; D'Ávila et al. 2016).

In some cases, pollinivory (i.e., the habit of feeding on pollen) may be incidental, for instance, when pollen is attached to other foods, such as nectar or honeydew. When this occurs, the amount of pollen ingested is very small (Duelli 1999; Lundgren 2009b). This type of ingestion differs from that of true pollinivory, in which insects consume large amounts of pollen during flower visitation (Lundgren 2009b). True pollinivory was observed under laboratory conditions in adults of *Chrysoperla externa* (Hagen) (Neuroptera: Chrysopidae) confined to cages containing flowers of Apiaceae (Resende et al. 2017).

The diet of the predatory ladybug *Coleomegilla maculata* DeGeer (Coleoptera: Coccinellidae) may consist of up to 50% pollen. The ladybug can produce fertile eggs while feeding solely on floral resources such as corn pollen (*Zea mays* L.,

Poaceae) (Hoffmann and Fordsham 1993; Shelton 2021). Lundgren et al. (2005) estimated that *C. maculata* larvae and adults consumed 3.30 and 13.15 mg of transgenic corn pollen, respectively, during anthesis in Illinois, USA. However, for some predatory ladybugs, pollen consumption is sufficient for survival but not for egg production, necessitating the intake of prey (usually aphids), as reported in *Hippodamia* spp. and *Antalya bipunctata* L. (Hodek and Evans 2012; He and Sigsgaard 2019).

Pollen grains differ in protein content and amino acid composition between and even within plant families. Moreover, some pollen grains may be toxic to entomophagous insects (Nielsen et al. 1955; Roulston et al. 2000; Orre et al. 2007; Lundgren 2009a; Nicolson and Human 2013). Such nutritional variability between pollen grains can exert different effects on the biology of predatory insects (Lundgren and Wiedenmann 2004; Venzon et al. 2006; D'Ávila et al. 2017; Resende et al. 2017; Melo et al. 2020). Predatory insects can therefore be habitual flower visitors (anthophilous), seeking pollen and sometimes nectar (Fig. 6.5), similar to pollinating insects, such as bees (Aguiar-Menezes et al. 2013).

In New Jersey, USA, Patt et al. (1997) observed adults of six species of predatory ladybugs (Coleoptera: Coccinellidae) visiting coriander (*Coriandrum sativum* L.) and dill (*Anethum graveolens* L.) flowers. *C. maculata*, *Hippodamia variegata* Goeze, and *Coccinella septempunctata* L. were the most abundant ladybug species. Lixa et al. (2010) found that different ladybug species frequently visited flowers of two aromatic plants and fennel (*Foeniculum vulgare* Mill.) in Brazil. Another

Fig. 6.5 Adult *Coleomegilla quadrifasciata* ladybug visiting flowers of coriander (*Coriandrum sativum*). (Photograph taken by Elen L. Aguiar-Menezes)

Brazilian study reported the same behavior of ladybugs with coriander flowers (Resende et al. 2011). Aguiar-Menezes et al. (2018) underscored the importance of these plants for the conservation of predatory ladybugs in agroecosystems.

Flowers of some plant species may be more attractive to pollinating insects than to entomophagous insects, and vice versa (Fiedler et al. 2007). Plants that supply pollen for insects that exert pest control are internationally known as beneficial insectary plants or simply insectary plants (Valenzuela 1994; Ambrosino 2005; Ambrosino et al. 2006; Hatt et al. 2019; Altieri et al. 2020). In Brazil, they are referred to as attractive plants for natural enemies (Aguiar-Menezes and Silva 2011; Barbosa et al. 2011). These plants may also attract alternative prey, produce nectar, and provide suitable microenvironments for shelter, mating, and oviposition (Lixa et al. 2010; Resende et al. 2011; Hatt et al. 2019).

The use of insectary plants in agroecosystems is an important strategy underlying conservation biological control (Ambrosino 2005; Aguiar-Menezes and Silva 2011; Altieri et al. 2020). Pollen is used not only as food by pollinating insects, which promotes the reproductive success of higher plants, but also as a reward for the indirect defense exerted by predatory insects and other biological agents that depend on pollen as food. Thus, competition for pollen among flower visitors may be inevitable. On an evolutionary scale, plants have developed several adaptive mechanisms to make the most of these harmonic interactions, as have insects that interact with them (Marques 2012). For instance, pollen may not be available to predatory insects in some flower species; access to pollen may depend in part on floral (e.g., petal opening and corolla depth) and insect (head shape and type of oral apparatus) morphology (Orre et al. 2007; Lundgren 2009a, b).

One of the challenges of selecting insectary plants is the identification of plant species that provide the floral resources needed by predatory insects. In the case of pollen, palynology and entomology have contributed to such knowledge (Medeiros et al. 2010; Lixa 2013; D'Ávila et al. 2016; Andrade et al. 2018; Souza et al. 2019). Nevertheless, palynological studies applied to entomophagous insects are still scarce, especially compared with studies on pollinating insects, such as bees (Jones and Jones 2001; Silva et al. 2020).

Medeiros et al. (2010) used palynological techniques for screening plants that could serve as a source of pollen for the predatory insects *C. externa* and *Hippodamia convergens* Guérin-Menéville (Coleoptera: Coccinellidae). Adult individuals were collected at an experimental site in Embrapa Hortaliças, Brasília, Federal District, Brazil. A total of 11,335 pollen grains distributed in 21 families were recovered from the digestive tract of 53 adults of *C. externa*, with greater abundance of the family Poaceae. A total of 46 pollen from 10 families were extracted from 43 adults of *H. convergens*, with greater abundance of Asteraceae pollen.

Pollen types consumed by adult ladybugs and chrysopids were identified in collaborative studies between the Laboratory of Palynology of the Botany Department of the National Museum, UFRJ, and the Department of Entomology and Phytopathology of UFRRJ (Lixa 2013; D'Ávila et al. 2016; Andrade et al. 2018; Souza et al. 2019). A summary of the results is presented below.

Lixa (2013) extracted pollen from adults of seven species of aphid-eating ladybugs [*Cycloneda sanguinea* (Linnaeus), *H. convergens*, *C. maculata*, *Coleomegilla quadrifasciata* (Schönherr), *Eriopis connexa* (Germar), *Harmonia axyridis* (Pallas), *Olla v-nigrum* (Mulsant)], collected at the Integrated System of Agroecological Production (SIPA), Seropédica, Rio de Janeiro, Brazil. Pollen grains of the following ten families were identified: Adoxaceae, Amaranthaceae, Apiaceae, Asteraceae, Fabaceae, Malvaceae, Myrtaceae, Oxalidaceae, Passifloraceae, and Poaceae. Five plant species were recorded: *Averrhoa carambola* L. (Oxalidaceae), *C. sativum* (Apiaceae), *Helianthus annuus* L. (Asteraceae), *Sambucus nigra* L. (Adoxaceae), and *Z. mays* (Fig. 6.6).

D'Ávila et al. (2016) observed that coriander, fennel, and dill have the same pollen type (i.e., stenopalynological). *C. maculata* larvae and adults ingested pollen from Apiaceae flowers under laboratory conditions. Of the total number of pollen grains recovered (*n* = 69,679), 22.3% were from coriander, 26.3% from fennel, and 51.4% from dill. Andrade et al. (2018) recovered 38,277 pollen grains distributed in 19 families of angiosperms from 345 adults of *C. externa* collected at different sampling sites in SIPA; 84.6% of the digestive tract of insects contained Poaceae pollen. However, 28.4% of insects fed on more than one pollen type (polyphagous behavior), suggesting that *C. externa* adults seek pollen in a variety of plant species. Three plant species and one botanical genus were identified: *Z. mays*, *Flemingia macrophylla* (Willd.) Alston (Fabaceae), *Coffea canephora* Pierre ex A.Froehner var. *kouilouensis* De Wild (Conilon) (Rubiaceae), and *Brachiaria* sp. (Poaceae). Souza et al. (2019) recovered pollen of 14 botanical families from adults of 3 ladybug species in an agroecological vegetable field. Most pollen samples were obtained from *C. sanguinea* adults (Table 6.1), whereas only three Asteraceae pollen types were recovered from *H. axyridis* adults and one pollen type from *C. quadrifasciata* adults; it was not possible to identify the pollen obtained from *C. quadrifasciata*.

The estimated global species diversity of angiosperms is about 300,000 species (Evert and Eichhorn 2014). However, according to Fiedler et al. (2008), only *C. sativum* (Apiaceae), *Fagopyrum esculentum* Moench (Polygonaceae), *Lobularia maritima* (L.) Desv. (Brassicaceae), and *Phacelia tanacetifolia* Benth. (Boraginaceae) are the most studied in terms of attractive potential to natural enemies. These species have been used for such purposes in agricultural production systems in the USA, Europe, and New Zealand. Considering that Brazil is home to 35,553 species of angiosperms (Flora do Brasil 2020), we have a wide range of native plants to be investigated for this purpose at our disposal.

6.1 Final Considerations

Palynology is a promising tool for the identification of plants that provide pollen for palynivore insects, contributing to the development of conservation biological control strategies applicable to Brazilian agroecosystems. Even if the use of such plants does not lead to an increase in the efficiency of biological pest control, it may

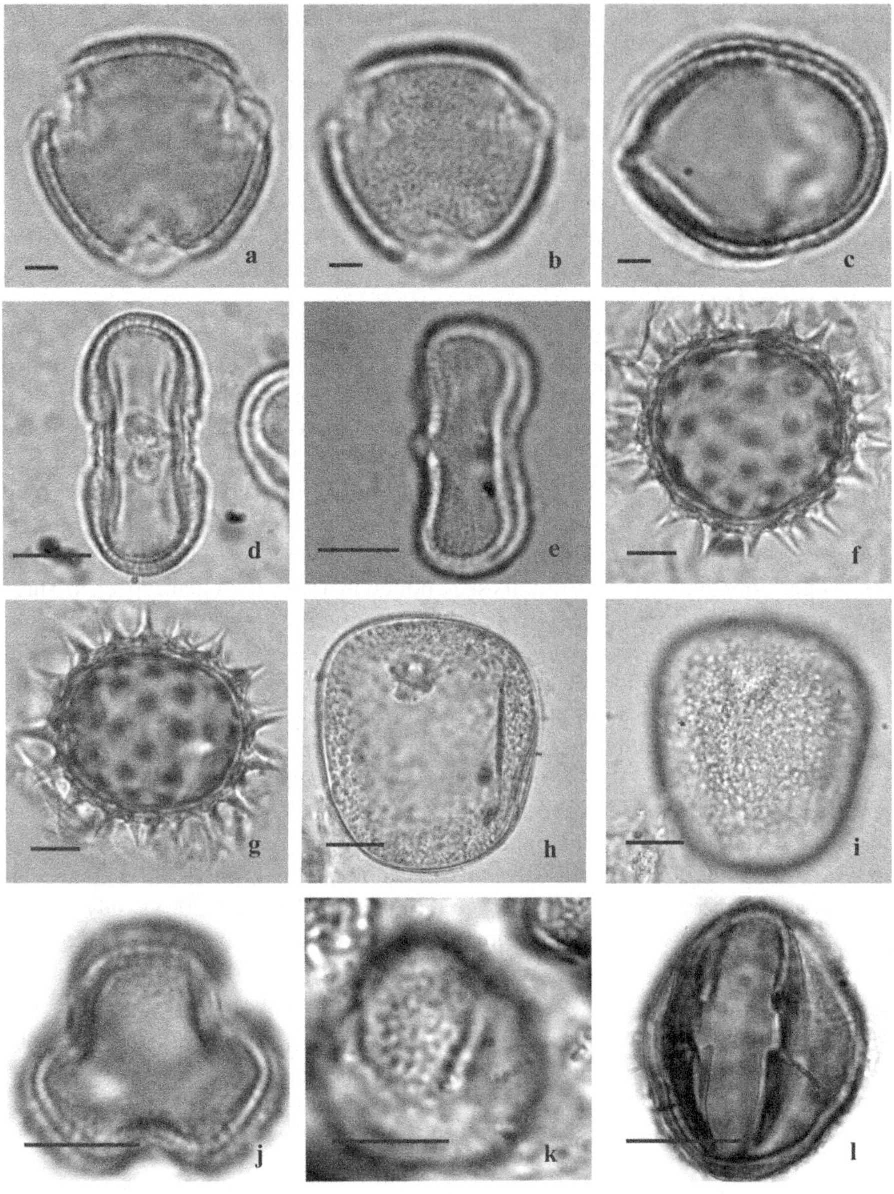

Fig. 6.6 Photomicrographs of pollen grains recovered from adult predatory ladybugs. *Averrhoa carambola*: (**a, b**) polar view, (**c**) equatorial view. *Coriandrum sativum*: (**d, e**) equatorial view. *Helianthus annuus*: (**f**) polar view, (**g**) equatorial view. *Zea mays*: (**h**) polar view, aperture, (**i**) surface. *Sambucus nigra*: polar view, (**j**) optical section, (**k**) surface; (**l**) equatorial view, aperture. (Lixa 2013)

Table 6.1 Pollen grains recovered by palynological techniques from the digestive tract of *Cycloneda sanguinea* adults sampled from December 21, 2018, to February 15, 2019, at the Integrated System of Agroecological Production, Seropédica, Rio de Janeiro, Brazil

Family	Number of pollen grains
Amaranthaceae	1
Apocynaceae	2
Arecaceae	2
Caesalpinioideae	2
Fabaceae	10
Leguminosae	1
Melastomataceae	820
Mimosaceae	2
Moraceae	3
Myrtaceae	2
Poaceae	30
Rutaceae	2
Solanaceae	7

Souza et al. (2019)

contribute to the conservation of native predatory insects. Various studies have investigated the feeding behavior of pollinating animals, providing important palynological data that can be used to define plants of interest, contributing to animal survival and bee production. Regarding flora, although pollen identification does not allow inferring whether effective pollination occurred, palynological studies generate large amounts of data, associating pollinators and floral visits. Such knowledge may guide more specific studies aiming to better understand the role of pollinators in the reproduction of a given plant species.

Acknowledgments We also thank the Rio de Janeiro Research Foundation (FAPERJ), the Brazilian National Council for Scientific and Technological Development (CNPq), and the Brazilian Federal Agency for Support and Evaluation of Graduate Education (CAPES) for the grants and financial support provided over the years.

References

Aguiar-Menezes EL, Menezes EB (2005) Bases ecológicas das interações insetos e plantas no manejo ecológico de pragas agrícolas. In: Aquino AM, Assis RL (eds) Agroecologia: princípios e técnicas para uma agricultura orgânica e sustentável. Embrapa Informação e Tecnologia, Brasília, pp 321–339

Aguiar-Menezes EL, Silva AC (2011) Plantas atrativas para inimigos naturais e sua contribuição para o controle biológico de pragas agrícolas. Embrapa, Agrobiologia, Seropédica

Aguiar-Menezes EL, Menezes EB, Lorenzon MCA et al (2013) Os insetos. In: Abboud ACS Introdução à agronomia. Interciência, Rio de Janeiro, pp 284–343

Aguiar-Menezes EL, Lixa AT, Resende ALS (2018) Joaninhas predadoras, as aliadas do produtor no combate às pragas. A Lavoura, Sociedade Nacional de Agricultura – SNA, Rio de Janeiro, pp 38–41

Albuquerque GS (2009) Crisopídeos (Neuroptera: Chrysopidae). In: Panizzi AR, Parra JRP (eds) Bioecologia e nutrição de insetos: base para o manejo integrado de pragas. Embrapa Informação Tecnológica, Brasília, pp 969–1022

Almeida LM, Ribeiro-Costa CS (2019) Coleópteros Predadores (Coccinellidae). In: Panizzi AR, Parra JRP (eds) Bioecologia e nutrição de insetos: base para o manejo integrado de pragas. Embrapa Informação Tecnológica, Brasília, pp 931–968

Altieri M, Nicholls CI, Fritz MA (2020) Manage insect on our farm: a guide to ecological strategies. Sustainable Agriculture Research & Education (SARE), College Park

Ambrosino MD (2005) Practical guidelines for establishing, maintaining and assessing the usefulness of insectary plantings on your farm. Oregon State University, Corvallis

Ambrosino MD, Luna JM, Jepson PC et al (2006) Relative frequencies of visits to selected insectary plants by predatory hoverflies (Diptera: Syrphidae), other beneficial insects, and herbivores. Environ Entomol 35(2):394–400

Andrade KA, Aguiar-Menezes EL, Gonçalves-Esteves V et al (2018) Pollen ingestion by Chrysoperla externa (Hagen) adults in a diversified organic agroecosystem. Neotrop Entomol 47:118–130

Baker HG, Hurd PD (1968) Intrafloral ecology. Annu Rev Entomol 13:385–414

Barbosa FS, Aguiar-Menezes EL, Arruda LN et al (2011) Potencial das flores na otimização do controle biológico de pragas para uma agricultura sustentável. Revista Brasileira de Agroecologia 6(2):101–110

Bastos M, Prado RM, Santiago AMA et al (2009) Estrutura econômica e organização sóciocultural e política. In: Bastos M, Callado CH (eds) O Ambiente da Ilha Grande. Ed. Universidade do Estado do Rio de Janeiro, Rio de Janeiro, pp 371–470

Bauermann SG (2006) Grãos de Pólen: uso e aplicações. Cadernos La Salle XI 2(1):99–107

Belina-Aldemita D, Fraberger V, Scheiner M et al (2019) Nutritional composition of pot-pollen produced by stingless bees (Tetragonula biroi Friese) from the Philippines. J Food Compos Anal 82:103215

CNAAA (2007) Dados obtidos através do Sistema de Meteorologia da Central Nuclear Almirante Álvaro Alberto – CNAAA e encaminhados pela área de meteorologia da Gerência de Meio Ambiente da Eletrobrás Termonuclear S.A.- Eletronuclear

D'Ávila VA, Aguiar-Menezes EL, Gonçalves-Esteves V et al (2016) Morphological characterization of pollens from three Apiaceae species and their ingestion by twelve-spotted lady beetle (Coleoptera: Coccinellidae). Braz J Biol 76(3):796–803

D'Ávila VA, Aguiar-Menezes EL, Pereira RN et al (2017) Effect of provision of apiaceous flowers associated to foods on the biology of Coleomegilla maculata. Phytoparasitica 45(4):471–484

Dórea MC, Santos FAR, Lima LCL et al (2009) Análise Polínica do Resíduo Pós-emergência de Ninhos de Centris tarsata Smith (Hymenoptera: Apidae, Centridini). Neotrop Entomol 38(2):197–202

Dórea MC, Aguiar CML, Figueroa LER et al (2010) Residual pollen in nestsof Centris analis (Hymenopera, Apidae, Centridini) in an area of Caatinga vegetation from Brazil. Oecologia Australis 14(1):232–237

Duelli P (1999) Honigtau und stumme gesänge: habitat- und partnersuche bei florfliegen (Neuroptera: Chrysopidae). Stapfia 60:35–48

Edwards J, Pardoe HS (2018) How palynology could have been paepalology: the naming of a discipline. Palynology 42(1):4–19

Erdtman G (1952) Pollen morphology and plant taxonomy-angiosperms. Almqvist and Wiksell, Stockholm

Evert RF, Eichhorn SE (2014) Biologia vegetal, 8th edn. Guanabara Koogan, Rio de Janeiro

Fiedler AK, Tuell J, Isaacs R et al (2007) Attracting beneficial insects with native flowering plants. Michigan State University Extension, East Lansing

Fiedler AK, Landis DA, Wratten SD (2008) Maximizing ecosystem services from conservation biological control: the role of habitat management. Biol Control 45:254–271

Flora do Brasil (2020) Jardim Botânico do Rio de Janeiro. http://floradobrasil.jbrj.gov.br/. Accessed 2 May 2021

Fonseca LCN (2012) Transporte de pólen de espécies orintófilas e energia disponível para beija-flores em área de Mata Atlântica, Núcleo Picinguaba/PESM, Ubatuba, SP. Tese, Unicamp

Fonseca LCN, Rech AR, Bergamo PJ et al (2016) Heterospecific pollen deposition among plants sharing hummingbird pollinators in the Brazilian Atlantic Forest. Rodriguésia 67(2):335–345

Freitas S (2002) O uso de crisopídeos no controle biológico de pragas. In: Parra JRP, Botelho PSM, Corrêa-Ferreira BS, Bento JMS (eds) Controle biológico no Brasil: parasitóides e predadores. Manole, São Paulo, pp 209–224

Gutiérrez BP, García DLQ (2007) Estudio Melisopalinológico de dos mieles de la porción Sur Del Valle de México. Polibotánica 23:57–75

Hatt S, Xu Q, Francis F et al (2019) Aromatic plants of East Asia to enhance natural enemies towards biological control of insect pests. A review. Entomologia Generalis 38(4):275–315

Hautequestt A, Depra M, Gonçalves-Esteves V et al (2020) Pollen load spectrum of tomato pollinators. Neotrop Entomol 49:491–500

He X, Sigsgaard L (2019) A floral diet increases the longevity of the coccinellid Adalia bipunctata but does not allow molting or reproduction. Front Ecol Evol 7(6):1–6

Hodek I, Evans EW (2012) Food relationships. In: Hodek I, Van Emden HF, Honěk A (eds) Ecology and behaviour of the ladybird beetles (Coccinellidae). Blackwell Publishing, Chichester, pp 141–274

Hoffmann MP, Fordsham AC (1993) Natural enemies of vegetable insect pests. Cornell University, Ithaca, Cornell Cooperative Extension

Jones GD, Jones SD (2001) The uses of pollen and its implication for entomology. Neotrop Entomol 30(3):341–350

Kiill LHP, Costa JG (2003) Biologia floral e sistema reprodutivo de Annona squamosa L. (Annonaceae) na região de Petrolina-PE. Ciência Rural 33(5):851–856

Lixa AT (2013) Bioecologia de joaninhas predadoras em sistema orgânico diversificado e efeitos de defensivos alternativos sobre Coleomegilla maculata (Coleoptera: Coccinellidae) em laboratório. Thesis, Universidade Federal Rural do Rio de Janeiro

Lixa AT, Campos JM, Resende ALS et al (2010) Diversidade de Coccinellidae (Coleoptera) em plantas aromáticas (Apiaceae) como sítios de sobrevivência e reprodução em sistema agroecológico. Neotrop Entomol 39:354–359

Lundgren JG (2009a) Pollen nutrition and defense. In: Lundgren JG (ed) Relationships of natural enemies and non-prey foods. Springer, Dordrecht, pp 127–139

Lundgren JG (2009b) Pollinivory. In: Lundgren JG (ed) Relationships of natural enemies and non-prey foods. Springer, Dordrecht, pp 85–116

Lundgren JG, Wiedenmann RN (2004) Nutritional suitability of corn pollen for the predator Coleomegilla maculata (Coleoptera: Coccinellidae). J Insect Physiol 50(6):567–575

Lundgren JG, Huber A, Wiedenmann RN (2005) Quantification of consumption of corn pollen by the predator Coleomegilla maculata (Coleoptera: Coccinellidae) during antesis in an Illinois cornfield. Agric For Entomol 7(1):53–60

Maciel NC, Araújo DSD, Magnanini A (1984) Reserva Biológica Estadual da Praia do Sul (Ilha Grande, Angra dos Reis, RJ). Boletim da FBCN 19:126–148

Marques RJ (2012) Uma abordagem geral das defesas das plantas contra a ação dos herbívoros. In: Del-Claro K, Torezan-Silingardi HM (orgs) Ecologia das interações plantas-animais: uma abordagem ecológico-evolutiva, 1st edn. Technical Books, Rio de Janeiro, pp 55–66

Medeiros MA, Ribeiro PA, Morais HC et al (2010) Identification of plant families associated with the predators Chrysoperla externa (Hagen) (Neuroptera: Chrysopidae) and Hippodamia convergens Guérin-Menéville (Coleoptera: Coccinellidae) using pollen grain as a natural marker. Braz J Biol 70(2):293–300

Menezes GB, Gonçalves-ESteves V, Bastos EMAF, Augusto SC, Gaglianone MC (2012) Nesting and use of pollen resources by Tetrapedia diversipes Klug (Apidae) in Atlantic Forest areas (Rio de Janeiro, Brazil) in different stages of regeneration. Revista Brasileira de Entomologia 56: 86–94. https://doi.org/10.1590/S0085-56262012000100014.

Melo GJB, Berber GCM, Aguiar-Menezes EL et al (2020) Parâmetros biológicos de adultos de Chrysoperla externa (Hagen, 1861) (Neuroptera: Chrysopidae) alimentados com pólen de Asteraceae em condições de laboratório. Sci Electron Archiv 13(2):43–50

Michener CD (1979) Biogeography of the bees. Ann Mo Bot Gard 66(3):277–347

Morgado LN, Barth OM (2011) The detection of pollen in royal jelly of honey bees (Apis mellifera). J Api Prod ApiMed Sci 3(3):137–139. https://doi.org/10.3896/IBRA.4.03.3.05

Morgado LN, Lorenzon MCA (2014) Recurso polínico utilizado por Tetragonisca angustula Latreille. In: Lorenzon MCA, Morado CN (orgs). A abelha jataí: Flora visitada na Mata Atlântica. Ed. Letras e Versos Gráficos e Editores, Rio de Janeiro, p 3-57

Morgado LN, Rocha CFD (2016) Diversity of wild bees and their mediated dispersal of pollen from the genus tillandsia (Bromeliaceae, Tillandsioideae) in an insular area. Athens J Sci 3(4):297–307

Morgado LN, Lima DD, Mendonça CBF et al (2018) Pollen Spectrum in Propolis of Tetragonisca angustula Latreille (Apidae: Trigonini) in an Island Area, Brazil. Bee World 95(3):84–88. https://doi.org/10.1080/0005772X.2018.1486277

Mouga DMDS, Dec E (2012) Catálogo polínico de plantas medicinais apícola. DIOESC, Florianópolis

Nicolson S, Human H (2013) Chemical composition of the 'low quality' pollen of sunflower (Helianthus annuus, Asteraceae). Apidologie 44:144–152

Nielsen N, Grommer J, Lunden R (1955) Investigations on the chemical composition of pollen from some plants. Acta Chem Scand 9:1100–1106

Oliveira PP, Santos FAR (2014) Prospecção palinológica em méis da Bahia. Print Mídia, Feira de Santana

Orre GUS, Tompkins JM, Jonsson M et al (2007) Provision of floral resources for biological control restoring an important ecosystem service. Function Ecosyst Commun 1(2):96–84

Patt JM, Hamilton GC, Lashomb JH (1997) Impact of strip insectary intercropping with flowers on conservation biological control of the Colorado potato beetle. Adv Hortic Sci 11:175–181

Pires O'brien MJ, O'brien CM (1995) Aspectos evolutivos da fenologia reprodutiva das árvores tropicais. FCAP, Serviço de Documentação e Informação, Belém

Resende ALS, Lixa AT, Santos CMA et al (2011) Comunidade de joaninhas (Coleoptera: Coccinellidae) em consórcio de couve (Brassica oleraceae var. acephala) com coentro (Coriandrum sativum) sob manejo orgânico. Revista Brasileira de Agroecologia 6(1):81–89

Resende ALS, Ferreira RB, Silveira LCP et al (2015) Desenvolvimento e reprodução de Eriopis connexa (Germar, 1824) (Coleoptera: Coccinellidae) alimentada com recursos florais de coentro (Coriandrum sativum L.). Entomotropica 30(2):12–19

Resende ALS, Souza B, Ferreira RB et al (2017) Flowers of Apiaceous species as sources of pollen for adults of Chrysoperla externa (Hagen) (Neuroptera). Biol Control 106:40–44

Roulston TH, Cane JH, Buchmann SL (2000) What governs protein content of pollen: pollinator preferences, pollen-pistil interactions, or phylogeny? Ecol Monogr 70:617–643

Shelton A (2021) Coleomegilla maculata (Coleoptera: Coccinellidae). In: Shelton A (org). Biological control: a guide to natural enemies in North America. Cornell University, Ithaca. https://biocontrol.entomology.cornell.edu/predators/Coleomegilla.php. Accessed 2 May 2021

Silva LS (2016) Diversidade polínica encontrada nas borbiculas de Tetragonisca angustula Latreille, Horto Botânico, Museu Nacional/UFRJ. Monografia, Universidade Veiga de Almeida

Silva CI, Arista M, Ortiz PL et al (2010) Catálogo polínico: palinologia aplicada em estudos de conservação de abelhas do gênero Xylocopa no Triângulo Mineiro. Universidade Federal de Uberlândia – EDUFU, Uberlândia

Silva CI, Imperatriz-Fonseca VL, Groppo M et al (2014) Catálogo polínico das plantas usadas por abelhas no campus da USP de Ribeirão Preto, 1st edn. Holos, Ribeirão Preto

Silva CI, Radaeski JN, Arena MVN et al (2020) Atlas of pollen and plants used by bees, 1st edn. Consultoria Inteligente em Serviços Ecossistêmicos (CISE), Rio Claro

Souza TS, Aguiar-Menezes EL, Mendonça CBF et al. (2019) Análise do potencial de diferentes subsistemas de um cultivo agroecológico de hortaliças como fonte de pólen para coccinelídeos. Paper presented at the CONGRESSO NACIONAL DE MEIO AMBIENTE, Poços de Caldas, 24–27 September 2019

Valenzuela HR (1994) Insectaries; the use of insectary plants as a reservoir for beneficials in vegetable agroecosystems. College of Tropical Agriculture & Human Resources, University of Hawaii, Manoa

Venzon M, Rosado MC, Euzébio DE et al (2006) Suitability of leguminous cover crop pollens as food source for the green lacewing Chrysoperla externa (Hagen) (Neuroptera: Chrysopidae). Neotrop Entomol 35:371–376

Wäckers FL (2005) Suitability of (extra-)floral nectar, pollen and honeydew as insect food source. In: Wäckers FL, Van Rijn PCJ, Bruin J (eds) Plant-provided food for carnivorous insects: a protective mutualism and its applications. Cambridge University Press, Cambridge, pp 17–73

Węglińska M, Szostak R, Kita A et al (2020) Determination of nutritional parameters of bee pollen by Raman and infrared spectroscopy. Talanta 212(1–4):120790

Part III
Ethnobotanical Aspects

This part will address ethnobotanical aspects as one approach within some of the *Aspects of Brazilian Floristic Diversity: From Botany to Traditional Communities*. Certain case studies to portray a scenario of the social diversity and its interaction with floristic diversity will be introduced.

This part will encompass historical treatments of this human-floristic interaction considering the first scientific expedition of the Brazilian Empire and the aesthetic representations in the art of gardening during the Second Empire of Brazil, both during the nineteenth century.

As a way to deepen the understanding of how these interactions occur, two *quilombola* communities will be taken as an inspiration for these reflections as representatives of traditional communities that interact with Brazilian floristic diversity.

As closure, but not signifying the finalization of crucial discussions on this theme, this section will recognize and highlight the sharing of knowledge with the participants of investigations about the biocultural heritage as a premise of ethical and legal conduct.

Chapter 7
Scientific Exploration Commission (1859–1861): Freire Allemão and the Invisible Network of Collaborators

Luiz José Soares Pinto and Luci de Senna-Valle

Abstract The Scientific Exploration Commission was the first Brazilian Empire's scientific expedition exclusively conducted by Brazilians, who went through the State of Ceará, including neighbor states regions, between February of 1859 and July of 1861. The object was to do an inventory of the natural resources from the North of Brazil's provinces. The Botanical Section was directed by Francisco Freire Allemão (1797–1874) and his adjunct and nephew Manoel Freire Allemão de Cysneiros (1832–1861). The botanical material collected was deposited in the National Museum's herbarium (R). All the materials related to the expedition were reunited besides the National Library's manuscripts. A total of 1794 exsiccates were gathered from this Expedition, comprising 1169 species, distributed in 137 botanical families. From the manuscripts 713 files were related to 695 species and 1300 quotes in their diaries. The exchange of cultural information sometimes is missed in scientific papers, in opposition to Freire Allemão that clearly mentioned his contributors; one of the famous ones was the Brazilian poet, Gonçalves Dias. Freire Allemão identified in his diaries 702 ethnospecies, of which 436 that are distributed in 102 are possible to be currently recognized. Noteworthy is that the *carnauba* palm stands out as the most quoted and indicated plant.

Keywords Botanical contributors · Ceará · Historical ethnobotanic · Useful plants

L. J. Soares Pinto (✉)
Postgraduate Program in Biological Sciences (Botany), National Museum, Federal University of Rio de Janeiro, Rio de Janeiro, RJ, Brazil
e-mail: ljspinto@gmail.com

L. de Senna-Valle
Departamento de Botânica, Museu Nacional, Universidade Federal do Rio de Janeiro (MN/UFRJ), Rio de Janeiro, RJ, Brazil

M. F. T. Medeiros, B. de Sá Haiad (eds.), *Aspects of Brazilian Floristic Diversity*,
https://doi.org/10.1007/978-3-031-07453-0_7

7.1 Introduction

The Portuguese government, in 1800, in a correspondence sent to the Brazilian colony, ordered the vigilance and prohibition of entrance for any foreigner, and it should be verified if any foreigner could contain tools, notes, and, overall, astronomical observations, which would be seized for being considered a threat to the security and sovereignty of the state (Lisboa 1999). With the arrival of the Portuguese royal family, in 1808, the Brazilian's harbor was reopened to the friendly nations, interrupting the injunction above the foreigners. After this event it can be observed the coming of many European naturalists interested in investigating the Brazilian lands, which contributed to a new "discovery" of Brazil (Lisboa 1997; Kury 1998; Senna-Valle and Sá 2007).

Between the foreign naturalists that were in Brazil, in this period, it is possible to stand out: two visits made by Georg Heinrich von *Langsdorff* (1803–1804; 1813–1820); Friedrich *Sellow* (1814–1831); Auguste de *Saint-Hilaire* (1816–1822); Carl Friedrich Philipp von *Martius* and Johann Baptist von *Spix* (1817–1821), besides other visitors after the Independence of Brazil in 1822, such as Charles Robert *Darwin* and George *Gardner* (1836–1841); Johannes Eugenius Bülow *Warming* (1863–1866); and Auguste François Marie *Glaziou* (1858–1897), among others (Lisboa 1999; Nogueira 2000; Brandão et al. 2008).

According to Brandão et al. (2008), the vast knowledge acquired by these naturalists, about the Brazilian flora, has an immeasurable value, where hundreds of new plants were discovered and described by them, based on these collected materials; besides relevant information about its use were also compiled during their stay, such as Auguste de Saint-Hilaire, who published in 1824 *Plantes Usuelles des Brésiliens* and *Histoire des plantes les plus remarquables du Brésil et du Paraguai*, and also Carl Friedrich Philipp von Martius in 1843 *Systema Materia Medicae Vegetabilis Brasiliensis* and in 1844 the work *Nature, illnesses, physic and medicines of Brazilian indigenous*, considered a mark to the ethnographic study and the indigenous medicine, besides the notes in the researcher's field diaries that are extraordinary sources of research and reference to the nature studies, medicines, ethnography, landform, and other areas of science.

Some Brazilian naturalists also did their studies during this period, with the subject of discovering and registering the knowledge about the local flora, like Pataca and Pinheiro (2005) who stand out, quoting *Frei* José Mariano da Conceição *Velloso* (1742–1811), Manoel *Arruda da Câmara* (1752–1810), Alexandre *Rodrigues Ferreira* (1756–1815), Francisco *Freire Allemão* (1797–1874), and *Manoel* Freire Allemão *de Cysneiros* (1825–1863), by their participations in the Scientific Exploration Commission, which was the first big Brazilian scientific expedition, that also played important populational and territorial observation that were essential to the elaboration of occupation and defense policies of the national territory.

7.2 The Brazilian Historical and Geographic Institute, Origin of the Scientific Exploration Commission

The construction of a national science made by Brazilians to Brazilians, between 1870 and 1939, according to Figueirôa (1998), presented the habits and behavior in the Empire's everyday life, with a strong development and promotion of the scientific activities and the challenges of building a national state as the mark of these efforts. Clearly expressed by nativism,[1] which permeates many areas of knowledge, such as literature, art, among others, where science took place in the construction of this expedition.

The Brazilian Historical and Geographic Institute (IHGB) was founded on October 21st of 1838, under the Auxiliary Society of National Industry's (SAIN) promotion and by an expressive group of politicians and intellectuals from the Court at Rio de Janeiro with the leadership of Marechal Raimundo da Cunha Matos and Cônego Januário da Cunha Barbosa; it was destined to gather, methodize, file, and publish the necessary files to the writing of History and Geography of Brazil (Guimarães 2007).

The Scientific Exploration Commission from the north provinces, the Scientific Commission of the Empire, the Imperial Scientific Exploration Commission or just Scientific Exploration Commission,[2] Ceará Commission, and even the Butterflies Commission[3] and the Deflowering Commission, all of them are the different titles for the same Expedition (Paiva 2002; Cavalcante 2012). According to Alves (2012), that was the first big scientific expedition created, during the Second Reign (1840 to 1889), composed just by Brazilians. That was part of a political project of the nation's integration, which had subjects such as knowledge, cataloging, quantifying, and registering the different regions of the empire. The emperor D. Pedro II supported it, and it was associated with the Brazilian scientific institutions, such as IHGB, the National Museum (MN), and the Auxiliary Society of National Industry (SAIN).

The Brazilian Historical and Geographic Institute (IHGB) had the responsibility of indicating the positions and the areas that were supposed to be in the Commission and its administration indicated, in the July 25th of 1856 session, the following components: *Francisco Freire Allemão* (1797–1874), director and president of the botanical section; *Guilherme Schüch de Capanema* (1824–1908), director of the geological and mineralogical sections; *Manoel Ferreira Lagos* (1816–1871), director of the zoological section; *Giácomo Raja Gabaglia* (1826–1872), director of the

[1] Nativism or nativist movement was so-called the isolated revolts that took place in the then Portuguese colony of Brazil, between the end of the seventeenth century and the beginning of the nineteenth century, in which the feeling of conflict between the "sons of the earth" and the "kingdoms" was expressed and that point to a picture of national feeling that was growing in several Latin territories (Souto Maior 1968).

[2] The name in which this Commission will be called in the text

[3] Pejorative title to this undertaking, formulated by newspapers and critics of the time

astronomical and geographical sections; and *Antônio Gonçalves Dias* (1823–1864), director of the ethnographic and traveling narrative sections.

A question hovered in the air: To where they should start the exploration? And which would be the first investigated provinces[4]? And who could give the foundation and ordering to this research? Capanema, during the IHGB 17th session, on November 28th of 1856, proposed a visit to the north provinces[5] that currently correspond to the northeast of Brazil: Maranhão e Ceará, based on possible economic resources and in particular the benefits that which section could find in these areas. However, there is a letter from Freire Allemão, deposited in the National Library, with the date of April 15th of 1849, asking to Antônio Paulino Nogueira,[6] a possible relative of João da Silva Feijó, biographical information about this naturalist. Ceará would become the start point for this journey, because its knowledge should be investigated, from this shortage of information that was already known of these lands.

Besides the provinces of the north being less known and distant from the Court, according to Alves (2012), the politics issues started to present special attention for having been the stage of contestatory movements against the central power since Brazil colony.

Upon arrival at Ceará, the commission was complete, with their kept directors since its idealization and complemented by its auxiliars. Worth the recognition merit were the counselor *Francisco Freire Allemão* (president and chief of botanical section) and his nephew and auxiliar, the young Dr. Manoel Freire Allemão de Cysneiros; *Guilherme Schüch de Capanema* (chief of geologic and mineralogic sections) and his auxiliar João Martins da Silva Coutinho; *Manoel Ferreira Lagos* (chief of zoological section) and his auxiliars, the brothers João Pedro Vila-Real and Lucas Antônio Vila-Real, preparing experts; *Giácomo Raja Gabaglia* (chief of astronomical and geographic sections) and his auxiliars, the Capitão Agostinho Victor de Borjes Castro and the First Lieutenants Antônio Alves dos Santos Souza, Francisco Carlos Lassance Cunha, João Soares Pinto, Caetano de Brito de Sousa Gayoso, and Basílio Antônio de Siqueira Barbedo; and *Antônio Gonçalves Dias* (chief of ethnographic and the trip narrative sections) and his auxiliar Francisco de Assis de Azevedo Guimarães, besides *José dos Reis Carvalho* that was responsible for the pictorial and photographic records (Venâncio Filho 1944; Damasceno 1961; Damasceno and Cunha 1961; Pereira 2006; Kury 2001a, 2009a).

[4] Province was how it was called, which we currently refer to as states of the federation.

[5] Northern provinces defined what we have today as parts of the North and Northeast regions of Brazil.

[6] Alemão FF (1849) Letter to Antônio Paulino Nogueira, asking for biographic information about a relative of his, the naturalist João da Silva Feijó. S.l.: s.n., 15 April 1849. 2 p. http://objdigital.bn.br/objdigital2/acervo_digital/div_manuscritos/mss1416119/mss1416119.pdf. Accessed 5 Jan 2017.

The itinerary to be carried[7] out by the Scientific Commission did not happen by chance; it has the start point at the province of Ceará and followed the few indications left by foreign naturalists that had visited the region during the nineteenth century, such as Henry Koster (1793–1820) and George Gardner (1810–1849) (Sá and Kury 2012). In addition to the possible potential that this region and its surroundings offered the expedition. After all, the northern provinces were already being scrutinized by naturalists who passed through the region, as pointed out by Alves (2012). The Scientific Exploration Commission specially the botanical section, chaired by Freire Allemão, gathered itself in a region of Ceará, as registered in the marked itineraries in his field diaries; however there were records of collections in Pernambuco, Rio Grande do Norte, Alagoas and Piauí.

By the end of the expedition, in 1861, the botanical section could collect a substantial number of samples and notes, that according to Braga (1962), gathered in an herbarium that was sent to the National Museum,[8] that counted with:

> that counted with 14000 of samples saved in tinplate welded boxes and covered in woods which arrived in very good condition.

Until that time, it was perhaps the greatest botanical contribution which had been added to the collection of the herbarium of the National Museum (R), where in a letter deposited at the National Library, Freire Allemão would have written to De Candolle:

> few phanerogamic plants and of a certain size escaped us.

About this collection, Freire Allemão wrote on *Works of the Scientific Exploration Commission*, 1862… that:

> ... from the balance that is taking place in the collections, we came to know the number of copies of the dry branches, which exceed 12 thousand, and we are missing a box that has not yet arrived from Ceará and which must contain a work of two thousand copies. Those who have already been counted and separated are included in 110 natural families. Care is now being taken to separate the genera and species…

The journalists and senators' expectations about the Commission to find wealth, such as silver and gold mines, weren't achieved; however, the subject of creating collections was fully reached, like Lopes (1996) points out. Besides the loss of part of the collected material on the Palpite shipwreck (Figueirôa 2008), the National Museum was the main destiny to the botanic, mineralogic, and zoologic collections and also receiving the instruments and materials to be used on the preparation of products and more than 1000 books that would be part of this institution Library.

---

[7] The script followed by each session of the Scientific Committee on Exploration was independent, especially due to the interests of each session, with some parts following other regions than Ceará; the same was seen in the botany section, which also visited Pernambuco, Rio Great North, Alagoas, and Piauí.

[8] National Museum referenced here, the botany section, that is, the Herbarium of this institution, which has the acronym R

Once the expedition was over, the work resulting from the organization, assembly, identification of plants and preparation of reports, pertinent to the publication of works on the recognition of vegetation and its uses, mainly timber and medicinal, were only completed in 1864.

7.3 Historical Ethnobotanics

The search for the knowledge about the plants and its use by men took Harsberger, in 1895, to propose the term ethnobotany, as a field of botany, which is the study about the various inductive processes of knowledge and the uses of vegetables, by different people and cultures throughout time. These processes point to an integration between forms of management, genetic improvements, bioprospecting, and conservation of useful species (Senna-Valle and Sá 2007).

The interdisciplinary character developed by ethnobotany research is the result of the interlocution between the natural sciences and the human sciences, the result of this form of investigation, which will observe the relationships between man and plants, in addition to the interaction with other disciplines, such as pharmacology, between others, who come to add their concepts and methodologies to help address issues about the relationship between man and the environment (Martin 1995).

Studies in ethnobotany cite several categories of use of plant resources, according to the worldview of the informants, members of the studied community. However, these categorizations pass through the observation point of the researcher who, in general, is in the role of interpreting or adapting this information, according to his own mental organization and worldview (Sá 2007). In this way, the data of the observed plants are informed in several emic categories during the fieldwork; these are reorganized by the ethnobotanists in other categories of use, with some adaptations, to adapt to the context of the community under study. Generally, the categories of uses raised in ethnobotanical studies are food, medicinal, ornamental, ritualistic, construction, fuel, and technology (Prance et al. 1987; Phillips and Gentry 1993; Albuquerque and Andrade 2002; Rios 2002; Sá 2007; Boscolo and Senna-Valle 2008).

Studies related to the use of plant species by different human cultures, whether extinct or not, have needed an approach to the discipline "history" as a fundamental instrument to help understand the synchronic and diachronic processes that involve this relationship (Heinrich et al. 2006). In this context, the ethnobotanist must be aware of the potential of historical documentation, which houses a rich collection of information (Noelli 1998).

Historical ethnobotany has as its central point observed by researchers the studies of the interrelationships between man and plants, through time, seeking answers from the perspective of botany, anthropology, ecology, and history, including the search for new or forgotten useful species (Medeiros 2009). Traditional knowledge can be preserved, in written documentation from distinct or even extinct communities, the role of history being important in consolidating the analyzed data and sources.

The historical-cultural exchange and biodiversity promoted by different cultures (Amerindians, Europeans and Africans) since the discovery of Brazil has been important for the understanding and use of natural resources, domestications, and exchange strategies between these cultures, thus promoting the development of knowledge about vegetation in the most different regions of the country (Herrera et al. 1998).

The recovery of pictorial records made by different naturalists, throughout the colonial and imperial periods, has been shown to be an important tool in understanding how plant resources were used by different elements of society, from time to time to the present day (Brandão et al. 2008). Ethnobotanical studies can emerge from this historical information, compiling data observed in the past that would be at risk of being lost if these studies were disregarded.

This study aimed to study the botanical collection of the Freire Allemão, deposited in the herbarium of the National Museum, and to compare it with their manuscripts, publications, and pictorial records, kept in the National Library, with special attention to the useful plants collected during the expedition of the Scientific Commission of Exploration, as well as recognizing in the material collected by this naturalist and in his manuscripts and diaries the existence of collaborators in the work of collecting and acquiring materials by the Freire Allemão.

7.4 Methods

In order to achieve the proposed objectives, the following methodologies were established.

7.4.1 Survey of Documental and Bibliographic Sources

It was made a survey in different bibliographic sources, searching for information about the botanist Francisco Freire Allemão, his nephew Manoel Freire Allemão de Cysneiros, and the Scientific Exploration Commission and about Ceará, both with regard to the circumscription of the former province and that of the current state.

It was visited the reference files and libraries: National Library (BN), National Museum (MN), Rio de Janeiro Botanical Garden Institute (JBRJ), Brazilian Historical and Geographical Institute (IHGB), National Archives (AN), National Historical Museum (MHN), Dom João VI Museum (MDJVI), and Historical Archive of the Imperial Museum of Petrópolis (AHMIP), in order to search for bibliographic and iconographic materials referring to this research. Analyses of handwritten, administrative documents, field diaries, travel diaries (Alemão 2006,

2007, 2011), collection books, correspondence, and maps were evaluated regarding indications of uses and methods of preparation of products of plant origin.

The interpretation of the visual representations, which aim to enrich the study, were carried out through observations of the diagnosis of the images, seeking a descriptive explanation of the content they carry, especially of species with resource citations of their uses, which may be contained in the collection of the Section of Manuscripts of the National Library on the registration of botanical studies.

7.4.2 Study of the Botanical Collection

A survey of specimens collected by Freire Allemão deposited in the herbarium of the National Museum (R) was carried out, with the aim of gathering and obtaining information contained in the exsiccate labels, with the process of restoration, reassembly, computerization, and digitization of the botanical material being carried out.

In order to validate and update, the scientific names of the taxa were consulted in the databases SciELO (2020), CRIA (2021), The Plant List (2021), IPNI (2020), APG IV (2016), and Flora do Brasil 2020 (2020) and reference works and whenever possibly assisted.

The information contained in the exsiccate tags were entered into spreadsheets in the Excel software; later, the data will migrate to the speciesLink website (CRIA 2021). The specimens collected by Freire Allemão and Manoel de Cysneiros, belonging to different categories of use, were verified, in addition to the citations of material sent to other naturalists.

7.4.3 Evaluation of Ethnobotanical Data

The species indicated with ethnobotanical were inventoried, according to the documental bases and publications left by Francisco Freire Allemão and Manoel de Cysneiros, such as in the second part of the works of the Scientific Commission for Exploration, from 1862, manuscripts, historical documents, and pictorial elements, which they mention the uses of these plants deposited in the National Library, in the lists of the Statistical Essay of the Province of Ceará (Brasil 1997) and of the Statistical Yearbooks of Ceará in the years 1922 and 1925, in addition to comparing and obtaining the scientific and popular names of the plants mentioned by Braga (1976) and Pio Corrêa (1926–1978), in updating the scientific names in these listings.

Works were carried out that addressed the various aspects of the uses of plants, bibliographic and electronic indexers, which were complemented by crossing references cited in specific works, verified in the articles on *Portal CAPES*, SciELO

(2020), and reference works. As for the analysis of medicinal indication, the databases of SciELO (2020) and Lorenzi and Matos (2002) were used.

The categories of use to be evaluated in this work will be timber (construction and joinery), medicinal, food, veterinary (forage "animal food" and animal medicine), technological (oils, gums and resins, fibers, dyes), and ornamental.

7.5 Results and Discussion

7.5.1 Ceará's Flora Between a Trip and a Mission

The scientific commission ran the most different regions of the northern provinces that currently correspond to the Northeast of Brazil, having a special focus, the then province of Ceará, not paying attention to what today circumscribes this state of the federation. The records of Freire Allemão's diaries point to the predominant permanence of his section, in the limits of Ceará, with small incursions to some bordering states, reaching the attempt to release an advance of this section to Piauí, Maranhão, and Pará, without success.

The documentation and scientific record collected during this endeavor generated numerous reports and publications. Upon returning to Rio de Janeiro, its members exalted having with them a precious set of 14,000 plant samples, which, according to Allemão (1862), were deposited in the collection of the herbarium of the National Museum (R), but with the sinking of the yacht Palpite, about 2000 samples, especially that of medicinal plants, were lost. This collection is not presented in a linear way, following the field trip and the records in the herbarium, as pointed out by Kury (2008) and Kury and Sá (2009), being sources for researchers from different areas. However, what we see today is the gathering of this collection, in a herbarium of the Scientific Commission for Exploration, which was initially distinct from the general collection and which over time was incorporated and distributed within this collection.

The collections carried out by Freire Allemão and Manoel de Cysneiros, during the period 1859–1861, could contribute to the formation of a checklist of the species collected by these naturalists during their performance in the botany session of the Scientific Exploration Commission and allowing the creation of a database, with historical value, on the flora of this region, in the middle of the nineteenth century. Generate subsidies for the identification of threatened species and indicate the priority species for conservation in these places. As Théberge (1898) said, almost 120 years ago:

> the present work has the merit of removing from the dust of the archive's writings of great scientific importance, whose copies are now very rare, and whose subjects are completely ignored, by many people who, in fact, are generally considered and considered by illustrated.

From the general collection of the herbarium of the Museu Nacional (R),[9] a total of 1965 exsiccates were gathered, from these 1794 samples, which are especially correlated to the Scientific Exploration Commission, to Ceará, and to the Freire Allemão, between the period 1859 and 1861, comprising a total of 1169 species, distributed in 137 plant families, which already demonstrates a considerable number of specimens of these collectors deposited in this herbarium, which in many cases corresponds to the gathering of many collections within the same herbarium record number, as you can observe in some exsiccates that have different location information and collection dates, but gathered under the same record.

Their collections/exsiccates were not restricted only to the herbarium of the National Museum, as over many years these materials, especially duplicates, served as exchange material or donation to other national and international institutions, where when searching in the sites, which keep the information in other collections and/or institutions, we can observe a very limited number of copies in these other collections, as can be seen in Fig. 7.1, proving that the largest collection of these

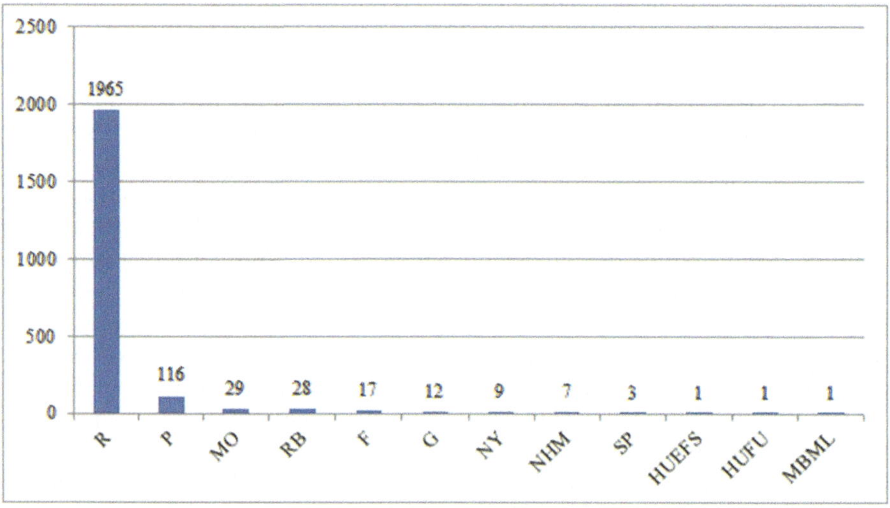

Fig. 7.1 Distribution of collections by Freire Allemão in scientific collections in Brazil and abroad. Acronyms: *BM* Natural History Museum of London, *F* Field Museum of Natural History, *G* Conservatoire et Jardin botaniques de la Ville de Genève, *HUEFS* Feira de Santana State University, *HUFU* Federal University of Uberlândia, *INPA* National Institute of Amazonian Research, *MBML* Mello Leitão Biology Museum, *MO* Missouri Botanical Garden, *NY* The New York Botanical Garden, *P* Muséum National d'Histoire Naturelle de Paris, *R* herbarium of the National Museum, *RB* Botanical Garden of the Rio de Janeiro, *SP* Institute of Botany

[9]The Herbarium of the National Museum, whose acronym is R, is the first founded in Brazil, dating from 1831, created by Ludwig Riedel. Its general collection comprises more than 550,000 specimens, of which around 6575 are nomenclatural types; its collection also includes historical collections by naturalists such as Glaziou, Freire Allemão from the Brazilian imperial family, and many others, safeguarding information on national and international flora (http://www.museunacional.ufrj.br/dptbot/herbario.html)

collectors is found in the herbarium of the National Museum (R). However, there was one species that caught our attention, *Ziziphus cotinifolia* Reissek, which presents duplicates of specimens distributed, especially among national collections and with a smaller number of samples.

As for the representativeness in terms of the botanical groups, the following were obtained: 1 species of bryophyte (1 collection); 16 species of ferns and lycophytes, distributed in 6 families; and 1152 species of angiosperms, distributed in 130 families, deposited in the herbarium of the National Museum (R). When comparing these numbers to that obtained in an analysis on the Flora do Brasil 2020 website (2020), the number of angiosperm families indicated for the current areas of Ceará is indicated, 157 families and 2604 species, which shows that the collection of these naturalists agrees in 44.81% similarity between them.

From the amount of species collected by the Freire Allemão, there was an expressive number of new species, which were described for science and are deposited in the herbarium of the National Museum (R); among these, 77 samples are registered as nomenclatural types, of which 36 species are distributed in 17 families.

When analyzing the 10 families with the highest number of species, Fabaceae (181 species) is the one with the greatest richness (Fig. 7.2), that is, 16.88% of the total material collected from these collectors, in agreement with several studies for the region, such as Lima (2012), Lima et al. (2009), and Araújo et al. (1999).

The Fabaceae family is the most diverse both in number of species (181) and in 262 collections. Among the genera, it has 93, the most numerous being *Mimosa* L. with 12 species, in agreement with Lima (2012), who points to 1 dominance of

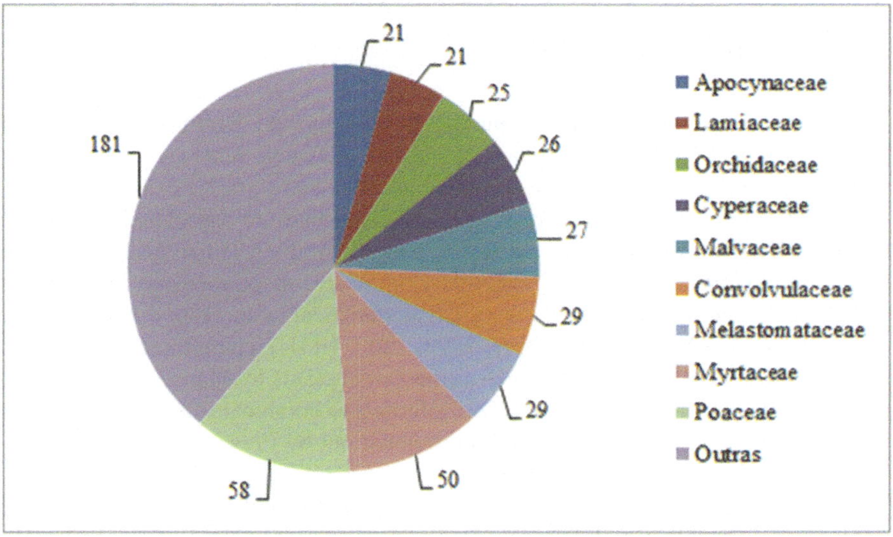

Fig. 7.2 The ten most diverse families from the Freire Allemão collection deposited in the National Museum (R) herbarium

the Mimosoideae subfamily in the study region. For the Caatinga we can highlight *Amburana cearensis* (Allemão) AC Sm. (*imburana-de-cheiro*, *cumarú*), *Copaifera duckei Dwyer* (*copaiba*, *pau-d'óleo*), *Libidibia ferrea* (Mart. ex Tul.) L.P. Queiroz (*jucá*, *pau-ferro*), *Mimosa caesalpiniifolia* Benth. (*sabiá*, *sabiá-da-serra*), *Mimosa tenuiflora* (Willd.) Poir (*acássia-jurema*, *jurema-preta*, *jurema*), *Myrospermum erythroxylum* Allemão (*bálsamo*), *Parkia pendula* (Willd.) Benth. ex Walp. (*angico*, *espinho-de-judeu*, *visgueiro*), and *Schinopsis brasiliensis* Engl. (*braúna-do-sertão*, *guaraúna-do-sertão*), among others.

The evaluation of Freire Allemão's manuscripts deposited in the National Library, represented by the collection of botanical studies referring to the Scientific Commission, which consist of 713 studies of plants from the northern provinces, carried out by Freire Allemão and his nephew, resulted in the classification of around 695 identifiable quotes. Of these studies, 18 were not possible to be categorized, depending on a deeper analysis and comparison with the materials in the National Museum collection. The identifications were possible, as these studies are very detailed, as they involve descriptions, designs, ecological, and use. From these 695 citations, we can identify 109 botanical families and about 650 species.

The collected materials and the data from the manuscripts may raise some doubts, but when we compare these studies to the botanical studies carried out by Freire Allemão and his nephew reported in this botanist's travel diaries (Alemão 2006, 2007, 2011), we can observe that there is a standard between exsiccates and manuscripts. However, often there is only the record of the Scientific Commission for Exploration or even just Freire Allemão and the collection dates when these are verified.

7.5.2 *Useful Plants of Ceará*

Among the objectives of this exploratory visit by the Scientific Exploration Commission to visit Ceará and its surroundings, between 1859 and 1861, was to survey the riches of plant origin, especially those that had some type of use. For this, 702 popular names or ethnospecies were collected that are associated with some type of indication of use indicated in the lists of Freire Allemão and Manoel de Cysneiros, in the works of the Scientific Committee, as pointed out by Allemão (1862), and in complementary works, such as statistical essay from the province of Ceará (Brasil 1997), Théberge (1897, 1898, 1900, 1901), where there are lists and citations for uses of Ceará plants, such as medicinal plants by Manoel de Cysneiros, in addition to consulting the manuscripts referring to botanical studies, deposited in the National Library, in addition to those pointed out by Freire Allemão in his diaries from 1859 to 1861.

Of these 702 popular names cited by Freire Allemão and Manoel de Cysneiros, 436 species were identified; however, 2 species were not, even if using works that indicate the popular names and uses of plants in the Northeast, especially in Ceará,

Braga (1976), or more general as by Pio Corrêa (1926–1978), Cruz (1992), Lorenzi (2000, 2008, 2009), Lorenzi and Mattos (2002), and Kinupp and Lorenzi (2014).

Of these 436 species evaluated with indication for use, 102 families were recognized, with the richest families in number of species, Fabaceae (61), Asteraceae (21), Malvaceae (17), Euphorbiaceae (16), Rutaceae and Myrtaceae (14), Solanaceae (13), Apocynaceae (12), Lamiaceae (11), and Cucurbitaceae (11), followed by the other families.

Indications for use were categorized according to reports contained in travel journals, manuscripts, and primary and secondary sources related to the Scientific Exploration Commission. The species were grouped into 6 categories, namely, timber (construction and joinery, 101 species), medicinal (291), food (116), veterinary (forage "animal feed" and animal medicine, 37), technological (oils, gums and resins, fibers, dyes, 63), and ornamentals (8).

7.5.3 Freire Allemão and the Network of the Invisible in the Scientific Exploration Commission

The development of natural sciences and national identity, during the nineteenth century, were the mainspring of transformation of both American and European countries, which fostered the applicability of different technologies, as this was a trend in various parts of the world at that time. Naturalists were no longer satisfied with the acquisition and assembly of cabinets of curiosities; they wanted to "see is to believe"; a period of searching and learning in loco begins (Kury 2001b). Many of the collections were assembled from specimens and/or materials, collected by merchants, navigators, and adventurers.[10]

From the travels made by Alexandre von Humboldt,[11] a golden period opened in the development, organization, and execution of these scientific endeavors in search of the new, the unexpected, and recognition. The "young naturalists" set out in search of adventures, as Kury (2001c) states:

> The journey is generally considered by natural history as one of the necessary steps for the transformation of nature into science, and which fostered colonial and territorial, which

[10] Süssekind (1990) "states that the look of these characters was different and did not have the focus of science. Diverse – epistles, diary, memoirs, reports – their accounts. Very diverse observers, however, of the traveling scientists also in transit at the time."

[11] These trips coincide with the new European expansion, since in this period at the end of the eighteenth century, the great discoveries of new territories are concluded. According to Pratt (1991), Europe becomes interested in the interior lands of these continents, seeking new riches. And Humboldt registers this moment of change:

> It is not sailing along the coast, [he says,] that we can discover the path of the mountain ranges and their geological constitution, their climate etc....

came under the responsibility of political, economic, and military interests that favored the work of naturalists, to know and explore our planet. (Antunes 2015)

After the arrival of the Portuguese Royal Family, in 1808, fleeing the Napoleonic expansion over Europe, which elevated and transformed Brazil from an exploration colony to a United Kingdom, the first measure was the opening of the Friendly Nations Ports, which facilitated commerce and the search for the new world that was unfolding for the sciences. A legion of Europeans landed in Brazilian lands, in search of this knowledge, making Brazil to be discovered again; this happened under the foreign gaze. Each of the incursions had particular objectives, destinations and resources, or even international interests involved (Kury 2009a):

> Despite the small differences in style, good travelers, since the eighteenth century, followed some common basic procedures, such as accurate notes in a diary, the collection and preservation of products and, more than that, they pursued the art of good to inquire. In the mid-nineteenth century, when there was already a well-developed specialization in natural history, the traveler continued to be the one who researched much more than his strict domain. (Kury 2009a)

The almost uncontrollable desire in the search for new beings, products, and knowledge and to open up new lands made Antunes (2015) highlight the concern with not transforming these naturalists into mere adventurers and pathfinders who sought to fight for survival and science in hostile environments. The discontent in relation to these reports made by some naturalists, led to the IHGB, the creation of a Scientific Commission formed by Brazilian scientists, to investigate our lands and consequently their resources, as pointed out by: Alemão (1862), Damasceno (1961), Braga (1962), Kury (2001a, 2009a,b), Sá and Kury (2012), Pereira (2006), Cavalcante (2012), and Alves (2012).

After its formation, the Scientific Exploration Commission took steps to assemble its guidelines and what each section should follow in its work; it took a long time to gather all the resources and personnel necessary for its execution (Veloso Junior 2011). Remember that the results obtained by these naturalists would not be the result solely of their intellects and personal abilities, but the set of actions and collaborations established throughout their endeavor, as we can see in the instructions, which should be followed by the botanical section (Alemão 1856), where it indicates that:

> He will inquire from the practical men of the place the indigenous and common name of each vegetable, and its popular uses… Each of these things, not being able to collect specimens or products, will seek to have someone from the place undertake to do so, indicating how to practice it, and to send it safely…

This demonstrates in a unique way that scientific activity is not solitary and depends on the exchange of information between different parts,[12] for the production and obtaining of consistent results. The analysis of collaboration between the forming

[12] For Lopes and Heizer (2011), the correspondence is also a source of information about the details of the trajectories taken by traveling naturalists and collectors in their naturalist, diplomatic, and commercial missions.

parts of science, especially in scientific trips, serves as a background for historiography, tracing the social profile of contributing partners, which often go unnoticed, but important in the development process of science. The study of the role of collaborators or the "network of invisibles" began to gain prominence in Camerini (1996), Moreira (2002), Fan (2003), and Raj (2007).

The resources for carrying out the work of the Scientific Exploration Commission may have started long before its actual proposal in 1856, given the IHGB's interest in systematically investigating and organizing matters relating to knowledge of the natural resources of the different provinces of Brazil at the time. They generated an interest in Freire Allemão, a member of this Institute, in obtaining information about explorers in the province of Ceará, in 1849, as recorded in a letter deposited at the National Library (Allemão 1849), where he requested information on the biography of João da Silva Feijó, to Antônio Paulino Nogueira, a supposed relative of this naturalist who was investigating this region.

According to Antunes (2015), the acquisition of contacts with local informants sometimes started long before the departure and execution of the expedition itself, and this information network was usually assembled from the contacts made through letters to authorities and/or representatives of the places to be visited, such as farmers, hunters, transporters, guides, and others. Until the naturalists hired woodsmen and acquired some slaves to accompany them on their journey and that they could count, on many occasions, on the support and knowledge of local communities and indigenous groups:

> From several of these writings, one can establish the main types of contributions of local personnel: identification, location, collection and nomenclature of animals and plants; preparation and preservation of specimens; discoveries of 'new' species; analysis of habits and uses of animals and plants; geographic, meteorological and animal and plant distribution knowledge; anthropological reports; indication of more favorable places for research; domestication of animals; and manufacture of instruments (including for capturing and preserving animals). (Moreira 2002)

One of the most interesting points of the Scientific Committee for Exploration was the memorable acquisition of camels and the hiring of Bedouins from Morocco, to carry out the transport of people and materials during the undertaking in Ceará. This fact was so remarkable that even after many years of the Commission in Ceará, it was remembered as the plot of a samba school in Rio de Janeiro.

The scientific commission consisted of five sections: botany, zoological, geological and mineralogical, astronomical and geographical, and ethnographic and travel narrative, which had their heads and assistants, as well as a painter to record pictorially and photographically (Venâncio Filho 1944; Damasceno 1961; Damasceno and Cunha 1961; Pereira 2006; Kury 2001a, 2009a). Freire Allemão was in charge of the Scientific Commission and the botany section, and as his assistant, nephew Manoel de Cysneiros, but they did not work in isolation, as shown in many records in their field diaries and letters exchanged between their members and other informants in the studied region (Fig. 7.3); these sometimes go unnoticed and even invisible in travel narratives, during the process of scientific dissemination (Camerini 1996).

Fig. 7.3 Scientific Exploration Commission of Ceará – author: José dos Reis Carvalho, "Rustic expedition camp," watercolor on paper, 1859, Ceará. (Source: Dom João VI Museum, available in public domain)

The formation of a network of assistants can be traced by reading the diaries left by Freire Allemão, which have recently gained publications (Alemão 2006, 2007, 2011), as well as letters and other documents kept in archives, from the National Library and from SEMEAR. A significant number of exchanges of information was observed, ranging from personal and family interest to official information and complementing data for the expedition.

When evaluating the records left in their diaries, the topic of botanical studies appears cited 68 times, which indicates a correlation with the volume of manuscripts deposited in the National Library, which consists of 9 volumes and 925 documents, referring to description, drawings, ecological information, and the use of many plants analyzed by Freire Allemão and his nephew, as well as information on the receipt of materials from other members of the Scientific Commission, people and informants, and collaborators in the investigation process.

Among this information, we can highlight Francisco Freire Allemão's attempts to transfer his position as head of the botany section to his nephew and deputy, Manoel de Cysneiros, as seen in the letter sent to D. Pedro II, on September 11, 1859, and who had health problems and could not perform the field visits satisfactorily:

> … My absence will not be felt in the works of my section. If V.M. do not want to send another naturalist, to replace me (in which case, I ask VM's permission, which in my opinion, would not be without inconvenience) here is my assistant, who has all the capacity to perform this task well…. (Allemão 1859)

A large number of manuscripts report that Manoel de Cysneiros and other members of the Commission, such as Capanema, Lagos, Dias, constantly sent him materials for his studies, in addition to other assistants, such as: Barreto, Dr. Pimentel, Dr. Pompeu, Juvenal, Captain Henrique Gonçalves da Justa, Captain Henrique Gonçalves dos Santos, Bernardo de Castro Freire Jucá and an unidentified woodcutter. But what most caught our attention was the record of the sending of two collections made by Gonçalves Dias, a Rubiaceae, *Hillia parasitica* Jacq. (BN: I.45, Pacatuba, Serra do Aratanha, 28/III/1859), and an Orchidaceae, *Epidendrum goncalvii* Barb. Rodr. (BN: II.56, Pacatuba, Serra do Aratanha, 12/VI/1859), synonymized as *E. saxatile* Lindl. (Fig. 7.4).

In his diaries, Freire Allemão highlighted several situations where these collaborators actively participate in the sending and study of plants and their respective usage information, such as the quotes where:

> Capanema (Guilherme Schüch de Capanema), indicates a plant as not being Linoceira (Caprifoliaceae) from China and yes *Corape guiarinensis*, called European madre-silva, but in the analysis of the referred material, it is a *Lonicera japonica* Thunb.

In another passage he mentions Dr. Théberge, who is invited to collect violetwood flowers (*Dalbergia cearensis Ducke) and braúna flowers (Melanoxylon brauna Schott) during his trip from Crato to Pacatuba. Like Barreto, he collected some samples sent from Fortaleza for his studies. There is also the record of two interesting facts, the first mentions the list of woods made by Francisco Félix, which were used on the top of the Serra do Meruoca, on February 3, 1861, and the other occurs in the Pacatuba diary, where there is talk of ordering 33 or 34 samples of wood from Mr. Captain Henrique Gonçalves da Justa, for the amount of 30$ (30,000 réis), and some dry plants, poorly processed, where only a few served and did not ask for pirauá* flowers (*Basiloxylon brasiliensis* (Allemão) K. Schum.), rabugens, and paus-d'arco amarelos to Mr. Juvenal, recorded in his journal of the Estada de Forteleza (23/V-27/VI/1860):

> After I got here I have done little or nothing in botany. The rains and the great and continuous heat, which is felt more in me by copious sweating, are part of the reason for this. Also the good one from Barroso has [been] lazy and has only been collecting some grasses and samples of trees or construction wood…. (Alemão 2011)

In one of his manuscripts, Allemão (1859) records the information passed on by Henrique Gonçalves da Justa, about the use of wood: *sabiá* (*Mimosa caesalpiniifolia* Benth.), *aroeira* (*Myracrodruon urundeuva* Allemão), and *carnauba* (*Copernicia prunifera* (Mill.) H.E. Moore), for different types of fences (caiçaras, corilio-fine, wattle and daub, rods, and fence posts). For the production of clogs, the *jenipapo* (*Genipa americana* L.) and, for the sticks of shotguns, carpentry, and tool handles, the *cauaçu* (*Coccoloba* sp.) were highlighted. In observations carried out on *catingueira* (*Caesalpinia pyramidalis* Tul.), the trunk was used by parrots and bees to build their nests.

Among the records of several conversations and observations about the uses of many plants, one that draws attention was recorded in a conversation with Father Sucupira, about medicinal plants, where the priest speaks of the virtues of the *angélica-brava* root (*Aristolochia* – Aristolochiaceae) and says there is a potato,

Fig. 7.4 *Epidendrum goncalvii* Barb. Rodr. Holotypus, deposited in the herbarium of the National Museum: currently synonymized as *E. saxatile* Lindl. (Source: creative commons)

which they call *batata-de-teiú*, *batata-de-cobra*, and, in *inhamuns*, *cabeça-de-negro* and *taiuia* (*teyia*), a vine, which was planted by Dr. Lacerda in Maranhão. This indicates that several records of Freire Allemão's collection deposited in the Museu Nacional may have come from regions other than Ceará and that they may have been improperly incorporated as being from Ceará.

In other passages, the informant remains anonymous, recording only the action, as in the case of citing the consumption of *ata fruits (Annonaceae), which cause "evacuações" (diarrhea), possibly referring to the different species that are consumed in this region and warns that excessive consumption causes this reaction. Or when he mentions the inerus roots that are consumed like manioc by the troops.*

Among the records referring to botanical studies and Freire Allemão's diaries, his assistant was a constant figure, in the role of collecting, sending materials, analyzing, and describing them. The role of interlocutor, played by Manoel de Cysneiros, among distant collaborators and Freire Allemão, favored the acquisition of interesting materials for collection, as in the case of an Araceae indicated as a parasite, sent by D. Maria Theophila, this being the only one registration of a woman participating in this Commission.

7.6 Final Considerations

The notoriety of the Scientific Exploration Committee for being the first major Brazilian scientific expedition to define: who we are and what we have, has been unquestionable in countless works resulting from these results. In particular, what led this group of scientists to investigate possible riches and thus help in the unity of the nation in the imperial period. With the approval of Emperor D. Pedro II, and the main research and science institutions in Brazil, at the time, the Brazilian Historical and Geographical Institute, the National Museum and the National Society for Industrial Assistance. This expedition obtained a portrait of Brazil, until then little known to Brazilians.

Among the scientists who made up this expedition, we highlight the importance of its head and director of the botany section, physician, and naturalist Francisco Freire Allemão and his nephew and assistant, also physician Manoel Freire Allemão de Cysneiros. The exploration of the most northern regions of Brazil, especially Ceará and its surroundings, was essential for the process of affirmation and contact between the central government and the more distant provinces.

This expedition lasted about 2.5 years (1859–1861). The botanical section covered the different regions and phytogeographic landscapes of the northeast of Ceará; these collectors claimed to have collected about 14,000 plant samples, which totaled about 110 known botanical families. In letters exchanged with foreign naturalists, they even extolled that they had collected almost all types of vegetables existing in that region.

When evaluating the existence of his collections deposited in the herbarium of the National Museum, the main point of deposit of his collection, we came across a large personal collection of this collector linked to this expedition, totaling, so far, 1794 samples, of which we can verify the existence of 1169 species, distributed in 137 botanical families. When compared with the most modern information about the Brazilian flora, they collected almost 45% of the species of the flora existing in this region, demonstrating a really huge sampling effort in such a short execution time and with all the tribulations suffered in this undertaking.

However, when comparing these quantities, quoted in the past and gathered in the present, the numbers do not add up. However, we can say that the material that we analyzed today is a portrait of the volume of material collected by the members of this expedition, because when evaluating these collections we came across multivariate information in terms of collection dates, studies and locations, enabling the assertion that specimens collected in different locations and dates were grouped in numbers subsequent to the actual fieldwork, thus demonstrating this duality in the numbers presented in the past and present, in addition to, of course, possible losses that occurred over time.

The representation of some families, in numerical terms, corresponds to what is still shown in many works on our flora today. As Fabaceae, the largest family evaluated, it groups 262 collections, distributed by 181 species in 93 genera. *Mimosa* is the most numerous genus, with 12 species. This high representation is corroborated by different floristic studies in the Brazilian northeast region.

The research guidelines outlined by Freire Allemão for the analysis of plant resources would be very similar to the most modern principles of inventorying the man/plant kingdom relationship, that is, it was completely in accordance with the nuances and guidelines followed in ethnobotanical works, long before the creation of this part of botanical science.

The records of Freire Allemão's diaries and other manuscripts demonstrate a great network of collaboration with different elements of Ceará society. Conversations with the most varied characters and their journeys. He can trace aspects of daily life, vegetation, climate, uses, and indications, as well as his own impressions about his visits and studies. Some doctors, politicians, and farmers were constantly cited in these notes. However, the fact that a woman has contributed even if discreetly is very interesting. The record of collections sent by Gonçalves Dias is a real novelty, a fact recorded in his diaries and botanical studies, the poet and rapporteur of the Commission actively participates in the contributions on the flora studied in this undertaking, and the herbarium record is even observed in some others, citing the fact.

We conclude that even discreet Manoel de Cysneiros was the great performer of manual and fieldwork, in addition to being the second director of the current Botany Department at the Museu Nacional. His uncle was the translator of plant data into detailed reports, studies, and drawings, which can now enable a reading of the flora of Ceará, which was inaccessible to the general public.

References

Albuquerque UP, Andrade LHC (2002) Uso de recursos vegetais da caatinga: o caso do agreste do estado de Pernambuco (nordeste do Brasil). Interciência 27:276–285

Alemão FF (1849) Carta para Antônio Paulino Nogueira, perguntando sobre informações biogeográficas relativas ao naturalista João da Silva Feijó. s.l.: s.n., 15 abr. 1849. 2p. http://objdigital.bn.br/objdigital2/acervo_digital/div_manuscritos/mss1416119/mss1416119.pdf. Accessed 5 Jan 2017

Alemão FF (1859) Informações sobre cercas e madeiras da região em geral, dadas por Henrique Gonçalves da Justa. Pacatuba, CE: [s.n.], 5 abr. 1859–6 abr. 1859. 6 p. http://objdigital.bn.br/objdigital2/acervo_digital/div_manuscritos/mss1449200/mss1449200.pdf. Accessed 20 Dec 2016

Alemão FF (2006) Diário de viagem de Francisco Freire Alemão: Fortaleza – Crato, 1859, vol I. Museu do Ceará, Fortaleza

Alemão FF (2007) Diário de viagem de Francisco Freire Alemão: Crato – Rio de Janeiro, 1859/1860, vol II. Museu do Ceará, Secretaria de Cultura do Estado do Ceará, Fortaleza

Alemão FF (2011) Diário de viagem de Francisco Freire Alemão (1859–1861). Fundação Waldemar Alcântara, Fortaleza

Allemão FF 17ª. (1856) Sessão em 28 de Novembro de 1856. Revista do Instituto Historico e Geographico do Brazil, Rio de Janeiro, t. 19, supl., p. 74–82

Allemão FF (1862) Trabalhos da Commissão Scientífica de Exploração: secção botânica. Tipografia do Imperial Instituto Artístico, Rio de Janeiro

Alves CJ (2012) Natureza e cultura nas ilustrações da Comissão Científica de Exploração, (1851–1861). Tesis, Instituto de Filosofia e Ciências Humanas da Universidade Estadual de Campinas

Antunes AP (2015) A rede dos invisíveis: uma análise dos auxiliares na expedição de Louis Agassiz ao Brasil (1865–1866). Dissertation, História das Ciências e da Saúde. Casa de Oswaldo Cruz, Fundação Oswaldo Cruz. http://www.ppghcs.coc.fiocruz.br/images/dissertacoes/dissertacao_anderson_antunes.pdf. Accessed 12 Jul 2016

APG IV (Angiosperm Phylogenetic Group) (2016) An update of the angiosperm phylogeny group classification for the orders and families of flowering plants: APG IV. Bot J Linnean Soc 181(1):1–20

Araújo FS, Martins FR, Shepherd GJ (1999) Variações estruturais e florísticas do carrasco no planalto da Ibiapaba, estado do Ceará. Acta Botânica Brasilica 13(1):1–13

Boscolo OH, Senna-Valle L (2008) Plantas de uso medicinal em Quissamã, Rio de Janeiro, Brasil. Iheringia Sér Bot 63:263–277

Braga R (1962) História da Comissão Científica de Exploração. Imprensa Univeritária do Ceará, Fortaleza

Braga R (1976) Plantas do Nordeste, especialmente do Ceará, 3rd edn. ESAM (Coleção Mossoroense v. 42), Mossoró

Brandão MGL, Zanetti NNS et al (2008) Brazilian medicinal plants described by 19th century European naturalists and in the official pharmacopeia. J Ethnopharmacol 120:141–148

Brasil TPS (1997) Ensaio Estatístico da Província do Ceará. Tomo I. 1. ed. 1863; Fundação Waldemar Alcântara, Fortaleza

Camerini JR (1996) Wallace in the field. Osiris, 2nd series. http://www2.warwick.ac.uk/fac/arts/history/students/modules/hi916/week5/camerini_wallace_in_the_field.pdf. Accessed 14 Oct 2016

Cavalcante FHB (2012) Ciência e civilização desvendam o sertão: História Cultura e Natureza nos relatos de viagem de Francisco Freire Alemão (1859–1861). In: Mollo HM (ed) Biografia e história das ciências debates com a historiografia. EDUFOP/PGHIS, Ouro Preto, pp 78–102

CRIA (2021) SpeciesLink network, specieslink.net/search Accessed 3 May 2021

Cruz GL (1992) Dicionário das Plantas Úteis do Brasil. Bertrand Brasil, Rio de Janeiro

Damasceno D (1961) O Botânico Freire Alemão. In: Damasceno D, Cunha W Os manuscritos do botânico Freire Alemão – catálogo e transcrição, vol 81. Anais da Biblioteca Nacional, pp 9–35
Damasceno D, Cunha W (1961) Os manuscritos do botânico Freire Alemão – catálogo e transcrição. Anais da Biblioteca Nacional 81:1–350
Fan F (2003) Science in a Chinese entrepôt: British naturalists and their Chinese associates in old Canton. Osiris, 2nd series. Science and the City 18:60–78
Figueirôa SFM (1998) Mundialização da ciência e respostas locais: sobre a institucionalização das ciências naturais no Brasil (de fins do século XVIII à transição ao século XX). Asclepio 50(2):107–123
Figueirôa SFM (2008) A Repartição Geral dos Telégrafos e o trabalho de Guilherme Schüch de Capanema (1824–1908) em Geociências. In: Almeida M, Vergara MR (eds) Ciência, história e historiografia. Via Lettera/MAST, São Paulo/Rio de Janeiro, pp 125–135
Flora do Brasil 2020 (2020) Jardim Botânico do Rio de Janeiro. http://floradobrasil.jbrj.gov.br/ Accessed 10 Oct 2020
Guimarães LMP (2007) Entre dois Mecenatos. In: Da escola Palestina ao Silogeu: Instituto Histórico e Geográfico Brasileiro (1889–1938). Museu da República, Rio de Janeiro, pp 21–37
Heinrich M, Kufer J et al (2006) Ethnobotany and ethnopharmacology – interdisciplinary links with the historical sciences. J Ethnopharmacol 104:387–406
Herrera T, Ortega MM et al (1998) Uma etnobotânica no México. Episteme 15:133–136
IPNI (2020) The International Plants Names Index. http://www.ipni.org/. Accessed 17 May 2020
Kinupp VF, Lorenzi H (2014) Plantas alimentícias não convencionais (PANC) no Brasil. Instituto Plantarum, São Paulo
Kury LB (1998) Les Instructions de Voyage Dans Les Expéditions Scientifiques (1750–1830). Revue d'histoire des sciences, Paris 51(1):65–92
Kury LB (2001a) A Comissão Científica de Exploração (1859–1861). A ciência imperial e a musa cabocla. In: Heizer A, AAP V (eds) Ciência, civilização e império nos trópicos. Access, Rio de Janeiro, pp 29–54
Kury LB (2001b) Viajantes-naturalistas no Brasil oitocentista: experiência, relato e imagem. História, Ciências, Saúde-Manguinhos, Rio de Janeiro 8(supl):863–880
Kury LB (2001c) A sereia amazônica dos Agassiz: zoologia e racismo na Viagem ao Brasil. Rev Bras Hist São Paulo 21(41):157–172. http://run.edu.ng/media/2566903943121.pdf. Accessed 24 Feb 2020
Kury LB (2008) As artes da imitação nas viagens científicas do século XIX. In: Almeida M, Vergara MR, Ciência, história e historiografia, Via Lettera/MAST, São Paulo/Rio de Janeiro 321–333
Kury LB (2009a) Francisco Freire Alemão, botânico e viajante. In: Kury L (ed) Comissão Científica do Império. Andrea Jakobsson Estúdio, Rio de Janeiro, pp 181–221
Kury LB (2009b) Explorar o Brasil: o Império, as Ciências e a Nação. In: Kury L (Org.). Comissão Científica do Império. Andrea Jakobsson Estúdio, Rio de Janeiro, 19–49
Kury LB, Sá MR (2009) Flora brasileira, um percurso histórico. In: Martins ACI (Org.). Flora Brasileira. História, Arte e Ciência. Casa da Palavra, Rio de Janeiro 18–57
Lima BG (2012) Caatinga: espécies lenhosas e herbáceas. EdUfersa, Mossoró
Lima JR, Sampaio EVSB et al (2009) Composição florística da floresta estacional decídua montana de Serra das Almas, Ceará. Acta Botânica Brasilica 23(3):756–763
Lisboa KM (1997) A nova Atlântida de Spix e Martius: natureza e civilização na viagem pelo Brasil (1817–1820). Editora HUCITEC. FAPESP, São Paulo
Lisboa KM (1999) Humboldt e os viajantes no Brasil na primeira metade do século XIX. In: Zea L & Magallón M. (orgs.) El mundo que encontró Humboldt. Instituto Panamericano de Geografia e História, Fondo de Cultura Econômica. México
Lopes MM (1996) Mais vale um jegue que me carregue que um camelo que me derrube… lá no Ceará. História Ciências, Saúde – Manguinhos. Fiocruz, Rio de Janeiro 3:50–64
Lopes MM, Heizer A (2011) Colecionismos, práticas de campo e representações. EDUEPB, Ciência & Sociedade collection. Campina Grande. https://static.scielo.org/scielobooks/rk6rq/pdf/lopes-9788578791179.pdf. Accessed 6 Jan 2020

Lorenzi H (2000) Plantas daninhas do Brasil, 3rd edn. Instituto Plantarum, Nova Odessa
Lorenzi H (2008) Árvores brasileiras: manual de identificação e cultivo de plantas arbóreas nativas do Brasil, 5th edn. Instituto Plantarum, Nova Odessa
Lorenzi H (2009) Árvores brasileiras: manual de identificação e cultivo de plantas arbóreas nativas do Brasil, 3rd edn. Instituto Plantarum, Nova Odessa
Lorenzi HE, Matos FJA (2002) Plantas medicinais no Brasil/Nativas e exóticas. Instituto Plantarum, Nova Odessa
Martin GJ (1995) Ethnobotany: a methods manual. Chapmman and Hall, London
Martius CFP (1843) Systema Materia Medicae Vegetabilis Brasiliensis. Lipsiae
Martius CFP (1844) Natureza, doenças, medicina e remédios dos índios brasileiros. Traduzido por Silva, P. 1939. Empresas Gráficas dos Tribunais. Série 5. Brasiliana. São Paulo
Medeiros MFT (2009) Etnobotânica histórica: princípios e procedimentos. Nupeea/Sociedade Brasileira de Etnobiologia e Etnoecologia, Recife
Moreira IC (2002) O escravo do naturalista. Ciência hoje 31(184):31. http://www.casadaciencia.ufrj.br/caminhosdedarwin/downloads/escravo.pdf. Accessed 6 Jan 2020
Noelli FS (1998) Múltiplos Usos de Espécies Vegetais Pela Farmacologia Guarani Através de Informações Históricas. Universidade Estadual de Feira de Santana. Diálogos, DHI/UEM 02:177–199
Nogueira E (2000) Uma história brasileira da botânica. Marco Zero, São Paulo
Paiva MP (2002) Os naturalistas e o Ceará. Instituto do Ceará, Fortaleza
Pataca EM, Pinheiro R (2005) Instruções de viagem para a investigação científica do território brasileiro. Rev Soc Bras Hist Cien 3(1):58–79
Pereira NA (2006) Francisco Freire Alemão – Cientista do Mendanha. Publit Soluções Editoriais, Rio de Janeiro
Phillips O, Gentry AH (1993) The useful plants of Tambopata, Peru: I. Statistical hypotheses tests with a new quantitative technique. Econ Bot 47:15–32
Pio Corrêa M (1926–1978). Dicionário das plantas úteis do Brasil e das exóticas cultivadas. Imprensa Nacional. Rio de Janeiro
Prance GT, Baleé W et al (1987) Quantitative ethnobotany and the case for conservation in Amazônia. Conserv Biol 1:296–310
Pratt ML (1991) Humboldt e a reinvenção da América. Estudos Históricos Rio de Janeiro 4(8):151–165
Raj K (2007) Surgeons, fakirs, merchants and craftsmen: making L'Empereur's Jardin in early modern South Asia. In: Raj, Kapil. Relocating modern science. New York: Palgrave Macmillan
Rios M (2002) La comunidad Benjamin Constant y las plantas útiles de la "capoeira": un enlace etnobotánico en la Región Bragantina, Pará, Amazonía Brasileña. Universidade Federal do Pará, Núcleo de Altos-Estudos Amazônicos, Belém
Sá, I. M. de 2007. Levantamento etnobotânico em Santo Antônio do Rio Grande, sul de Minas, Brasil. Dissertation, Museu Nacional/Universidade Federal do Rio de Janeiro/Programa de Pós-graduação em Ciências Biológicas (Botânica)
Sá MR, Kury LB (2012) Naturalista europeus nas caatingas. In: Kury LB (ed) Sertões adentro: viagens nas caatingas, séculos XVI a XIX, vol 1. Andrea Jakobsson Estúdio Editorial, Rio de Janeiro, pp 204–257
Santos Junior NA (Eds.). A botânica no Brasil: pesquisa, ensino e políticas públicas ambientais. Sociedade Botânica do Brasil, 68°, São Paulo. p: 558–561
SciELO (2020) The Scientific Electronic Library Online. http://www.scielo.br/. Accessed 17 May 2020
Senna-Valle L, Sá IM (2007) Dos naturalistas à etnobotânica contemporânea: um panorama dos estudos de plantas no Brasil. In: Botânica Brasileira futuro e compromissos. 60° congresso nacional de botânica. Sociedade Botânica do Brasil. EDUNEB: BA. 1083–1087
Souto Maior A (1968) Unidade X: O Sentimento Nativista. In: Companhia Editora Nacional. História do Brasil. 6 edição, São Paulo. 181–200
Süssekind F (1990) O Brasil não é longe daqui. Companhia das Letras, São Paulo

THE PLANT LIST (2021) www.theplantlist.org. Accessed 17 May 2016
Théberge H (1897) Fauna e flora Cearense. Rev Acad Cear 2:189–199
Théberge H (1898) Fauna e flora Cearense. Rev Acad Cear 3:9–32
Théberge H (1900) Fauna e flora Cearense. Rev Acad Cear 5:92–124
Théberge H (1901) Fauna e flora Cearense. Rev Acad Cear 6:145–170
Veloso Junior C (2011) Ciência e nação no Brasil do século XIX: anotações de Freire-Allemão na expedição ao Ceará (1859–1861). In: Anais do XXVI Simpósio Nacional de História – ANPUH. São Paulo 1–19
Venâncio Filho F (1944) Freire Allemão Revista do Museu Nacional 1(1):22–24

Chapter 8
The Former Imperial Plant Nursery of *Quinta da Boa Vista*

Mariana Reis de Brito, Luiz Fernando Dias Duarte, and Luci de Senna-Valle

Abstract The nineteenth century brought with it a series of renovations in different scientific areas and, also, in aesthetic representations. An increasing desire to be close to nature, to examine it, to unravel it, to systematize it, and to portray it, contributed to the existence of a new dialogue between nature and civilization. The appreciation of plant resources influenced the development of botany and the arising of a new grammar in the art of elaborating gardens. The first "artificial landscape" elaborated in nineteenth-century Brazil were creations of the French landscaper Auguste François Marie Glaziou, who stood out not only for leaving deep marks on the urban landscape of the capital of the Second Empire but also as a scholar of the natural world. Through the analysis of documents of the time, it was possible to catalog the plants that were present in the former imperial garden of *Quinta da Boa Vista*, conceived by Glaziou, and thus to reconstitute some of the history of this important horticultural laboratory, providing information of botanical, historical, cultural, and artistic importance.

Keywords Botanical practices · Gardens · Nineteenth century · Ornamental plants · Plant nurseries

M. R. de Brito (✉)
Departamento de Biologia, CCBS-Centro de Ciências Biológicas e da Saúde, Pontifícia Universidade Católica do Rio de Janeiro (PUC-Rio), Rio de Janeiro, RJ, Brazil
e-mail: marianareis2002@puc-rio.br

L. F. D. Duarte
Departamento de Antropologia Social, Museu Nacional, Universidade Federal do Rio de Janeiro (MN/UFRJ), Rio de Janeiro, RJ, Brazil

L. de Senna-Valle
Departamento de Botânica, Museu Nacional, Universidade Federal do Rio de Janeiro (MN/UFRJ), Rio de Janeiro, RJ, Brazil

M. F. T. Medeiros, B. de Sá Haiad (eds.), *Aspects of Brazilian Floristic Diversity*,
https://doi.org/10.1007/978-3-031-07453-0_8

8.1 Auguste François Marie Glaziou: The Constructor of Landscapes

In Brazil there was not a long tradition in the elaboration of large gardens; only late in the eighteenth century emerged the concern of idealizing a public garden in the middle of a colonial city (Terra 2000). The creation of the first public garden in Brazil, the so-called *Passeio Público* of Rio de Janeiro, commissioned by viceroy Luís de Vasconcelos, in 1779, is justified by the need to combat putrid environments that, according to miasma theory, were favorable places for the development of diseases. However, the insertion of a public garden in the heart of the colonial city also sought the creation of a European "scenario," imprinting notions of "prosperity" and "modernity" and manifesting the emergence of a new ideology that imposed its values through the appropriation and manipulation of its landscape (Macedo 2012).

In the mid-nineteenth century, the process of urban change in the Brazilian Empire was intensified, through a strong feeling of overcoming the ills of colonial heritage and desire to reach the same stage of development of the most fortunate European nations. Thus, seeking inspiration in the prosperous French capital, which radiated its model of urban environment integrated to green spaces, one can observe the implementation of a metropolitan set of gardens, squares, and parks[1] as also afforestation programs in several Brazilian cities (Dourado 2011). At that time, the services of French landscapers were constantly requested to assist in the enhancement of this art, according to both the characteristic typologies and the strategies adopted in Paris (Terra 2000). Thus, upon arriving in Brazil, the Frenchman, gardener, and horticulturist Auguste François Marie Glaziou, driven by the desire to know the vegetation of this tropical country closely, would find the ideal conditions to launch himself and stand out in the careers of landscaper and horticulturist and – when the time came – even as a botanical researcher.

François Marie Glaziou, called by Auguste,[2] was born on August 30, 1828, in France, more precisely in Lannion, Côtes-d'Armor, Brittany. During his childhood and youth, several regions of the country underwent considerable social and landscape transformations inspired by the heritage of the Enlightenment movement and by the urban and technological development (Hetzel 2011).

From a humble family, Glaziou began his learning in gardening and horticulture with his father. Bureau (1908), his first biographer, says that at the age of 16, after a

[1] In this chapter, the terms gardens, parks, and squares have the same meaning, tied to the idea of "cultural landscape," in which the perception of environment comprises its most current perspective: a dynamic system of mutual relationships and interferences, encompassing natural, social, cultural, economic, ethical, and political aspects. In 2009, the *IPHAN* (National Historical and Artistic Heritage Institute) conceived the expression "Brazilian cultural landscape" as a particular portion of the national territory, representative of the interactive process between man and the natural environment, to which life and human science printed its marks or attributed its values.

[2] On August 30, 1858, Glaziou presented himself for the first time under the forename Auguste when signing his passport application. Some believe that the attribution of the name Auguste is a tribute to his great inspirer Auguste de Saint-Hilaire (Bériac 2009).

paternal punishment, François Marie left his hometown to begin what invariably crowned then the qualification of young workers: his *Tour de France*. He worked in Nantes, Angers, Bordeaux, etc. He even attended courses taught by renowned botanists and naturalists, such as Brongniart and Decaisne, at the *Jardin des Plantes*, in Paris, as well as that one of Durieu de Maisonneuve, in the Botanical Garden of Bordeaux, with whom he learned the art of acclimatizing plant species in new latitudes (Mérian 2009). The obstinate Breton took all the opportunities he was offered to hone his horticultural skill during his pilgrimage through France.

Encouraged by travel books, newspaper news, and illustrated magazines, in addition to extensive studies on the nature of the Brazilian territory,[3] at the age of 30, already married and a family father, he landed in Brazil to try his luck, moved by his very living taste for botany, by the spirit of adventure and, above all, by the desire to follow the example of Auguste de Saint-Hilaire, his main inspirer (Mérian 2009). In his passport he declares himself a cooper and agriculturist[4] (Bériac 2009). Glaziou would be just another skilled foreign technician to arrive in Rio de Janeiro, among hundreds of other young immigrant workers, if he did not bring in his professional luggage a great differential: his excellent knowledge in horticulture, management of parks and gardens, and botany, acquired in the free courses he attended in the main scientific societies of his country of origin and with horticulturists and gardeners with whom he had contact during his pilgrimages (Hetzel 2011).

Without any title, prestigious diploma, or official contract and with almost insignificant resources, the beginning of his life in the new territory was quite difficult (Hetzel 2011). But in 1860, 2 years after landing in the imperial capital, Glaziou embraced the opportunity to join Congressman Francisco Fialho in the reform of the *Passeio Público* of Rio de Janeiro. This would be the beginning of the trajectory of a protagonist of French landscape activity in Brazil in the second half of the nineteenth century. In his remodeling of the garden, it was possible to observe the compositional grammar of Parisian public spaces, as informed by modern logic. The

[3] Countless publications about the Brazilian territory came to light since the arrival of the Royal Family in 1808. Among them, we can mention the *Voyage Pittoresque et Historique au Brésil* (Picturesque and Historical Trip to Brasil) by Debret, published in Paris between 1834 and 1839. The work is composed of 153 boards accompanied by texts that elucidate each portrait, documenting aspects of nature, man, and Brazilian society at the beginning of the nineteenth century. Also the famous French monthly magazine *Revue des deux mondes* (Journal of two worlds), created in 1829, regularly published articles on South America, and, in 1851, under the signature of Emile Adet, it was possible to read an article entitled *L'Empire du Brésil et la societé Brésilienne en 1850* (The Empire of Brazil and Brazilian society in 1850).

[4] In order to trace Glaziou's professional profile, several publications on the character attribute to him the titles of civil engineer and botanist. It is interesting to observe how a mythology is built to make this man more admirable. A past is created with university degrees that he never attended, perhaps because his main professional activities of horticulturist, gardener, and cooper are not in vogue in the contemporary world.

French landscaper broke with the rectilinear traces of Master Valentim (1745–1813),[5] an expression of the eighteenth-century late Baroque garden, and introduced winding paths that hid picturesque nooks and favored different points of view, among lakes, caves, bridges, bowers, kiosks, and viewpoints that inspired the contemplation of the produced scenario (Segawa 1996).

With the completion of the work of the *Passeio Público,*[6] in1862, Glaziou assumed the direction of the garden and, in 1869, began to occupy, simultaneously, the post of General Director of Forests and Gardens of the Imperial House (Cardozo and Azevedo 2009). He also accumulated the position of Inspector of Municipal Gardens, besides being part of the Brazilian Association of Acclimatization (Noronha Santos 1945). The simple and unknown immigrant, agriculturist, and cooper who arrived with his bare hands in Rio de Janeiro in 1885 now united the work as a designer to public positions of direction and administration of green spaces in the imperial capital.

Glaziou continued to design several public and private gardens in the city of Rio de Janeiro and its surroundings, being responsible for the production of new urban scenarios, impregnated with natural elements, which expressed a portrait of modernity[7] presentable to the civilized world and promoted a positive image of the imperial capital (Cunha 2007). He became known as "the Emperor's landscaper" (Cunha 2007).

Among his most important achievements in the city of Rio de Janeiro, stand out, in addition to the reform of the *Passeio Público*, the idealization of the gardens of the Imperial Palace of São Cristóvão, in *Quinta da Boa Vista*, and the park of *Campo da Aclamação*, later known as *Campo de Santana*, his largest and most ambitious project (Cardozo and Azevedo 2009).

[5] Several studies that refer to the history of the *Passeio Público* corroborate that the garden was created by order of the viceroy of Brazil, D. Luís de Vasconcelos. Valentim da Fonseca e Silva, better known as Master Valentim, was one of the most important artists in Brazil colony. Recognized for his work as a sculptor, architect, and urban planner, he was chosen to design and execute the first landscaped enclosure of the colony that would have the role of serving the inhabitants of the city. The execution of the park, which followed the rectilinear molds of the Enlightenment representations, took place between 1779 and 1783 (Segawa 1996).

[6] The renovation of *Passeio Público* was Glaziou's first relevant performance on Brazilian soil and gave him great visibility. The proposal not only equated the situation of the slouch that spread over the oldest public garden in Brazil but also replaced the classic layout of Master Valentim (1745–1813) with a composition based on the vocabulary of the Romantic landscape garden, a pioneer of that style in Brazilian public spaces (Dourado 2011).

[7] Modernity is understood here as a first expressive movement enabling a new order presided over by the values of progress and civilization. The intention was to break the colonial backwardness to bring Brazil closer to a "civilized European world."

8.2 The Creation of the Horticultural Laboratory of *Quinta da Boa Vista*

In a letter written by Glaziou to Baron de Nogueira da Gama, member of His Majesty's Council, on May 6, 1873,[8] less than a month after the beginning of the gardening works of the *Campo de Santana,* which began on February 17, 1873,[9] the French landscaper, who already had the post of General Director of Forests and Gardens of the Imperial House since 1869, requests the authorization for the use of some wasteland existing in *Quinta da Boa Vista* for the creation of a plant nursery, which should be managed entirely by the State:

> Entrusted by the Imperial Government to transform the former *Campo de Santa Anna* into a landscape garden and, wanting to carry out this task with the confidence that was given to me, I come to ask *V. Exa.* [Your Excellency] for your valuable support in order to save the State some "contos de réis" [Brazilian currency used at the time]. There is in *Quinta Imperial da Boa Vista,* and mainly in the lower part, some plots of land, accompanied by shade and water, where I would like to start the preparation of the numerous plants that would be needed in the planting of said garden; and as these lands are now unproductive, I beg *V. Exa.* [Your Excellency] to grant me permission to use, in the next eighteen months [...] one or other of these places in favor of the public and in the interest of the Imperial Government.

And at the request of Auguste F. M. Glaziou, the Counselor of His Majesty writes a letter to Emperor Pedro II, on May 13, 1873,[10] requesting consent to transform those rough lands into a rich plant nursery:

> His Majesty the Emperor, appreciating the zeal with which *V. Sa.* [Your Lordship] is dedicated to the public service and seeks to save money of His House, and those of the State, in the important work of which you were entrusted, the landscaping of *Campo de Aclimação,* he decided to authorize me [...] permission to use some unproductive lands existing at *Quinta da Boa Vista,* converting them into land nurseries of the numerous plants that must be planted in the garden of the same Campo [...].

As informed by the correspondence sent by Glaziou to the Director of the First Directorate of the State Secretariat for Administration of the Empire, Manoel Jesuíno Ferreira, on November 9, 1878,[11] the application for the constitution of the plant nursery to be installed at *Quinta da Boa Vista* was met by Pedro II:

> [...]for the cultivation of plants that exist in the plant nurseries of State, established with the authorization of His Majesty the Emperor, in the grounds of the *Quinta Imperial da Boa*

[8] Correspondence belonging to the collection of the National Archives – *Série interior – Administração IJJ²*.

[9] Date reported in the letter sent by Glaziou to the Director of the First Directorate of the State Secretariat for Administration of the Empire, Manoel Jesuíno Ferreira, on November 9, 1878. This document is deposited in the National Archives – *Série interior – Administração IJJ²*.

[10] Document belonging to the collection of the National Archives – *Série interior – Administração IJJ²*.

[11] Document belonging to the collection of the National Archives – *Série interior – Administração IJJ²*.

> *Vista*, in *São Cristóvão*, which require the assiduous service of 5 workers and 1 master gardener [...].

The creation of the plant nursery of the *Quinta da Boa Vista* reminds us of the radical change that occurred in the horticultural sector as a consequence of the admirable profusion of green areas seen in Paris during the Second Empire under the rule of Napoleon III. The considerable multiplication of parks and urban public gardens required a huge amount of plants many of which were little known or totally unknown to the French horticulturists at the time. In addition the prices of the species supplied by local horticultural companies were extremely high. Such inconveniences caused Baron Haussmannand his team to feel the necessity to create a series of establishments for cultivation germination development and even acclimatization of all kinds of plants needed to supply the numerous landscape creations in the French imperial capital. And from 1854 onward several agricultural facilities[12] began to appear which through an exemplary organization provided the new Parisian gardens with their raw material their main ornament thus constituting the essence of the functioning of Haussmann's projects (Limido 2002)

Following the Parisian example, these new horticultural laboratories, together with the culture of green landscapes insertion within urban centers, spread to most cities in France, such as Rennes, which inaugurated, in1868, its first plant nursery that would take care of the creation of public gardens, and Grenoble in the 1890s (Limido 2002).

The city of Bourdeaux underwent a revitalization program, approved in1853, covering a very substantial sector related to the inclusion of several parks and other green corners in the city (Bériac 2012). Both Alphand and Barillet-Deschamps, the "Men of Haussmann," resided in Bordeaux at that time and participated in the elaboration of the plan for the transformation of the capital of the *Gironde* department, until the time when they were invited to be part of the team of the program of promenades and gardens of Paris, the first in 1854 and the second in 1858. Barillet-Deschamps, like Glaziou, was the son of a gardener and became a self-taught horticulturist (Bériac 2012). Member of the Horticultural Society of Gironde, he received several awards, and his private nursery, which had large greenhouses, was a symbol of the greatest horticultural progress. In 1850, he opened his establishment to the public, appearing as one of the favorite walking places of the people of Bordeaux (Bériac 2012). Being an articulated man, with broad participation in regional events, in the activities of the Horticultural Society and, particularly, for sharing the same interests of Glaziou, everything leads to believe that these two characters knew each other and, even, that our protagonist attended the gardens of Barillet-Deschamps (Bériac 2012).

Considering that Glaziou spent practically his entire childhood in a plant nursery and horticulture was his first training, through the work carried out with his father,

[12] The most famous and important plant nursery in Paris was *le Fleuriste de la Muette*, considered a modern establishment, which had a large number of greenhouses and was valued for its extreme order (Moncan 2009).

who also knew the departmental gardens that served the public gardens and who probably had the habit of visiting the establishment of Barillet-Deschamps, it is easy to understand all his engagement for the implementation of a space like this in Rio de Janeiro, able to provide a great quantity of plants necessary for the execution of his projects. Glaziou knew the importance of the horticultural spaces administered by the State and the real financial economy they would represent to the coffers of the Crown. This would be the first initiative to bring to Brazil a tradition of creating plant nurseries that would be managed by the State, which had started in France.

8.3 Information About the Cast of Species Present in the Garden of *Quinta da Boa Vista*

From the analysis of the catalogue organized by Glaziou, found in the *Seção de Memória e Arquivo do Museu Nacional/UFRJ – SEMEAR* (Memory and Archive Section of the National Museum) – an archive unfortunately destroyed in the tragic fire of the National Museum that occurred on September 2, 2018 – which showed the list of all the plants that were present in the botanical garden of the *Quinta da Boa Vista*, one could verify a total of 656 taxa, which means that all these species were available for the supply of the public gardens of Glaziou. These species are distributed in 96 botanical families, the most representative ones being Araceae (70 species), Arecaceae (69 species), and Bromeliaceae (66 species). These species were responsible for the formation of a wide set extremely attractive for its chromatic effects and formal patterns, attributes that aroused the interest of the nineteenth-century public (Brito et al. 2015).

Of all the taxa surveyed, 414 species are exotic and 242 are native, indicating that, although Glaziou favored native plants in his gardens, he gave preference to the cultivation of exotic species in the nursery he was in charge of.

After examining five reports submitted by different Ministers and Secretaries of State of Agriculture, Commerce and Public Works, sent to the General Assembly at the first session of the nineteenth legislature, referring to the years 1875, 1877, 1882, 1884, and 1885,[13] and the correspondence sent by Glaziou to Antônio da Costa Pinto e Silva, Minister and Secretary of State for Administration of the Empire, on October 29, 1877,[14] it was possible to inventory a total of 1.080 plant species present in the gardens of *Campo de Santana* and *Passeio Público*. It was also verified that 208 species listed in these documents also occurred in the garden

[13] In 1875, the report was presented by the Minister and Secretary of State of Agriculture, Commerce and Public Works José Fernandes da Costa Pereira Júnior; in 1877, by the Minister and Secretary Thomaz José Coelho de Almeida; in 1882, it was sent by Manoel Alves de Araújo; in 1884, by Affonso Augusto Moreira Penna; and in 1885, by João Ferreira de Moura. These reports were found in the collection of the National Archives.

[14] Document belonging to the collection of the National Archives – *Série interior – Administração IJJ²*.

of *Quinta da Boa Vista*, suggesting that the nursery had provided them for the parks of Glaziou.

However, according to the manuscripts analyzed both in the *Passeio Público* and the *Campo de Santana*, a total of 449 taxa registered in the imperial garden catalogue was not found. It is not easy to understand why Glaziou did not insert some of these species into two of his most relevant projects, since many of these plants presented all the characteristics which were desired and appreciated both in his landscaping, as verified by Brito et al. (2015), and in the European parks of the nineteenth century, as indicated by Dourado (2011). To exemplify, we can mention the absence of individuals from the families Bromeliaceae and Cactaceae, whose vast majority was native to Brazil and which would delight the eyes of visitors by displaying their showy inflorescences and quite peculiar shapes, as well as the species of Orchidaceae available in the garden of *Quinta da Boa Vista*, also native for the most part and which have blooms of extraordinary beauty. However, it is interesting to point out that representatives of these families were selected and taken by Glaziou to the Paris Universal Exhibition of 1889, as an expression of the autochthonous nature of Brazil, singularized by its richness, beauty, and potentialities (Heizer 2009). No historical document was found regarding the survey of plant species that were present in the gardens of *Quinta da Boa Vista*, also designed by Glaziou. It may be that these plants were part of the floristic composition of these spaces, since they were available for use in the botanical garden located within the same property as the leisure gardens. Currently, Bromeliaceae, Cactaceae, and Orchidaceae native to Brazil have extreme relevance in nurseries throughout the country and are highly valued for ornamentation of national landscape nooks (Lorenzi and Souza 2001). It is possible to observe an expressive number of specimens from the botanical families mentioned above in the works of Roberto Burle Marx, plastic artist and one of the most renowned and important Brazilian landscapers of the twentieth century. Burle Marx favored, like Glaziou, the use of native plants as a symbol of Brazilianness (Tabacow 2004; Dourado 2009). However, there was also in the former a tone related to environmental issues, which was not seen in the work of Glaziou. It is probable that Burle Marx himself was responsible for the current popularization of the ornamental use of these Brazilian plants (Duarte 2011).

If 208 taxa inventoried in the gardens of *Passeio Público* and *Campo de Santana* came from the nursery of *Quinta da Boa Vista*, where were the other 872 present in these green spaces obtained? The diversity of species and the number of seedlings needed to express the intention of Glaziou's landscape projects were too bulky. While he sought to obey the grammatical syntax of the Romantic landscape garden, the French aspired to a fantastic reproduction of the exaltation and the various charms of Brazilian nature. The correspondence written by Glaziou, on November 1878,[15] to the Director of the First Directorate of the State Secretariat for Administration of the Empire, Manoel Jesuíno Ferreira, reveals that the number of

[15] Document belonging to the collection of the National Archives – *Série interior – Administração IJJ*[2].

plants cultivated in the park of *Aclamação*, the later *Campo de Santana*, had already exceeded 46,000. The plan proposed by Glaziou for this ambitious work said that "[…] the general plantation of the garden […] will exceed fifty thousand plants" (Noronha Santos 1945 p.141), and, with the completion of the park, in 1880, it was verified that "in the woods and groups that stand out from the lawns, there were about sixty- six thousand plants" (Noronha Santos 1945 p.162).

Noronha Santos (1945 p.149), in his detailed study on the history of the *Campo de Santana*, provide us with relevant information related to the garden of *Quinta da Boa Vista*:

> […][in] a land granted by the stewardship of the Imperial House was created a plant nursery […], whose number exceeds 25000, being largely of hardwood trees and shrubs and herbs interesting mainly for their industrial and medicinal applications.

Based on the above report, it is suspected that the availability of only this botanical garden to supply all the demand for plants indispensable for the compositions of Glaziou's works would not be sufficient. An alternative way to solve this issue had to be found. A clue as to how Glaziou managed to increase the supply of plants for the elaboration of his gardens was found in a correspondence sent by the French landscaper to the Minister and Secretary of the State for Administration of the Empire, João Alfredo Corrêa de Oliveira, on July 8, 1873[16]:

> Finding myself ready to start preparing plants to help the planting of the park of *Aclamação*, I come to ask *V. Exa.* [Your Excellency] the favour of interceding for me with the *Exm°. Snr.* [Honorable] Minister of Agriculture, the favor of giving the orders to those responsible for the conservation of the forests of *Tijuca* and *Corcovado* […] in order that they grant me seedlings and seeds of the trees which they have available, and that they do not harm their crops in any way; in this regard, and after the warning, I will get along directly with those concerned.

It was also verified, through the analysis of eight receipts of purchases of plants from different horticultural establishments of Rio de Janeiro, comprising the years of 1876, 1877, and 1878,[17] that 73 species destined to the *Campo de Santana* were acquired in private horticultural establishments. This fact draws attention to the new and promising market of the imperial capital, that of commercialization of vegetation for gardens.

The consolidation of a garden culture in Rio de Janeiro, which spread throughout Brazil in the second half of the nineteenth century, refers not only to the various green spaces implanted in public areas but also to the new habits of the population that, inspired by the ways of life of the European aristocracy, began to take pleasure in preparing the land, practicing gardening, and cultivating domestic gardens. This new trend was not exclusive to the Brazilian elite, but even reached the middle classes, as observed by the Swiss couple of scientists Agassiz and Agassiz (1937)

[16] Document belonging to the collection of the National Archives – *Série interior – Administração IJJ²*.

[17] Document belonging to the collection of the National Archives – *Série interior – Administração IJJ²*.

who, when crossing the country between 1865 and 1866, mentioned the beautiful gardens of the houses in the *Laranjeiras* neighborhood,[18] full of European flowers that exhibited their colors among greening shrubs and indigenous trees, and, as also informed by the eminent writer Machado de Assis (1906 p.1), in his play *Lição de Botânica* (Lesson of Botany), when quoting the phrase:

There is a lot of garden throughout this *Andaraí*[19] to examine.

The new urban lifestyle, marked by social interest in ornamental plants, the activity of cultivating residential gardens, and the yearning for a life closer to nature, directly influenced the expansion of the offer of products aimed at gardening. In 1880, the *Almanaque Laemmert* (Laemmert Almanac), one of the main commercial dissemination vehicles of the city of Rio de Janeiro, announced a total of 23 establishments and professionals specialized in the production and commercialization of seedlings and seeds of rare and coveted species at the time, to be introduced in the various landscape nooks (Laemmert and Laemmert 1880). There was also the circulation of specialized publications in horticulture, such as the *Revista de Horticultura* (Magazine of Horticulture) — *Jornal de Agricultura e Horticultura Prática* (Journal of Agriculture and Practical Horticulture), launched in 1876, by the *Typografia Universal de E. & H. Laemmert* (Universal Typography of E. & H. Laemmert), in Rio de Janeiro. This monthly Brazilian publication was conceived by the horticulturist Frederico de Albuquerque and had the collaboration of important nineteenth-century botanists and naturalists, such as Guilherme Schüch de Capanema and João Barbosa Rodrigues (Dourado 2011).

The purchase receipts of the plants found and analyzed in the present study refer to three horticulturists or horticultural establishments: Emilio Wittig, Figueiredo & Velloso, and Lourenço Hoyer & Cia. Consulting the *Almanaques Laemmert* referring to the corresponding years of the receipts examined here, of those three suppliers of plants indicated, only the announcement of the nurserymen Figueiredo & Silva was found in the 1876 edition. Emilio Wittig and Lourenço Hoyer & Cia are not present among the related horticultural establishments in the Laemmert brothers' annual catalogues of 1876, 1877, and 1888 (Laemmert and Laemmert 1876, 1877, 1878). However, in the *Revista de Horticultura* (Magazine of Horticulture) of February 1876, Frederico de Albuquerque published the news that the well-known horticulturist of the imperial capital, Mr. Lourenço Hoyer, was awarded a major gold medal on January 22 of the same year at the *Exposição de Horticultura de Petrópolis* (Petrópolis Horticulture Exhibition) for his splendid ornamental collection of Araceae.

Of the 73 species acquired in these 3 horticultural establishments, 43 were exotic and 30 were native. Like the garden of *Quinta da Boa Vista*, administered by Glaziou, it seems that the private nurseries were also represented by a majority of non-indigenous plants. Dourado (2011) comments that although it is difficult to

[18] Located in the South Zone of Rio de Janeiro, *Laranjeiras* is one of the oldest residential and middle-class neighborhoods in the city, with occupation started in the seventeenth century.

[19] Residential neighborhood of the North Zone of Rio de Janeiro, inhabited by the middle class. It began to be occupied in the sixteenth century by Jesuit priests for the cultivation of sugarcane.

recover the full extent of the plant list cultivated by these horticulturists, it is possible to extract interesting information from the announcements of the *Almanaque Laemmert*. In the 1877 and 1878 editions, the advertisements of Frederico de Albuquerque's horticultural establishment appear on full-page posters, which mention several ornamental plants commercialized by him, mostly foreign, such as rosebushes, *Chrysanthemum* spp., *Dahlia* spp., *Phlox* spp., also *Rhododendron* spp., *Yucca* spp., and rare and famous Cycadaceae.

Examining the catalogue of plants that existed in the former botanical garden of *Quinta da Boa Vista*, projected by Glaziou, it was found a list of 124 cultivars of *Rosa* and 59 of *Caladium*, indicating that these plants were possibly in fashion or that even a craze had been triggered by these plants. Observing Figs. 8.1 and 8.2, it is possible to notice that many of these cultivated varieties earned designations of illustrious people of the time, with the aim of honoring them, having as examples *Rosa Empereur du Brésil* and *Caladium Prince Albert.* Unfortunately, it was not found as evidence, material referring to the names present in the catalogue, which makes it difficult to specify whether the different epithets reported by Glaziou for *Rosa* and *Caladium* corresponded to a single species or if they were more than one species of these genera.

Roses are among the oldest flowers cultivated over the world. In Greece, in 1800 B.C., they already appeared depicted in the various forms of art, which recorded their planting for ornamental purposes. The genus *Rosa* is native to the temperate regions of the Northern Hemisphere, and Western Asia is considered its main center of dispersion. With about eight species frequently used in horticulture, it is recognized for its taxonomic complexity, due, in part, to the selective crossings conducted by men over the centuries, with the objective of selecting the most desired characteristics, such as number, size, and coloration of the petals, quantity of thorns, intensity of perfume, flowering season, and even resistance to pathogens (Barbieri and Stumpf 2005).

Admired for the beauty and perfume of their flowers, roses have always had a variety of historically complex codes involving their symbolic use.[20] In the West, when they appeared in private gardens, they were seen as an emblem of luxury, wealth, and power (Goody 1993). The Empress of France, Josephine de Beauharnais,[21] first wife of Napoleon Bonaparte and, therefore, the most influential woman during the First Empire, possessed an admirable collection of roses, which included all the 2.562 varieties known at the time, and was responsible for the spread of interest in this plant at the beginning of the nineteenth century (Barbieri and Stumpf 2005). During the nineteenth century, the *Rosa* became a favorite of gardeners and horticulturists, who increasingly specialized in the production of new hybrids and cultivars. Currently, more than 16.000 varieties are estimated, all descendants of wild roses from Europe, Asia, and North America (Bill 2013).

[20] The great variation of meanings that roses have in the West, throughout history, is beyond the scope of this work. To learn more about this theme, see Goody (1993).

[21] Her reign lasted from May 26, 1804, to January 10, 1810.

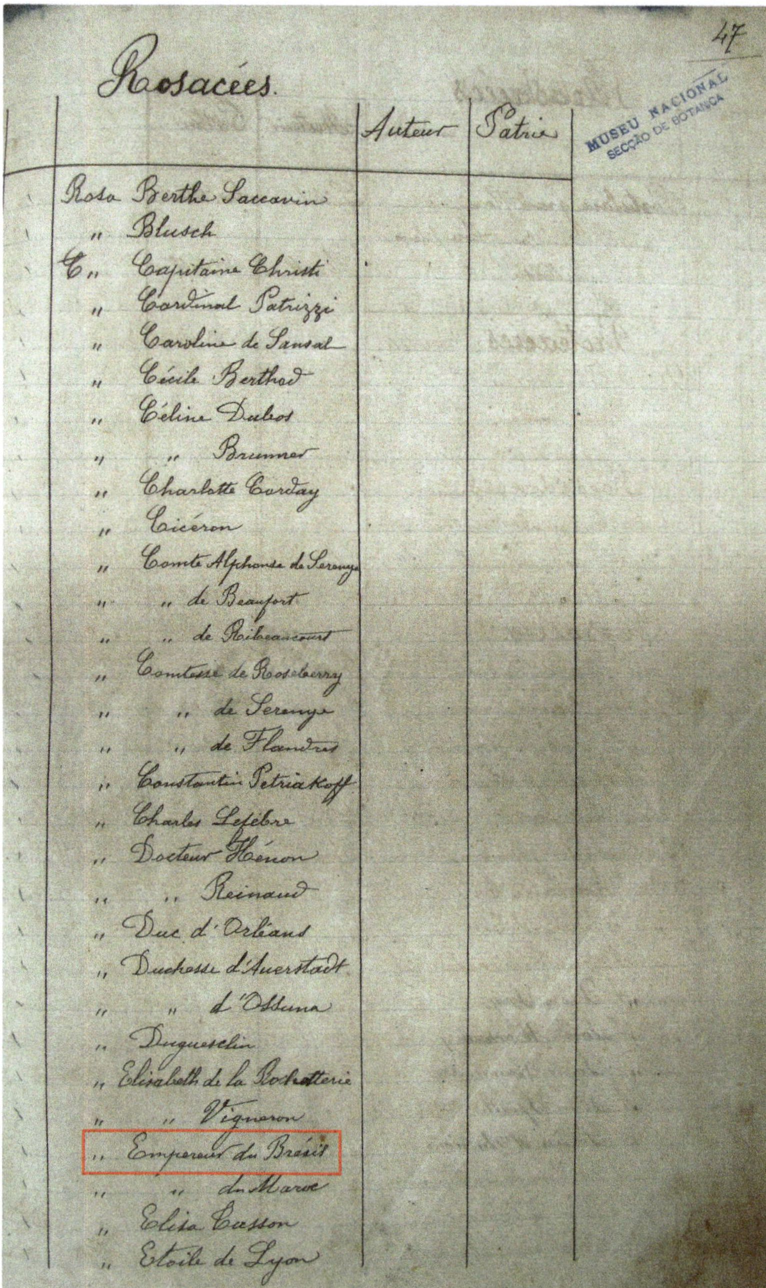

Rosacées.

	Auteur	Patria
Rosa Berthe Saccavin		
„ Blusch		
C. „ Capitaine Christi		
„ Cardinal Patrizzi		
„ Caroline de Sansal		
„ Cécile Berthod		
„ Céline Dubos		
„ „ Brunner		
„ Charlotte Corday		
„ Cicéron		
„ Comte Alphonse de Serenye		
„ „ de Beaufort		
„ „ de Ribeaucourt		
„ Comtesse de Roseberry		
„ „ de Serenye		
„ „ de Flandres		
„ Constantin Petriakoff		
„ Charles Lefébre		
„ Docteur Hénon		
„ „ Reinaud		
„ Duc d'Orléans		
„ Duchesse d'Auerstadt		
„ „ d'Ossuna		
„ Duguesclin		
„ Elisabeth de la Rochetterie		
„ „ Vigneron		
„ Empereur du Brésil		
„ „ du Maroc		
„ Elisa Boëlle		
„ Etoile de Lyon		

47

Fig. 8.1 Names given to different cultivars of *Rosa*. (Source: SEMEAR)

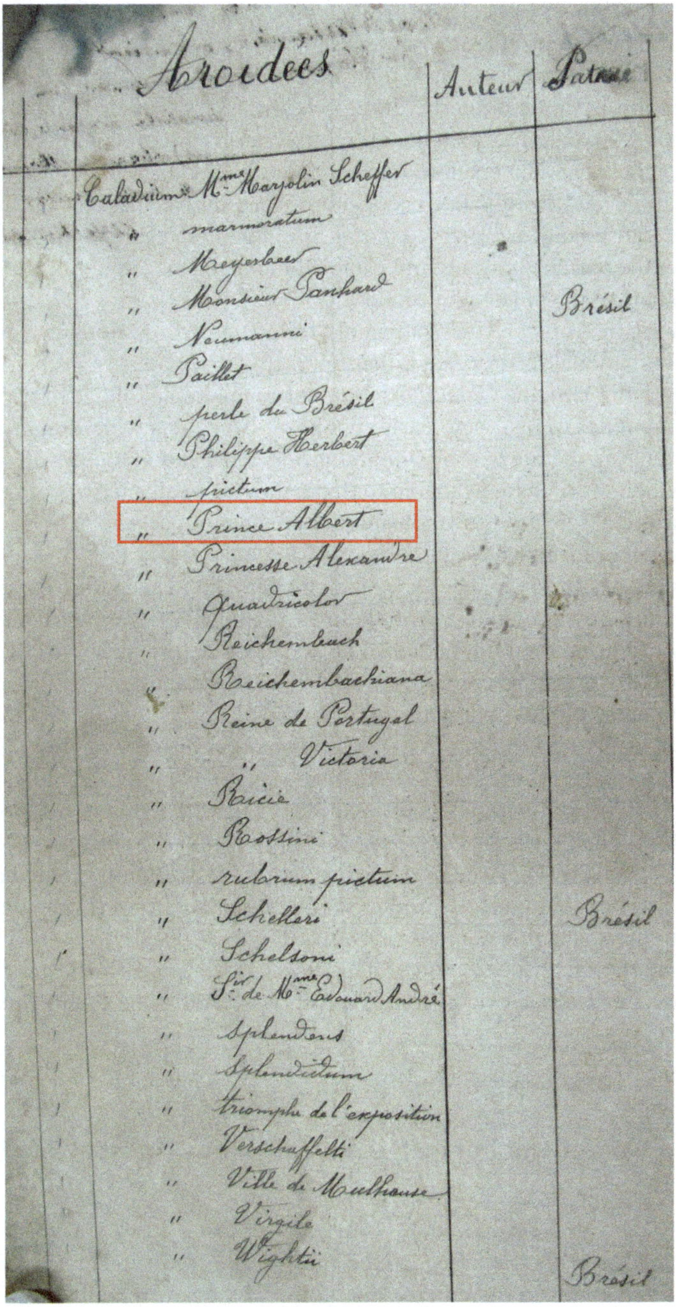

Aroidées	Auteur	Patrie
Caladium Mme Margolin Scheffer		
" marmoratum		
" Meyerbeer		
" Monsieur Panhard		Brésil
" Neumanni		
" Paillet		
" perle du Brésil		
" Philippe Herbert		
" pictum		
" Prince Albert		
" Princesse Alexandre		
" Quadricolor		
" Reichembach		
" Reichembachiana		
" Reine de Portugal		
" " Victoria		
" Ricii		
" Rossini		
" rubrum pictum		
" Schelleri		Brésil
" Schelsoni		
" Sr. de Mme Edouard André		
" Splendens		
" Splendidum		
" triomphe de l'exposition		
" Verschaffelti		
" Ville de Mulhouse		
" Virgile		
" Wightii		Brésil

Fig. 8.2 Names given to different cultivars of *Caladium*. (Source: SEMEAR)

Despite its origins from cold climate regions, the roses have adapted very well to Brazilian climates of mild temperatures, such as the south of the country or in the subtropical altitude locations, and, today, Brazil is a major producer of roses for cutting, serving both the domestic and foreign markets (Lorenzi and Souza 2001; Novaro 2005). We can infer that the current scenario of high commercialization of roses in the country is the result of many years of experience in intense cultivation, acclimatization, and testing in genetic improvements of these beautiful flowers, which were carried out, as observed in this research, in the garden of *Quinta da Boa Vista* and in the various horticultural establishments of the capital of the Empire in the second half of the nineteenth century.

Originating in Tropical America, mainly from Brazil, presenting extensive distribution in the national territory and, therefore, found very easily in their wild state (Flora do Brasil 2020), the *Caladium*, with their remarkable foliage of varied colors, forming splendid designs, were early recognized for their wide ornamental vocation. Cultivated in the nursery of *Quinta da Boa Vista* and in the private gardens of the time, which made them available to trade, they were also targets of tireless processes of artificial selections, responsible for the creation of new hybrids, extremely elegant and decorative.

The variety of *Caladium* that were taken by Glaziou to his public parks probably had the intention of provoking, stimulating Brazilians, of making them marvel at this beauty of the national flora that had always been before their eyes. If that really was his goal, it worked! The different cultivars of this genus gained full acceptance from the amateur public, who came to know them, buy them, and cultivate them widely in their gardens, tuning in to Glaziou's action in favor of Brazilian vegetation. The landscaper Burle Marx, born in the early twentieth century, said that, among his childhood memories, a very strong one was that that of his mother cultivating with skill and love, the countless *Caladium* which she collected in her greenhouse (Tabacow 2004). Even today, this large horticultural group has a high ornamental value, being widely produced and commercialized (Lorenzi and Souza 2001), a circumstance that demonstrates that this facet of Glaziou, in which he put his whole heart, of the valorization of national plants, was crowned by an extraordinary success, extending over a long period, which even he could probably not imagine.

8.4 The End of the Former Imperial Plant Nursery of *Quinta da Boa Vista*

The proclamation of the Republic, which took place on November 15, 1889, put an end to the reign of Pedro II. With the institution of the new republican government, the Brazilian imperial family went into exile in Europe, and the *Paço de São Cristóvão*, in *Quinta da Boa Vista*, where the Emperor grew up, was educated, and lived for almost all his life, began to host the works of the National Assembly responsible for the Brazilian Constitution of 1891. In 1892, there was the transfer of the National Museum from its premises at the *Campo de Santana* to the former imperial residence, with all its collections, laboratories, and researchers (Miranda 2009). At that time, the Museum was under the direction of Ladislau de Souza Mello e Netto, who was considered an example in administrative efficiency and attention to the scientific sector. His mandate was highlighted by the creation of a quarterly scientific journal, the *Archivos do Museu Nacional*[22] (Archives of the National Museum), for the major changes in the physical structure of the building to better accommodate its collection, following the European standards of conservation of scientific collections at the time, inauguration of new laboratories, and increase in the salary of employees. There was also an increase in the number of scientific expeditions, conference achievements, and international exhibitions, among other important events that gave great visibility to the institution (Duarte 2019). The National Museum was at its full apogee!

At that time, the Botanical Garden of Rio de Janeiro, which had been founded in 1808 as an acclimatization institution, was also undergoing a major restructuring promoted by the naturalist João Barbosa Rodrigues,[23] marked by the increase in botanical studies, reorganization, and creation of new collections of living plants and inauguration of the herbarium, museum, and library (Miranda 2009).

As recommended in a letter sent by the Minister of Agriculture, Commerce and Public Works, Antão Gonçalves de Faria, to the first section of the Directorate of Agriculture, on February 6, 1892,[24] the Republican government did not provide financial resources to bear the necessary expenses for the maintenance of the former imperial garden. And for this reason, everything indicates that Glaziou, the creator and responsible for the nursery, requested the sending of this to the Botanical Garden of Rio de Janeiro:

> You are authorized to deliver to the Directorate of the Botanical Garden the nursery of *Quinta da Boa Vista*, in *São Cristóvão*, with a numerical and nominal list of the plants that exist in it, since in the current budget there is no fund to finance it, which without prejudice to its purposes will be done by the Botanical Garden.

Regarding this episode, two issues are beyond our understanding. The first of these refers to the reason why Glaziou would have requested the authorization to deliver the garden of *Quinta da Boa Vista* to the care of Barbosa Rodrigues, the then

[22] Created in 1876, the *Archivos do Museu Nacional* (Archives of the National Museum) was the first journal specialized in Natural Sciences in Brazil – and the oldest Brazilian scientific journal still in circulation. The edition of this journal allowed the National Museum to expand its network of communication and exchanges with foreign museums, in tune with the international scientific community (Lopes 2000; Lopes and Muriello 2005).

[23] In 1890, the Brazilian naturalist João Barbosa Rodrigues took over the direction of the Botanical Garden of Rio de Janeiro. Presenting a plan to fulfill its traditional aspirations, he became the most prestigious leader in the history of the institution until then (Sá 2001).

[24] Correspondence found in the *Seção de Memória e Arquivo do Museu Nacional/UFRJ (SEMEAR)* (Memory and Archive Section of the National Museum). This document, unfortunately, no longer exists due to the fire of the National Museum on September 2, 2018.

Director of the Botanical Garden, since the French showed no admiration for him. In the correspondence sent to William Thiselton-Dyer, Director of the Royal Botanic Gardens, Kew, on September 14, 1891,[25] Glaziou complains about Barbosa Rodrigues' lack of scientific rigor. And also, in a letter of January 8, 1892,[26] also sent to Thiselton-Dyer, the indignation of Glaziou concerning the ethics and scientific practices of the naturalist can be observed:

> The box of living plants that you were kind enough to announce me on the date of November11 last year arrived here after a long time, but, by an unfortunate error of address, it is in the hands of Barbosa who keeps it without any scruple [...].
>
> I believe that your waiting for Barbosa's materials is useless, since he only gathered [...] some trifles in jars of barbers [...].

The second incomprehensible point, in our view, was the fact that the National Museum did not take care of the garden of *Quinta da Boa Vista.* If the importance of that establishment was evident both for the supply of plants destined to public parks in Rio de Janeiro and for the advances in horticulture research, why this institution, which was installed in the same area of the garden and also was in a period of full apogee with the excellent administration of the botanist Ladislau Netto,[27] did not appropriate the respectable horticultural laboratory? As Glaziou was a corresponding member of the National Museum and had a strong relationship of affinity and respect for Ladislau Netto, why did he demand the transfer of the former imperial nursery to the Botanical Garden of Rio de Janeiro? It is true that the Botanical Garden was in a totally favorable condition at the time to receive the garden of *Quinta da Boa Vis*ta. Numerous factors converged, notwithstanding, for it to be incorporated by the National Museum. Such questions will probably remain unanswered, but the fact is that this event represented the end of the former imperial nursery of *Quinta da Boa Vista.*

References

Adet E (1851) L'Empire du Brésil et la société brésilienne en 1850. Revue des Deux Mondes, pp 1083–1105

Agassiz L, Agassiz EC (1937) Viagem ao Brasil 1865–1866. Nacional, São Paulo

Albuquerque FG (ed) (1876) Revista de Horticultura, vol 1. E. & H. Laemmert, Rio de Janeiro

Assis M (1906) Lição de Botânica. In: Assis M (ed) (1982) Teatro completo de Machado de Assis. Funarte, Rio de Janeiro, pp 350–365

Barbieri R, Stumpf ERT (2005) Origem, evolução e história das rosas cultivadas. R Bras Agrociência 11(3):267–271

[25] Document in possession of the collection of the Royal Botanic Gardens, Kew: Archives: Director's correspondence.

[26] Document in possession of the collection of the Royal Botanic Gardens, Kew: Archives: Director's correspondence.

[27] Ladislau Netto had a long-standing public disagreement with Barbosa Rodrigues. On this subject, see Sá (2001).

Bériac JP (2009) Auguste Glaziou e seus mestres franceses em Bordeaux. In: Glaziou e os jardins sinuosos. Catálogo da exposição realizada no Jardim Botânico do Rio de Janeiro. Dantes Editora, Rio de Janeiro, pp 20–53

Bériac JP (2012) Auguste Glaziou, um paysagiste entre Bordeaux et Rio de Janeiro. Revuearchéologique de Bordeaux 103:231–262

Bill L (2013) 50 plantas que mudaram o rumo da história. Sextante, Rio de Janeiro

Brito MR, Nic Lughadha E, Duarte LFD, Senna-Valle L (2015) Exchange of useful plants between Brazil and England in the second half of the nineteenth century: Glaziou and the botanists of the Royal Botanic Gardens, Kew. Kew Bull 70:4. https://doi.org/10.1007/s12225-014-9553-6

Bureau É (1908) Notice historique sur F. M. Glaziou. Société Botanique de France, Paris

Cardozo FS, Azevedo MNS (2009) Um francês no Brasil Imperial do século XIX: Auguste François-Marie Glaziou. In: Vidal L, Luca, TR (orgs) Franceses no Brasil: séculos XIX–XX. São Paulo: Editora Unesp, p 387–396

Cunha MG (2007) O extraordinário Glaziou. In: Terra CG, Andrade R, Trindade J (orgs) Leituras paisagísticas: teoria e práxis 2. Rio de Janeiro: EBA/UFRJ, p 46–59

Debret JB (1834-39) Voyage pittoresque et historique au Brésil, v 3. Firmin Didot Frères, Paris

Dourado GM (2009) Modernidade verde: jardins de Burle Marx. Senac, São Paulo

Dourado GM (2011) Belle époque dos jardins. Senac, São Paulo

Duarte LFD (2011) Damascus in Dahlem: art and nature in Burle Marx tropical landscape design. Vibrant 8(1):495–510

Duarte LFD (2019) O Museu Nacional: ciência e educação numa história institucional brasileira. Horiz Antropol 53:359–384

Flora do Brasil (2020) Jardim Botânico do Rio de Janeiro. http://floradobrasil.jbrj.gov.br/. Accessed 26 Feb 2021

Goody J (1993) The culture of flowers. Cambridge University Press, Cambridge

Heizer A (2009) Os Jardins de Glaziou na Exposição de Paris de 1889. In: Glaziou e os jardins sinuosos. Catálogo da exposição realizada no Jardim Botânico do Rio de Janeiro. Dantes Editora, Rio de Janeiro, pp 54–59

Hetzel B (2011) Raízes na França, plenitude no Brasil. In: Hetzel B, Negreiros S (eds) Glaziou e as raízes do paisagismo no Brasil. Manati, Rio de Janeiro, pp 18–20

Instituto do Patrimônio Histórico e Artístico Nacional (IPHAN) (2009) Portaria n. 127, de 30 de abril de 2009. Diário Oficial da União

Laemmert E, Laemmert H (1876) Almanak Administrativo, Mercantil e Industrial da Corte e Província do Rio de Janeiro para o ano de 1862. Rio de Janeiro

Laemmert E, Laemmert H (1877) Almanak Administrativo, Mercantil e Industrial da Corte e Província do Rio de Janeiro para o ano de 1862. Rio de Janeiro

Laemmert E, Laemmert H (1878) Almanak Administrativo, Mercantil e Industrial da Corte e Província do Rio de Janeiro para o ano de 1862. Rio de Janeiro

Laemmert E, Laemmert H (1880) Almanak Administrativo, Mercantil e Industrial da Corte e Província do Rio de Janeiro para o ano de 1862. Rio de Janeiro

Lopes MML (2000) Cooperação Científica na América Latina no final do Século XIX: os intercâmbios dos museus de ciências naturais. Interciência 25(5):228–233

Lopes MM, Muriello SE (2005) Ciências e Educação em Museus no final do século XIX História, Ciências, Saúde. Manguinhos 12:13–30

Lorenzi H, Souza HM (2001) *Plantas Ornamentais no Brasil*: arbustivas, herbáceas e trepadeiras. Instituto Plantarum, Nova Odessa, São Paulo

Limido L (2002) L'Art des jardins sous le Second Empire: Jean-Pierre Barillet-Deschamps (1873–1924). Editions Champ Vallon, Ceyzérieu

Macedo J (2012) Modelos conceituais e análises arqueológicas na compreensão da paisagem urbana: a pesquisa arqueológica do Passeio Público do Rio de Janeiro. In: Macedo J, Andrade R, Terra C (eds) Arqueologia na paisagem: novos valores, dilemas e instrumentais. Rio Books, Rio de Janeiro, pp 74–75

Mérian JY (2009) A atualidade de Glaziou. In: Glaziou e os jardins sinuosos. Catálogo da exposição realizada no Jardim Botânico do Rio de Janeiro. Dantes Editora, Rio de Janeiro, pp 10–19
Moncan P (2009) Paris les jardins d'Haussmann. Les Éditions du Mécène, Paris
Miranda EE (2009) Jardins Botânicos do Brasil. Metalivros, São Paulo
Noronha Santos (1945) O parque da praça da República, antigo da Aclamação, notícia histórica do Campo de Santana. Revista do Serviço do Patrimônio Histórico e Artístico Nacional 8:102–163
Novaro N (2005) Breeders rights and Brazilian roses. Flora Culture International, Heiloo 15(4):32
Sá MR (2001) O botânico e o mecenas: João Barbosa Rodrigues e a ciência no Brasil na segunda metade do século XIX. História, Ciências, Saúde. Manguinhos 8:899–924
Segawa H (1996) Ao amor do público: jardins no Brasil. Studio Nobel: Fapesp, São Paulo
Tabacow J (ed) (2004) Arte e paisagem: Roberto Burle Marx. Studio Nobel, São Paulo
Terra CG (2000) Os jardins no Brasil no século XIX: Glaziou revisitado. EBA/UFRJ, Rio de Janeiro

Chapter 9
Medicinal Plants Used in *Quilombola* Communities in Piranga, State of Minas Gerais, Brazil

Renata Barreto Tostes and Luci de Senna-Valle

Abstract The purpose of this chapter was to analyse knowledge and use of medicinal plants in two *quilombola* communities, within the Atlantic Rainforest. This data was obtained through semistructured interviews that took place between 2017 and 2019. In Santo Antônio dos Quilombolas, 55 interviewees mentioned 132 species for medicinal use, while in Santo Antônio do Guiné, the 76 interviewees mentioned 142 different species. The botanic families with the greatest number of species in Santo Antônio dos Quilombolas were Asteraceae, Lamiaceae, Fabaceae, Solanaceae, Rosaceae and Rutaceae, while in the case of Santo Antônio do Guiné, they were the Asteraceae, Lamiaceae, Rutaceae, Apiaceae, Poaceae and Rosaceae. The data shows that the *quilombola* communities in Piranga make great use of a diversity of medicinal plants and that many other studies are necessary for the establishment of the efficiency of these plants, which opens new possibilities of pharmacological studies.

Keywords Atlantic Rainforest · Ethnobotany · Medicinal plants · Quilombola

Among the many traditional communities that occur in the Atlantic Rainforest, *quilombola* communities, of African inspiration, appear as a source of resistance to the slavery-based system of the time. This resistance was not only through slaves escaping en masse but also through the organisation of family units that sought a system of free and autonomous production, based on cooperation, and representing a new style of life and political organisation (Carvalho and Lima 2013).

The quilombos established in Brazil were copies of those in Africa, whose structure and organisation started in the *Mbunda* tradition (from Angola), whose main

R. B. Tostes (✉)
Universidade do Estado de Minas Gerais, Ubá, MG, Brazil
e-mail: renata.tostes@uemg.br

L. de Senna-Valle
Departamento de Botânica, Museu Nacional, Universidade Federal do Rio de Janeiro (MN/UFRJ), Rio de Janeiro, RJ, Brazil

M. F. T. Medeiros, B. de Sá Haiad (eds.), *Aspects of Brazilian Floristic Diversity*,
https://doi.org/10.1007/978-3-031-07453-0_9

goal was opposition to a slavery-based structure, and for the implementation of a political composition where all the oppressed could be found (Leite 2000). Many *quilombos* were started through donation of land by the *senhores de engenho* (slaveowners), as a payment for services provided to the State, purchase of land and the mere stay on the land that they occupied and cultivated inside large estates (Rothenburg 2008).

Throughout the domain of the Atlantic Rainforest, there are some 370 *quilombola* communities. Most of these communities sought shelter in Brazil's forest frontiers, far from urban centres, thus becoming established as rural Black communities who normally established lifestyles integrated with natural systems, their territories and also their ancestrality, traditions and cultural practices (da Silva 2013). According to Almeida (2002), *quilombola* identity is connected to its territory, and it is in this space that the communities uphold a direct connection with ancestrality, culture and traditions, as well as the common use of the land, with the family unit as an essential element, as political and ethnic affirmation.

The *quilombola* communities established within the domain of the Atlantic Rainforest have, in the backyards of their own homes, a social and cultural space where the families keep a wide variety of plants, using such plants in a sustainable way to make sure of their preservation. These backyards are considered agroforest systems that are practices of land usage where woody and herbaceous species are integrated with agricultural cultivations and also the rearing of small animals, in a sequential and integrated manner (Depommier 2003). According to Machado and Santilli (2008), forestry systems, such as yards, family farms and secondary agriculture, are areas that have preserved the agricultural and crop biodiversity throughout the tropical strips of the world, as well as being an important space for maintenance and exchange of traditional knowledge.

Even though there are many *quilombola* communities along the Atlantic Rainforest, few have had investigation of their cure practices (Mota and Dias 2012; Ferreira et al. 2014; Rogério 2014; de Siqueira 2014; Zank et al. 2016; dos Lisboa et al. 2017; Machado and Dória 2017). A range of authors mention that this part of the Brazilian population has kept up its traditions and knowledge through oral communication (Ki-Zerbo 2010; Schiffler and Nascimento 2012; de Almeida and Miranda 2015). According to de Almeida and Miranda (2015), the older members of the community use their memory and oral communication in conversation circles or in their daily activities to teach the younger people about the ways of seeing and living the world, thus seeking to preserve their local customs.

Studies carried out about the use of medicinal plants within the domain of the Atlantic Rainforest are concentrated in coastal areas and/or areas close to big cities, and this has left some gaps in the knowledge of the use of such plants in extensive countryside areas, within the Atlantic Rainforest (Liporacci et al. 2017). In this scenario, it is essential that there be the execution of ethnobotany research in *quilombola* communities in countryside regions, as they can contain important knowledge about medicinal plants.

The communities focused on this work, namely, Santo Antônio dos Quilombolas and Santo Antônio do Guiné, both lie within the Municipality of Piranga, in the countryside of the State of Minas Gerais. This region is part of the Atlantic Rainforest, and the prevailing phytophysiognomy being the semideciduous seasonal forest (Veloso et al. 1991). The municipality has a total area of 658,812 km^2, and it stands some 166 km (103 miles) from the state capital, Belo Horizonte (IBGE 2018).

The State of Minas Gerais was the Brazilian state with the largest enslaved population during the colonial period. It was originally occupied with the discovery of gold and diamonds in the Espinhaço mountain range,[1] leading to a gold rush of millions of people who relocated to Minas to work with mining, bringing with them an enormous number of slaves to carry out heavy manual labour in the mines (CEDEFES 2008).

The region of the present Municipality of Piranga was one of the first to be occupied in Minas Gerais for the exploitation of gold. According to Venâncio (1997), the growth of the city was so significant that, in 1724, the settlement was raised to the status of *freguesia* (parish), and then the discovery of new mineral lands allowed this location to keep 1193 African and *crioulo* slaves that same year.

With the demise of mining activities, a process that started at the beginning of the nineteenth century, the region around Piranga started to be a point of subsistence mercantile activity, which is the production of basic and staple foods for self-consumption and also for the internal market, inside and outside the Province (now State), and the slaveowner farmers were regularly inserted in these relationships (de Oliveira 2006; Silva 2012). A study conducted by do Silva (2014) shows that, even after the economic change that swept across Piranga, this region still had a significant number of slaves. Indeed, in 1831 the city had a population of 12,393 people, of which 4088 people (32.99%) were enslaved.

Santo Antônio dos Quilombolas was the first quilombola community to be certified by the Palmares Cultural Foundation as being a remnant of a quilombo in the Municipality of Piranga. According to the local residents, this community was set up by freed slaves in Taquaruçu Farm (probably the oldest in the region) that formerly belonged to Francisco Antônio de Souza. This story can be confirmed by the inventory of Francisco Antônio de Souza, dated 1883, belonging to the current owner of the Taquaruçú Farm. This inventory sets out the donation of land to the slaves, José Philomena and Antônio, but throughout the document another eight names are mentioned: Raymundo, Antônio, Fermino, Manoel, Ceriaca, Maria, Mara Crioula and Manoel, of African origin.

The community has 96 houses built in stone, as can be seen in the photograph below, organised mainly around the Catholic church present in the community. The church represents a large part of the sociability of the community, as this is where the religious festivities take place, from putting up the stalls for selling food through to the holding of raffles to raise funds for the maintenance of the churches. The community also has a municipal school, which offers basic education only, and a

[1] A large mountain range that cuts its way through the States of Minas Gerais and Bahia.

health centre which has a doctor every morning on weekdays and a community health agent.

Some of the residents, mainly the women, work harvesting coffee, in the neighbouring city of Mariana, between May and September. Many of the men in the community work on rural properties, or in the building trade, also in neighbouring municipalities such as Diogo Vasconcelos and Mariana, spending the whole week outside the community and only returning at weekends.

Santo Antônio do Guiné was the second *quilombola* community to have received certification granted by the Palmares Cultural Foundation, on 5 May 2009. *Território de Guiné*, as it is known by the local residents, comprises a host of small settlements: Santo Antônio do Guiné (16 houses) as seen in the following photograph, Coelho (2 houses), Catanho (33 houses), Chácara (2 houses), Baú (7 houses), Acabasacu (6 houses), Pedreira (12 houses), Palmeiras (7 houses), Passarinheiro (8 houses), Tenda (7 houses) and Congonhas (7 houses), giving a total of 107 houses in the Guiné sector.

According to stories told in Piranga, in 1721 Colonel João Amaro Maciel Parente freed his Guinean African slaves, who then founded the quilombo of Santo Antônio do Guiné. To mark off their land and commemorate freedom, the freed slaves planted a *jequitibá* (*Cariniana brasiliensis*) on the settlement of Palmeiras, where the quilombo of Guiné-Piranga had been founded. The *jequitibá* at Palmares exists to this day and is protected by municipal law.

It is in the hamlet of Santo Antônio do Guiné, as can be seen in the following photograph, where the school is located, as also the health centre and the church, which is used by the whole population of this sector. The municipal school only has primary education, and the health centre offers medical services 3 days a week together with a health agent who belongs to the community.

The church is also the main point of sociability of the quilombo, and this is where the association meetings are held, also being the venue for hosting the community parties. To raise funds and take care of the church, the local residents normally organise raffles a few months in advance.

Most of the residents in Santo Antônio do Guiné work in Piranga or are small rural farmers, mainly producing beans and corn. Eucalyptus plantations are also very common in the hamlets of Catanho for the production of plant-based coal.

For the study of medicinal plants used by the *quilombola* communities at Piranga, semistructured interviews were carried out, accordingly with Viertler (2002), Alexiades (1996) and Albuquerque et al. (2010), during 2017, 2018 and 2019, respectively. The residents' houses were numbered and based on a table of random numbers, and the interviewees were selected according to the number that corresponds to the numbering of the house (Albuquerque et al. 2010).

Interviews were conducted with only one resident of each house selected, the interviewee being the person who self-declared himself or herself as the person in the region with best knowledge of medicinal plants or the person who was nominated as interviewee by the members of the family. On starting the interview, the free and informed consent form was read out to the interviewees (informants), seeking to make sure of free, volunteered and informed participation of the interviewee.

For the interviews themselves, a special form was drawn up, asking about the species of medicinal plants used, their vernacular name, the purpose of their use, the method of preparation and the quantity used, how it is used, the duration of the treatment and the place of collection. After the interviews, the plants mentioned were located and collected by means of guided tours through the spaces as previously established and by common agreement with the interviewee, as in yards, wooded areas, roadsides and so on (Alexiades 1996; Albuquerque et al. 2010). Later these were incorporated into the herbarium of the National Museum.

The taxonomic identification of the botanical materials was made with the use of analytic keys and specific bibliography, as well as consultations with specialists, comparison with exsiccates housed in the herbarium of the National Museum and online herbaria such as the Reflora Virtual Herbarium, the Virtual Herbarium of the Emílio Goeldi Pará State Museum and the Missouri Botanical Garden (MOBOT). The system of classification that has been adopted here was that of the Angiosperm Phylogeny Group (APG IV 2016), with nomenclature updates of plant species and their respective families having been carried out on www.theplantlist.org.br.

To compare the wealth of species between the different *quilombola* communities, there has been an estimate of curves of rarefaction and extrapolation, based on the number of interviews, using the first Hill number, $q = 0$ (Chao et al. 2014). The extrapolations were made based on the presence or absence of species during the interview (Colwell 2013). These estimates were obtained using the "iNEXT" package (Hsieh et al. 2016). The Hill number was estimated as the mean of 100 repetitions of bootstrapping, to estimate confidence intervals of 95% (e.g. Rodrigues et al. 2019). Similarly, we always assume that there is no overlap of confidence intervals of 95% and that the number of species has shown a statistically significant difference, with $p < 0.05$ (Colwell et al. 2013).

In all, 59 residents were interviewed at the community of Santo Antônio dos Quilombolas: 46 women and 13 men. Out of this sample, 93.22% (55) were born in the community, while 6.77% (4) came after they married residents of the quilombo, having hailed either from Mariana or other neighbouring rural communities. In the case of the community of Santo Antônio do Guiné, 76 were the interviewees: 67 women and 9 men. Out of this group, 94.73% (72) were born within the community, with only 6.77% being born elsewhere, such as Piranga (seat of local government) and Ouro Preto.

The higher number of women interviewed in both *quilombola* communities took place due to them being mentioned by other residents in the house as the people with best understanding of medicinal plants. According to Viu et al. (2010), women are responsible for taking care of the home and for curing illnesses that occurred in children, through the use of medicinal plants. This fact can be confirmed by noting that women are also the people who best understand medicinal plants in most of the studies carried out (Vendruscolo and Mentz 2006; Gomes and de Bandeira 2012; da Silva et al. 2012; Ferrão et al. 2014; Ferreira et al. 2014; Ferreira et al. 2016; dos Lisboa et al. 2017).

Women's occupation with domestic tasks and plantation, within the community of Santo Antônio dos Quilombolas, was also observed by Santos (2012), when there was the development of a study in the aforementioned community about the interface between food and the *quilombola* culture. The author reports that the women in the community are responsible for taking care of children and the plantations in the yard and that they cultivate a wide range of medicinal plants. Cupertino (2012), on carrying out a study on the young people at the communities of Santo Antônio dos Quilombolas and Guiné, also reported on the traditional consumption of medicinal teas by the *quilombola* families.

The ages of the interviewees ranged from 23 to 94 years, in the community of Santo Antônio dos Quilombolas. For female interviewees the youngest was 34 years old and the interviewee of most advanced age was 94, while for the male gender, the youngest was 34 and the most chronologically advantaged was 76. The most common age brackets within the sample were 36–45 years old and 46–55 years old, both corresponding to 25.42% of the interviewees.

In the community of Santo Antônio do Guiné, the age of the interviewees ranged from 23 to 87 years. For the female gender, the youngest was 23 and the most chronologically advantaged was 87, while for males the youngest was 37 and the interviewee of most advanced age was 67. In this case, the most common age bracket among the interviewees was between 46 and 55 years old, which worked out at 30.26% of the interviewees.

The fact that there were very few young interviewees, between 23 and 35 years old, was because this is the age bracket that most leaves the communities to work outside the home, at present. According to Brumer (2013), the exodus of young people, leaving rural areas, is also due to the search for something more modern. The city is considered a place where things happen, somewhere fun and full of opportunity. This migration of young people from rural to urban locations is selective, being more common in the case of women (de Castro 2005).

This reality can best be seen in the work of Cupertino (2012) which was carried out with young people of these same two communities (Santo Antônio dos Quilombolas and Santo Antônio do Guiné), where it was observed that most young people wish to work in areas not related to agricultural production and that many people also want to study at university. The same author draws attention to the fact that immigration is very common in these communities, as 92% of the interviewees have at least one relative living away from the communities.

The occupations of the interviewees in the community of Santo Antônio dos Quilombolas are homemaker (54%), agricultural worker (25%), retired person (15%), rural producer (2%), shopkeeper (2%), school helper (1%) and joiner (1%). Even though most of the women are homemakers, it is important to mention that some of these interviewees have worked 5 months of the year harvesting coffee in rural properties of the region, this being the main job opportunity mentioned at these interviews.

The main occupations seen in the community of Santo Antônio do Guiné are no different from those found in Santo Antônio dos Quilombolas, the most mentioned being homemaker (50%), agricultural worker (33%) and retired person (7%). Other jobs were also mentioned, each with 1%: these were school helper, security guard, health agent, teacher, shopkeeper, artisan and coal merchant. It is also worth mentioning that these latter occupations are developed in the community itself, except for those of security guard and teacher, these being carried out in central Piranga and in a neighbouring community, respectively.

The main occupations found in the two *quilombola* communities are homemaker, agricultural worker and retired person. These same occupations were the most commonly found in studies carried out at other *quilombola* communities (de Siqueira 2014; de Nunes 2016). The occupational data as found in the two communities studied reinforce the importance of land for the development of subsistence activities on the part of the quilombolas.

Most of the interviewees at the community of Santo Antônio dos Quilombolas (55, 96.4%) said that they used medicinal plants for treating health problems; only 4 respondents said they did not use plants for this purpose. Out of the people who did not use plants, three are young women aged 35 years, who justified this by saying they did not have time to cultivate the plants and that they trust medication prescribed by the doctor at the health service. The other interviewee that did not use plants was a chronologically advantaged man, who lives alone.

Out of the cohort making use of plants, 91.23% learnt how to use and recognise medicinal plants, or 'tea plants', through their mothers. The grandmother was also mentioned by 7.02% of the interviewees, while only one of the interviewees (a man) said that he learnt how to use medicinal plants through his father, a *raizeiro*[2] who had already passed away many years before, in the community. This prevalence of the woman as the transmitter of knowledge about medicinal plants adds weight to the view that they are the people who best understand this class of use for medicinal plants.

My mother made teas from many plants picked in the yard and in the forest, to cure many diseases; everything was treated with plants. When we were children, we didn't even go to the doctor (female interviewee, 52 years old).

This reality is also confirmed in the community of Santo Antônio do Guiné, where most of the interviewees, 75 (98.68%), use medicinal plants for the treatment of health problems. There is just one person, a female resident aged 64, who does not use such teas any more. This lady says that, now her children have got married and moved away from Piranga, she did not keep up the habit of cultivating many of these plants in the vegetable garden for this purpose, with the plants that do exist being used as seasonings.

Out of the group of interviewees who make use of medicinal plants, the large majority (92.10%) learnt how to use such medicinal plants with their mothers, 6.57% with their grandmothers and only one single case (1.31%) interviewed declared learning of use through an uncle. During the interviews with the residents of Santo Antônio do Guiné, the role of the mother as owner and transmitter of this knowledge was also reinforced.

---

[2]A medicine man who makes medical treatments using roots of plants (Minas Gerais terminology).

My mother knew about many plants that cured many diseases. She cured all her health problems with plants that she would pick from the vegetable garden in the yard (female interviewee, 57 years old).

The recommendation of medicinal plants to other residents in the community is made by most people who participated in the survey at both *quilombola* communities, making recommendations mainly to family, neighbours and friends. This same exchange between family, friends and neighbours is also observed when they are asked about if they seek someone's help when they need to use plants as treatment for some illness. According to Zank et al. (2016), the picking of plants in neighbours' and family members' yards shows the importance of social relations and exchange, in the use of medicinal plants.

The residents aged between 76 and 85, at the community of Santo Antônio dos Quilombolas, are the people who have the best knowledge about medicinal plants, mentioning 14 species on average. A similar result was also found in the community of Santo Antônio do Guiné, where the interviewees in the age bracket between 66 and 75 were those with best knowledge, having mentioned an average of 13 species. Other work projects in ethnobotany and using medicinal plants also found greater knowledge among people aged over 60 (Gomes and de Bandeira 2012; da Silva et al. 2012; Linhares 2015).

There were a lower number of mentions of medicinal species by residents within the age bracket from 23 to 35 years old, in both communities, the mean being seven species within the community of Santo Antônio dos Quilombolas and eight at the community of Santo Antônio do Guiné, which shows that the transmission of knowledge of the usage of these plants has been declining. Amorozo (2002) reports that many factors have helped to bring about this loss of information about plant species with therapeutic value, including the new options for health care and the devaluation of local culture, which mainly affects the younger population, who abandon the traditional practices.

The perception of loss of knowledge of medicinal plants in both communities can be perceived during the interviews, when the people of more advanced age say that the communities use less teas than in the past and that these changes have been brought about by two key factors: younger people no longer have interest in learning about the plants, and medical care is more readily available.

The 55 interviewees at the community of Santo Antônio dos Quilombolas mentioned 132 species of plants for medical use. Out of these, 126 are identified at species level and 5 at genus level only. In the case of Santo Antônio do Guiné, there were 142 species identified with 5 being identified at genus level only.

In both communities, it is possible to observe how the curve of rarefaction stabilises to show that the number of interviews was sufficient to enable awareness of the wealth of medicinal plants as used by the two *quilombola* communities in Piranga (Fig. 9.1). A similar diversity of medicinal species was also found in other communities studied in the Atlantic Rainforest (Ferreira et al. 2014; de Siqueira 2014; Crepaldi and Peixoto 2010).

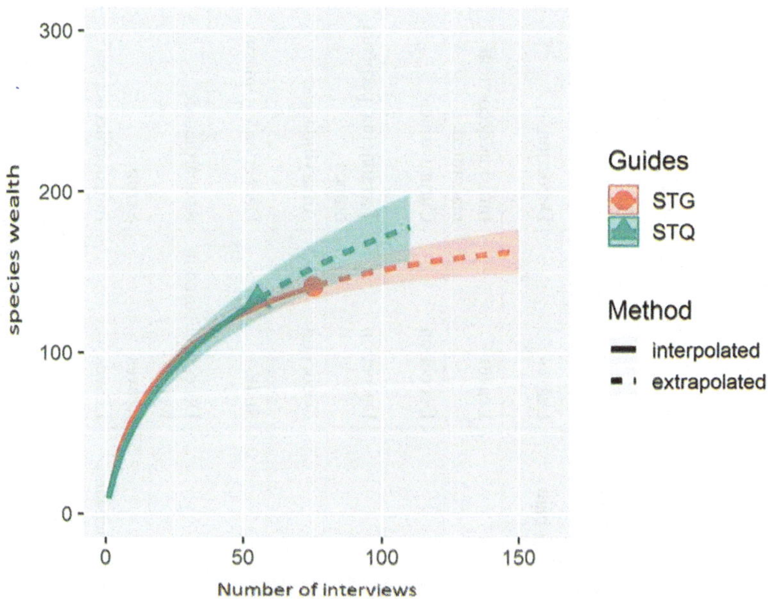

Fig. 9.1 Rarefaction curve (accumulation) of medicinal species cited by interviewees in Santo Antônio dos Quilombolas (*n = 55) and in Santo Antônio Guiné (n = 76), Piranga, Minas Gerais*

The medicinal species used by the community of Santo Antônio dos Quilombolas belong to 129 genera and 52 botanical families, according to Table 9.1. The most representative genera were *Mentha*, *Citrus* and *Solanum* with four species each; *Ocimum* with three species; and *Alpinia*, *Allium*, *Artemisia*, *Baccharis*, *Erythrina*, *Plectranthus* and *Rubus* each with two species. The other 118 genera had only 1 species mentioned, which represented 89% of the total number of species.

The 142 medicinal species used by the community of Santo de Guiné represent 123 genera and 54 families: the genus *Mentha* showed 6 species, *Citrus* 4 species, and then 2 genera showed 3 species each, *Ocimum* and *Artemisia*, and 5 genera with only 2 species apiece: *Allium*, *Baccharis*, *Apium*, *Rosa* and *Piper*. The other 114 genera had only 1 species mentioned, representing 92% of the total.

The prevalence of genera with only one species mentioned in research into the use of medicinal plants has been recorded in most of the academic works published (Amorozo 2002; Pinto et al. 2006; Silva and Proença 2008; Messias et al. 2015; de Nunes 2016; dos Lisboa et al. 2017). The genera that stood out in the two *quilombola* communities are mainly genera that have medicinal species that are widely used by popular medicine, most having been naturalised or cultivated within Brazilian territory (Forzza et al. 2010).

The most representative biological families in the community of Santo Antônio dos Quilombolas were Asteraceae with 21 species (15.90%); Lamiaceae with 19 species (14.39%); Fabaceae with 7 species (5.30%); Solanaceae, Rosaceae and

Table 9.1 Plants cited in interviews in the communities of Santo Antônio dos Quilombolas and Santo Antônio do Guiné, Piranga, Minas Gerais

Family	Vernacular name	Scientific name	Recommended for	Part used	Method of preparation
Acanthaceae	Sapatinho de judeu	*Thunbergia mysorensis* (Wight) T. Anderson	Fever	Leaves/root	Infusion/macerated
Achariaceae	Sapucaia	*Carpotroche brasiliensis* (Raddi) A.Gray	Wounds	Fruit	Cooked in water
Adoxaceae	Sabugueiro	*Sambucus australis* Cham. & Schltdl.	Colds/emagrecimento/catapora/alergia no corpo/influenza	Flower	Infusion/decoction
Alismataceae	Chapéu de couro	*Echinodorus grandiflorus* (Cham. & Schltdl.) Micheli	Cleanses the blood/body pain	Leaves	Decoction
Amaranthaceae	Santa maria	*Dysphania ambrosioides* (L.) Mosyakin & Clemants	Influenza/worms	Leaves	Infusion/decoction/macerated
Pérpetua	*Gomphrena arborescens* L. f.	Diarrhoea/influenza/coughs	Flower	Decoction	Cururu arnica
	Amaranthus sp.	Intestinal infection	Leaves/root	Decoction	Terramicina, perpétua branca
	Alternanthera brasiliana (L.) Kuntze	Infection/influenza/sore throat	Leaves	Decoction	Amaryllidaceae
	Cebola	*Allium cepa* L.	Influenza/wounds	Stalk	Decoction/cataplasm
Alho antigo	*Allium sativum* L.	Influenza/coughs/worms/hypertension	Stalk	Decoction/cured in water	Anacardiaceae
	Manga	*Mangifera indica* L.	Influenza	Leaves	Syrup
Apiaceae	Funcho	*Foeniculum vulgare* Mill.	Influenza/sedative/stomachache/hypertension/colds/coughs/hoarseness	Leaves	Infusion/decoction/syrup

Aipo	*Apium graveolens* L.	Chest pains/flatulence	Leaves	Macerated/infusion	Cheiro de coentro
	Eryngium foetidum L.	Influenza/swollen abdomen	Leaves	Decoction/cured in olive oil	Agrião
	Apium nodiflorum (L.) Lag.	Fever/influenza	Leaves/stalk	Decoction/in its natural state	Salsa
	Petroselinum crispum (Mill.) Fuss	Influenza/infection/kidney infection/cleans the womb/menstrual colic/reduces the flow of blood during menstruation	Root/leaves	Decoction/beaten in the liquefier with water	Araliaceae
	Capitão	*Hydrocotyle bonariensis* Comm. ex Lam.	Wounds/cutaneous lesions on the skin	Leaves	Decoction
Asparagaceae	Salsaparrilha	*Herreria salsaparrilha* Mart.	Cracked hands/rheumatism/infection	Root	Decoction/cured in *pinga* (Brazilian firewater)
Milindro	*Asparagus officinalis* L.	Chest pain	Leaves	Decoction	Asteraceae
	Boldo de árvore	*Acmella ciliata* (Kunth) Cass.	Headache/stomach/liver	Leaves	Macerated/decoction
Picão	*Bidens pilosa* L.	Influenza/hookworm disease/hepatitis/urine infection/jaundice/stomachache/anaemia	Leaves/root	Decoction	Mata pasto
	Acanthospermum australe (Loefl.) Kuntze	Worms/depression/anaemia	Root	Decoction	Mentrasto
	Ageratum conyzoides (L.) L.	Dysentery/infection	Leaves	Infusion	
	Serralha	*Sonchus oleraceus* (L.) L.	Diabetes/flatulence/dysentery	Leaves	Stewed/beaten in the liquefier with water

(continued)

Table 9.1 (continued)

Family	Vernacular name	Scientific name	Recommended for	Part used	Method of preparation
Marcelão/novalgina	*Achillea millefolium* L.	Headache/influenza/fever/diarrhoea	Leaves	Decoction/infusion/syrup	Alecrim do mato
	Baccharis dracunculifolia DC.	Depression/bronchitis	Leaves/stalk	Infusion/decoction	Camomila
	Coreopsis grandiflora Hogg ex Sweet	Influenza/fever/stomachache/dysentery/headache/hypertension/sedative/wounds	Flower/leaves	Infusion/macerated	Dente de leão
	Taraxacum campylodes G.E.Haglund	Dysentery/cleansing the blood	Leaves	Infusion/in natura	Carqueja
	Baccharis trimera (Less.) DC.	Heart/stomach/cholesterol/wounds/liver/diabetes	Leaves	Decoction/infusion	Marcela de travesseiro
	Achyrocline satureioides (Lam.) DC.	Sedative	Leaves	Infusion	Marcela da horta
	Chamaemelum nobile (L.) All.	Stomachache/influenza/hypertension/headache/coughs/fever/stomach	Leaves	Infusion/macerated	Losna
	Artemisia absinthium L.	Worms/menstrual colic/diabetes/stomach pain/influenza	Leaves	Infusion/macerated	Timijio
	Tanacetum parthenium (L.) Sch.Bip.	Fever/influenza/headache/menstrual colic	Leaves/flower	Infusion/decoction	Batata da diabetes
	Smallanthus sonchifolius (Poepp.) H.Rob.	Diabetes	Stalk/leaves	In natura/macerated	Insulina
	Tithonia diversifolia (Hemsl.) A.Gray	Diabetes	Leaves	Decoction/macerated	Guaco
	Mikania glomerata Spreng.	Influenza/coughs/bronchitis	Leaves	Decoction/infusion	Cânfora, canforada

	Artemisia alba Turra	Sinusitis/body pain/ shortness of breath	Leaves	Cured in alcohol	
	Assapeixe	*Vernonanthura phosphorica* (Vell.) H.Rob.	Chest pain/wounds/influenza/ coughs	Leaves	Cured in *pinga* (Brazilian firewater)/ cured in water/ cataplasm
Voadeira	*Tagetes minuta* L.	Worms	Leaves	Decoction	Ponta livre
	Artemisia vulgaris L.	Stomach twinge	Leaves	Infusion	Quitoco
	Pluchea sagittalis Less.	Coughs/colds	Leaves	Decoction	Cravo de defunto
	Tagetes erecta L.	Influenza/coughs	Leaves	Decoction	Cardo santo
	Centaurea benedicta (L.) L.	Headache	Leaves	Infusion	Bardana
	Arctium minus (Hill) Bernh.	Stomach/cholesterol/ hypertension	Leaves	Beaten in the liquefier with water	Cambará
	Chromolaena squalida (DC.) R.M.King & H.Rob.	Influenza/coughs	Leaves	Decoction	Fumo de cachorro
	Elephantopus mollis Kunth	Influenza	Leaves	Decoction	Arnica
	Solidago chilensis Meyen	Body pain	Leaves	Cured in *cachaça* (Brazilian firewater)	Bignoniaceae
	Cinco folhas	*Sparattosperma leucanthum* (Vell.) K. Schum.	Wounds/cleansing the blood/ throat/infection	Leaves	Decoction
Bixaceae	Urucum	*Bixa orellana* L.	High cholesterol	Seed	Decoction
Boraginaceae	Confrei	*Symphytum officinale* L.	Cleansing the blood/body pain/ healing of wounds/dysentery/ inflamed throat/coughs	Leaves	Infusion/decoction/ macerated

(continued)

Table 9.1 (continued)

Family	Vernacular name	Scientific name	Recommended for	Part used	Method of preparation
Marmelinho	*Cordia axillaris* I.M.Johnst.	Infection	Leaves	Decoction	Borragem
	Borago officinalis L.	Influenza/stomachache	Flower	Decoction	Brassicaceae
	Menstruce	*Lepidium didymum* L.	Influenza/coughs/bronchitis	Stalk and leaves	Decoction
Couve	*Brassica oleracea* L.	Ulcers/reflux	Leaves	Beaten in the liquefier with water	Agrião
	Nasturtium officinale R.Br.	Fever/influenza	Leaves, stalk	Decoction/syrup	Bromeliaceae
	Abacaxi ananás	*Ananas comosus* (L.) Merr.	Catarrh	Fruit	Syrup
Cactaceae	Ora pro nóbis	*Pereskia grandiflora* Pfeiff.	Any kind of illness/diabetes	Leaves	Beaten in the liquefier with water
Mandacaru	*Cereus jamacaru* DC.	Bronchitis	Stalk	Syrup	Caricaceae
	Mamão	*Carica papaya* L.	Headache/body itching/fever/indigestion/liver/dysentery	Flower/leaves	Infusion/decoction
Caryophyllaceae	Trevo branco	*Drymaria cordata* (L.) Willd. ex Schult.	Dysentery/stomachache	Leaves/stalk	Decoction
Combretaceae	Capitão do mato	*Terminalia argentea* Mart.	Cleansing the blood	Leaves	Decoction
Commelinaceae	Trapoerava	*Commelina erecta* L.	Kidney stones	Leaves	Infusion
Convolvulaceae	Batata doce	*Ipomoea batatas* (L.) Lam.	Toothache	Leaves	Decoction
Costaceae	Caninha do brejo/canela de macaco	*Costus* spicatus (Jacq.) Sw.	Kidney stones/urinary infection/bladder inflammation	Leaves and stalk	Decoction
Crassulaceae	Saião	*Kalanchoe gastonis-bonnieri* Raym.-Hamet & H. Perrier	Burning in the stomach/coughs	Leaves	Infusion/beaten in the liquefier with milk

Basto	*Sedum dendroideum* Moc. & Sessé ex DC.	Stomach pain/gastritis/ wounds	Leaves	Beaten in the liquefier with water/ in its natural state/ cataplasm	Fortuna
	Bryophyllum pinnatum (Lam.) Oken	Wounds	Leaves	Cataplasm	Saião
	Kalanchoe laciniata (L.) DC.	Sore throat	Leaves	Macerated	Cucurbitaceae
	Melão de são caetano	*Momordica charantia* L.	Influenza/headache/ill health	Leaves	Infusion/decoction
Abóbora d´anta	*Cayaponia podantha* Cogn.	Rheumatism	Root	Decoction	Chuchu
	Sechium edule (Jacq.) Sw.	Hypertension/sedative	Leaves	Decoction	Ebenaceae
	Caqui	*Diospyros kaki* L.f.	Insomnia	Leaves	Infusion
Equisetaceae	Cavalinha	*Equisetum giganteum* L.	Hypertension/cancer/healing wounds/cleansing the blood/urine infection	Stalk	Decoction
Euphorbiaceae	Metiolate	*Jatropha multifida* L.	Wounds	Leaves	Cataplasm
Sangra-dágua	*Croton urucurana* Baill.	Wounds	Bark	Cataplasm	Mandioca
	Manihot esculenta Crantz	Diarrhoea	Root	Cured in water	Mamona
	Ricinus communis L.	Wounds	Leaves	Cataplasm	Fabaceae
	Carrapichinho	*Desmodium adscendens* (Sw.) DC.	Kidney stones/infection	Leaves	Infusion

(continued)

Table 9.1 (continued)

Family	Vernacular name	Scientific name	Recommended for	Part used	Method of preparation
Barbatimão	*Stryphnodendron adstringens* (Mart.) Coville	Wounds/removes excess liquid from the body	Bark	Decoction	Copaíba
	Copaifera langsdorffii Desf.	Healing agent	Fruit	Decoction	Jatobá
	Hymenaea courbaril L.	Illnesses of the digestive system	Bark	Decoction	Samaduba
	Erythrina verna Vell.	Cleansing agent	Bark	Decoction	Olho de boi
	Dioclea violacea Benth	Stroke	Seed	Macerated	Angico
	Anadenanthera peregrina (L.) Speg.	Influenza, throat	Bark	Decoction	Bardana
	Erythrina speciosa Andrew	Cleansing agent	Bark	Cooking the bark	Pata de vaca
	Bauhinia forficata Link	Diabetes	Leaves	Infusion	Lamiaceae
	Boldo	*Plectranthus barbatus* Andrews	Headache/dysentery/liver/influenza/hypertension	Leaves	Macerated
	Boldo	*Plectranthus amboinicus* (Lour.) Spreng.	Stomach ailments/headache	Leaves	Macerated
	Zé serafim, mata pasto, mané magro	*Leonurus sibiricus* L.	Dysentery/stomachache/hypertension/headache/influenza/thinning the blood/stroke/wounds/diabetes/coughs	Leaves/flower	Macerated/cataplasm/infusion

Alfavaca	*Ocimum carnosum* (Spreng.) Link & Otto ex Benth.	Influenza/hoarseness/sedative/coughs	Leaves	Infusion/decoction	Cravo
	Ocimum gratissimum L.	Coughs/influenza/sedative/hoarseness/colds	Leaves	Infusion/decoction/syrup	Hortelã roxo/hortelã liso/hortelã verde/hortelã da horta
	Mentha x villosa Huds.	Influenza/stomachache/worms/inflammation/sedative/infection/coughs/throat inflammation/wounds	Leaves	Infusion/decoction/macerated	Hortelã menta/hortelã pimenta
	Mentha arvensis L.	Influenza/sore throat/worms/coughs/sedative/indigestion	Leaves	Infusion	Hortelã branco
	Mentha suaveolens Ehrh.	Worms/influenza/swollen feet/stomachache/inflammation/sedative/throat inflammation/coughs/infection/hypertension/toothache/wounds	Leaves	Infusion/decoction	Hortelã balinha
	Mentha x piperita L.	Influenza/sedative/inflamed throat/stomachache	Leaves	Decoction	Poejo
	Mentha pulegium L.	Influenza/rashes in children/bronchitis/coughs/stomachache/sedative	Leaves	Infusion/decoction	Levante
	Mentha sp.	Influenza/body pain/sedative/bronchitis/colds/stomachache/coughs/influenza	Leaves	Infusion/decoction	Levante

(continued)

Table 9.1 (continued)

Family	Vernacular name	Scientific name	Recommended for	Part used	Method of preparation
	Mentha spicata L.	Earache	Leaves	Toasted	Manjerona
	Origanum vulgare L.	Influenza/stomachache/sedative	Leaves	Infusion/decoction	Hortelã do mato
	Peltodon radicans Pohl	Wounds/infection/influenza/coughs	Leaves	Infusion	
	Salvia	*Salvia officinalis* L.	Body pain/sedative/chest pain/toothache/influenza/kidney stones	Leaves	Infusion/decoction/macerated
Alecrim da horta	*Rosmarinus officinalis* L.	Chest pain/depression/sadness/sedative/hypertension/menstrual colic/nervousness/headache/coughs/heart	Leaves/stalk	Infusion/cured in water/macerated/decoction	Erva terrestre
	Glechoma hederacea L.	Coughs/influenza/fever/colds/diarrhoea	Leaves	Infusion/decoction/toasted	Erva cidreira da horta
	Melissa officinalis L.	Sedative/influenza/hypertension/menstrual colic	Leaves	Infusion/decoction	Manjericão
	Ocimum basilicum L.	Influenza/depression/nervousness/sedative/coughs	Leaves	Infusion/decoction	Papagaio
	Aegiphila integrifolia (Jacq.) B.D.Jacks.	Burns	Fruit	Decoction	Cordão de frade
	Leonotis nepetifolia (L.) R. Br.	Headache/haemorrhage/lowers uric acid/stomachache/coughs/pneumonia/influenza	Leaves	Infusion/decoction	Alfazema
	Lavandula angustifólia Mill.	Stomachache in babies/heart/sedative	Leaves	Infusion/decoction	Lauraceae
	Abacate	*Persea americana* Mill.	Coughs/kidney stones/urinary infection/hypertension	Leaves	Decoction/syrup

Loranthaceae	Erva de passarinho	*Struthanthus* sp.	Pain/suffocation/bronchitis/influenza	Leaves	Decoction/macerated
Erva de passarinho	*Struthanthus* sp.	Worms	Leaves	Decoction	Lythraceae
	Romã	*Punica granatum* L.	Throat infection	Fruit	Decoction
Malpighiaceae	Douradinha	*Heteropterys macrostachya* A.Juss.	Kidney stones	Leaves	Infusion
Malvaceae	Algodão	*Gossypium barbadense* L.	Infection/cleanses the uterus/falls/healing of wounds	Leaves/flower	Infusion/decoction
Açoita-cavalo	*Luehea grandiflora* Mart.	Cleansing the blood	Leaves	Infusion	Malva lisa
	Malva sp.	Coughs/throat infection	Leaves	Infusion	Meliaceae
	Cutieira	*Guarea guidonia* (L.) Sleumer	Laxative/cleanses the blood	Fruit	In its natural state
Menispermaceae	Buta	*Abuta* sp.	Rheumatism	Root	Decoction
Moraceae	Amora preta	*Morus nigra* L.	Hypertension/hormone replacement/headache/heat flashes characteristic of menopause	Leaves	Infusion/decoction
Myrtaceae	Eucalipto	*Eucalyptus* sp.	Influenza/sinusitis	Leaves	Syrup/cured in alcohol
Goiaba	*Psidium guajava* L.	Diarrhoea/toothache	Leaves/bark	Infusion/decoction	Pitanga
	Eugenia uniflora L.	Stomachache	Leaves	Decoction	Musaceae
	Banana	*Musa paradisiaca* L.	Influenza/diarrhoea	Navel/leaves/stalk	Decoction/in its natural state
Nyctaginaceae	Erva tostão	*Boerhavia diffusa* L.	Kidney stones/liver	Root	Decoction
Bonina	*Mirabilis jalapa* L.	Wounds/infection	Leaves	Beat in the liquefier with aloe	Oxalidaceae
	Carambola	*Averrhoa carambola* L.	Diabetes	Leaves	Decoction
Passifloraceae	Maracujá	*Passiflora alata* Curtis	Sedative	Leaves	Decoction
Phyllanthaceae	Quebra-pedra	*Phyllanthus niruri* L.	Urinary infection	Root	Decoction

(continued)

Table 9.1 (continued)

Family	Vernacular name	Scientific name	Recommended for	Part used	Method of preparation
Phytolaccaceae	Guiné	*Petiveria alliacea* L.	Headache/toothache/influenza	Leaves	Infusion/bath
Piperaceae	Jamborandi	*Piper aduncum* L.	Hair	Leaves	Decoction
Capeva	*Piper umbellatum* L.	Stuck catarrh/influenza/ stomach	Leaves	Cataplasm	Plantaginaceae
	Tansagem	*Plantago major* L.	Urinary infection/throat infection/ sedative/influenza/expectoration/ stomachache	Leaves/root	Infusion/decoction
Tansagem	*Plantago australis* Lam.	Urinary infection/throat infection	Leaves/root	Infusion/decoction	Vassorinha de são Pedro
	Scoparia dulcis L.	Kidney stones	Leaves	Decoction	Poaceae
	Milho	*Zea mays* L.	Urinary infection/detox for the kidneys	Flower	Decoction
Pé de galinha	*Eleusine indica* (L.) Gaertn.	Kidney stones/bronchitis/ influenza	Root/leaves/root	Decoction	Cana miúda
	Saccharum officinarum L.	Hypertension	Leaves	Decoction	Grama miúda
	Cynodon dactylon (L.) Pers.	Kidney stones	Leaves	Decoction	Lágrima de nossa senhora
	Coix lacryma-jobi L.	Fever/body pain/urine infection	Leaves, root	Decoction/bath	Capim gordura
	Melinis sp.	Hypertension	Leaves	Infusion	Capim cidreira
	Cymbopogon citratus (DC.) Stapf	Sedative/hypertension/ coughs	Leaves	Infusion/decoction	Polygalaceae
	Encarangada	*Diclidanthera laurifolia* Mart.	Body itch	Leaves	Decoction

Polygonaceae	Erva de bicho	*Persicaria punctata* (Elliott) Small	Dandruff	Leaves	Decoction
Solitária	*Homalocladium platycladum* (F.Muell.) L.H. Bailey	Worms	Leaves	Decoction	Pteridaceae
	Avenca	*Adiantum* sp.	Fatigue/influenza	Leaves	Decoction
Rosaceae	Ameixa	*Eriobotrya japonica* (Thunb.) Lindl.	Hypertension	Leaves	Decoction
Amora vermelha do mato	*Rubus sellowii* Cham. & Schltdl.	Sore throat	Leaves	Syrup	Amora
	Rubus rosifolius Sm.	Throat problems	Leaves	Syrup	Rosa branca
	Rosa alba L.	Wounds/sores/infection	Flower	Cooked in milk	Fragaria
	Fragaria vesca L.	Influenza/stomachache	Leaves	Decoction/infusion	Rosa verde
	Rosa chinensis cv. Viridiflora	Haemorrhage	Leaves	Decoction	Grimonia
	Agrimonia eupatoria L.	Lack of appetite/diarrhoea/headache/worms	Leaves	Decoction	Rubiaceae
	Cura-tombo	*Chiococca alba* (L.) Hitchc.	Falls/body pain	Leaves	Decoction
Mária preta	*Palicourea tetraphylla* Cham. & Schtdl.	Kidney stones	Leaves	Infusion	Rutaceae
	Mexirica	*Citrus reticulata* Blanco	Sedative/influenza/depression/coughs	Leaves	Decoction

(continued)

Table 9.1 (continued)

Family	Vernacular name	Scientific name	Recommended for	Part used	Method of preparation
Limão	*Citrus limon* (L.) Osbeck	Influenza/stomachache/ thinning the blood/wounds/ diarrhoea	Leaves/fruit	Toasted/decoction/ juice	Laranja
	Citrus sinensis (L.) Osbeck	Influenza/coughs/headache/ sedative/hypertension/lack of appetite/heart/ constipation	Leaves/fruit	Toasted/decoction/ infusion	Arruda
	Ruta graveolens L.	Headache/body pain/ infection in the uterus/ swollen eye/fever/ migraine/colds	Leaves	Infusion/macerated/ bath/cured in water	Laranja da terra
	Citrus aurantium L.	Hypertension/sedative	Leaves	Infusion	Paratudo
	Hortia arborea Engl.	Diabetes	Bark	Decoction	Scrophulariaceae
	Barbaçu	*Buddleja stachyoides* Cham. & Schltdl.	Influenza	Flower	Decoction
Siparunaceae	Negramina, folha santa	*Siparuna guianensis* Aubl.	Influenza/fever	Leaves	Decoction
Solanaceae	Fruta de lobo	*Solanum lycocarpum* A. St.-Hil.	Influenza	Flower	Decoction

Berinjela	*Solanum melongena* L.	Cholesterol	Fruit	Cured in water	Erva moura, amora dormideira
	Solanum americanum Mill.	Sedative	Leaves and fruit	Infusion/in natura	Tomatinho
	Lycopersicon pimpinellifolium L.	Wounds/tooth infection/ influenza/coughs/ hoarseness	Leaves/root	Decoction	Fumo
	Nicotiana tabacum L.	Wounds	Leaves	Cataplasm	Jurubeba
	Solanum paniculatum L.	Worms	Fruit	Decoction	Meladoma
	Brugmansia suaveolens (Humb. & Bonpl. Ex Willd.) Bercht. & J.Presl	Sedative	Leaves	Infusion	Costa branca
	Solanum cernuum Vell.	Pain	Leaves	Decoction	Tropaeolaceae
	Chaga	*Tropaeolum majus* L.	Influenza/intestinal ailments/ cleansing the blood/infection/ mycosis	Flower	Decoction/cooked in milk/cataplasm
Urticaceae	Urtiga	*Urtica dioica* L.	Hair	Leaves	Decoction
Embaúba vermelha	*Cecropia glaziovii* Snethl.	Spinal pain	Leaves	Bath	Verbenaceae
	Mal-me-quer	*Lantana camara* L.	Influenza/allergy/throat inflammation	Leaves	Infusion
Erva cidreira de árvore	*Lippia alba* (Mill.) N.E.Br. ex Britton & P. Wilson	Coughs/influenza/sedative/ symptoms of the menopause	Leaves and stalk	Infusion/decoction/ beaten in the liquefier with water	Jurubão
	Stachytarpheta cayennensis (Rich.) Vahl	Wounds/kidney stones/ joint pain	Leaves	Cataplasm/decoction	Violaceae
	Violeta	*Viola odorata* L.	Dizziness	Leaves	Decoction
	Piraguaia	*Anchietea pyrifolia* (Mart.) G.Don	Cleansing the blood	Bark	Cured in *cachaça* (Brazilian firewater)

(continued)

Table 9.1 (continued)

Vitaceae	Insulina	*Cissus verticillata* (L.) Nicolson & C.E. Jarvis	Diabetes	Leaves	Infusion/decoction
Xanthorrhoeaceae	Babosa	*Aloe vera* (L.) Burm.f.	Influenza/growth of hair/wounds/cancer/burns/infection	Leaves	Decoction/cataplasm
Zingiberaceae	Pacová	*Alpinia zerumbet* (Pers.) B.L.Burtt. & R.M.Sm.	Hypertension	Leaves	Decoction
Lírio do brejo	*Alpinia purpurata* (Vieill.) K. Schum.	Wounds	Leaves/stalk	Cataplasm	Açafrão
	Curcuma zedoaria (Christm.) Roscoe	Cancer/leukaemia/cholesterol	Stalk	Powder	Mariazinha
	Hedychium coronarium J.Koenig	Colic pain	Leaves	Decoction/bath	Gengibre
	Zingiber officinale Roscoe	Sore throat/hoarseness/influenza	Stalk	Decoction	

Rutaceae with 5 species each (3.78% each); and Poaceae, Brassicaceae and Apiaceae with 4 species each (3.03% each). These families account for 56.02% of the species mentioned by the community. The other 42 families had only 1, 2 or 3 species mentioned.

The most representative families in the community at Santo Antônio do Guiné were Asteraceae with 25 species (17.73%); Lamiaceae with 19 species (13.47%); Rutaceae with 6 species (4.54%); Apiaceae, Poaceae and Rosaceae with 5 species each (3.78% each); and the Fabaceae with 4 species (3.03%). These families together accounted for just over half (50.11%) the species mentioned for the community as a whole. The remaining 46 families had 1, 2 or 3 species mentioned at the interviews.

Many different projects of ethnobotany in different parts of Brazil have shown the Asteraceae and the Lamiaceae as the most important families, showing that the use of its species is well consolidated in Brazil's popular pharmacopoeia for the treatment of several diseases and ailments (da Silva et al. 2012; Battisti et al. 2013; Brito and Senna-Valle 2011; Oliveira and Menini-Neto 2012; do Oliveira and de Gondim 2013; Neto et al. 2014; Cunha et al. 2015; Flor and Barbosa 2015; Messias et al. 2015). These two families are also widely used in other parts of the world and always rank among the most important (Tefera and Kim 2019; Vinagre et al. 2019).

The ten medicinal species most mentioned by the interviewees in the community of Santo Antônio dos Quilombolas were *Mentha* x *villosa* Huds. (65.45%), *Foeniculum vulgare* Mill. (50.90%), *Leonurus sibiricus* L. (49.09%), *Lippia alba* (Mill.) N.E.Br. ex Britton & P. Wilson (43.63%), *Plantago major* L. (40.0%), *Coreopsis grandiflora* Hogg ex Sweet (21.81%), *Chamaemelum nobile* (L.) All. (27.27%), *Mentha pulegium* L. (25.45%), *Rosmarinus officinalis* L. (25.45%) and *Plectranthus barbatus* Andrews (23.63%).

The medicinal species most mentioned by interviewees in the community of Santo Antônio do Guiné were *Mentha* x *villosa* (65.33%), *Leonurus sibiricus* (56%), *F. vulgare* (50.66%), *M. pulegium* (46.66%), *Mentha suaveolens* Ehrh (38.66%), *Mentha* sp. (36%), *P. major* (34.66%), *C. nobile* (33.33%), *Melissa officinalis* L. (28%), *Glechoma hederacea* L. (25.33%) and *R. officinalis* (25.33%).

Most of the medicinal plants mentioned and used by the community of Santo Antônio dos Quilombolas are collected in backyards (84.09%), in fragments of forest (7.57%), in grazing land (4.54%) and in swampy areas throughout the community (3.78%). In the community of Santo Guiné, the backyards have also been the most important sources of resources involving medicinal plants, and here 86.52% of the plants were collected in backyards, 9.21% in fragments of forest, 2.83% in pastures and 1.41% in swampy areas.

This prevalence of medicinal plants collected in yards is justified in the two communities, as women are the people who know these plants better. Many medicinal species are cultivated close to kitchens or at the sides of houses, making the plants more accessible to women. A larger number of medicinal plants collected in yards back up what has been found in other works on ethnobotany carried out in *quilombola* communities in Brazil (da Silva et al. 2012; Ferreira et al. 2014; Rogério 2014; de Siqueira 2014; de Nunes 2016).

The two *quilombola* communities in Piranga have the leaf as the most used element for preparation of medicines; this prevalence has been constant in the works on ethnobotany in Brazil (Di Stasi and Hiruma-lima 2002; Freitas et al. 2012; da Silva et al. 2012; Alves and Povh 2013; Rogério 2014; de Siqueira 2014). da Silva et al. (2012) say that the predominance of the use of leaves is due to two main factors: the fact that it is available the whole year round and the fact that it is easy to collect. Castellucci (2000) also stresses that most of the active ingredients are in the leaves.

The method of preparation of medicinal plants in Santo Antônio dos Quilombolas was very varied, with tea being the main method, being 67.2% prepared mainly by cooking (decoction) (48.33%) or in a 'scalded' form (infusion) (18.88%), followed by maceration (17.7%), syrup (4.4%), beaten in the liquefier (3.33%), cataplasm (2.2%), juice (1.6%) and baths (1.1%).

In Santo Antônio do Guiné, the most common way of preparation was also that of a tea (74.8%), this being prepared mainly by decoction (47.3%), with the second most common method being infusion (27.53%). The other methods of preparation found in the community were maceration (11.1%), beaten in the liquefier (3.8%), cataplasm (3.8%), syrup (2.7%), in natural form (2.2%), juice (0.96%), bath (0.48%) and in powder form (0.48%).

Many ethnobotanical studies have suggested that decoction of medicinal plants is the method of preparation most used by the population studies. Here we highlight the studies made by Pinto et al. (2006), Merétika et al. (2010), Liporacci and Simão (2013), Alves and Povh (2013), Araújo and Lemos (2015) and Palheta (2015).

In Santo Antônio dos Quilombolas, most of the plants were recommended as treatment of diseases of the respiratory tract (29%), the next largest use being that of treatment of diseases related to symptoms or signs in general (24%), such as headache, fever, inflammation and body pain and sedative (for nervousness), followed then by infectious and parasite diseases (19%), such as worms, infection and stomachache (diarrhoea) and disorders of the circulatory system (14%), such as hypertension, and for cleansing the blood.

In Santo Antônio do Guiné, plants were mainly prescribed for illnesses involving symptoms or general signs (50%), including headache, fever, sedative (nervousness), throat inflammations and body pain, followed by diseases of the respiratory tract (47%), influenzas, colds, bronchitis and sinusitis; infectious and parasite diseases (26%) such as worms, infection and stomachache (diarrhoea); and diseases of the circulatory system (19%) such as hypertension and for cleansing the blood.

According to studies carried out in other *quilombola* communities in Brazil, diseases related to the respiratory tract have also stood out (Massarotto 2009; de Siqueira 2014; Rogério 2014; Linhares 2015). Illnesses related to the respiratory tract are among the most common ailments for treatment with medicinal plants, in several other academic papers on ethnobotany (Freitas et al. 2012; Aguiar and Barros 2012; Liporacci and Simão 2013; Rodrigues and Andrade 2014).

Among the plants that have been mentioned for treatment of ailments affecting the respiratory tract in Santo Antônio dos Quilombolas are *Leonurus sibiricus*, *Achillea millefolium*, *Mikania glomerata*, *Vernonanthura phosphorica*, *Mentha* x villosa, *Plectranthus barbatus*, *Ocimum carnosum*, *Ocimum gratissimum*, *Artemisia absinthium*, *Origanum vulgare*, *Glechoma hederacea*, *Ocimum basilicum*, *Leonotis nepetifolia*, *Lantana camara*, *Citrus sinensis* and *Eucalyptus* sp.

In Santo Antônio do Guiné, the species mentioned include *Bidens pilosa*, *A. millefolium*, *Coreopsis grandiflora*, *Chamaemelum nobile*, *Tanacetum parthenium*, *M. glomerata*, *V. phosphorica*, *Tagetes erecta*, *Chromolaena squalida*, *Elephantopus mollis*, *O. carnosum*, *O. gratissimum*, *Mentha x villosa*, *Mentha* sp., *Salvia officinalis*, *G. hederacea*, *Melissa officinalis* and *Buddleja stachyoides*.

In both communities, medicinal plants have been reported as being the first option for treatment of diseases of the respiratory tract, meaning that they play an important preventive role within public health, with even the people interviewed having access to the government health system through health centres in the communities. Factors that should be taken into account, for these communities, on treating such illnesses almost exclusively with plants, lie in the efficiency of the treatment and in the lack of financial means for acquisition of medications, many of which are missing in the state health system.

The results presented in this research study show that the *quilombola* communities of Piranga make use of a wide range of medicinal plants. This result draws attention the need for preservation of knowledge about medicinal plants into the *quilombola* communities in Santo Antônio dos Quilombolas and Santo Antônio do Guiné, especially among the younger generations; for this reason, measures need to be taken, especially by the public sector, with the aim of giving value to this knowledge within the communities and publicising the rational use of such plants.

References

Aguiar LCG, Barros RFM (2012) Plantas medicinais cultivadas em quintais de comunidades rurais no domínio do cerrado piauiense (Município de Demerval Lobão, Piauí, Brasil). Revista Brasileira Plantas Medicinais 14(3):419–434

Albuquerque UP, Lucena RFP, da Cunha LVFC (eds) (2010) Métodos e técnicas na pesquisa etnobiológica e etnoecológica. NUPEEA, Recife

Alexiades MN (1996) Collecting ethnobotanical data: an introduction to basic concepts and techniques. In: Sheldon JW (ed) Selected guidelines for ethnobotanical research: a field manual, Advances in economic botany, vol 10. The New York Botanical Garden, New York, pp 53–94

Almeida AWB (2002) Os quilombos e as novas etnias. In: O'Dwyer EC (ed) Quilombos: identidade étnica e territorialidade. Editora FGV, Rio de Janeiro, pp 43–81

Alves GSP, Povh JA (2013) Estudo etnobotânico de plantas medicinais na comunidade de Santa Rita, Ituiutaba, Minas Gerais. Brazil Biotemas 26(3):231–242

Amorozo MCM (2002) Uso e diversidade de plantas medicinais em Santo Antônio do Laverger, Mato Grosso, Brasil. Acta Bot Bras 16(2):189–203

APG IV (2016) An update of the Angiosperm Phylogeny Group classification for the orders and families of flowering plants: APG III. Bot J Linn Soc 181:1–20

Araújo JL, Lemos JR (2015) Estudo etnobotânico sobre plantas medicinais na comunidade de Curral Velho, Luiz Correia, Piauí, Brasil. Revista Biotemas 28(2):125–136

Battisti C, Garlet TMB et al (2013) Plantas medicinais utilizadas no município de Palmeira das Missões, Rio Grande do Sul, Brasil. Revista Brasileira de Biociências 11(3):338–348

Brito MR, Senna-Valle L (2011) Plantas medicinais utilizadas na comunidade caiçara da Praia do Sono, Paraty, RJ. Acta Bot Bras 25(2):363–372

Brumer A (2013) A problemática dos jovens na pós-modernidade. In: Carneiro MJ, Castro EG (eds) Juventude rural em perspectiva, 1st edn. Mauad, Rio de Janeiro, pp 35–51

Carvalho RMA, da Lima GFC (2013) Comunidades quilombolas, territorialidade e a legislação no Brasil: uma análise histórica. Revista de Ciências Sociais 39:329–346

Castellucci S, Lima MIS, Nordi N, Marques JGW (2000) Plantas medicinais relatadas pela comunidade residente na estação ecológica de Jataí, município de Luís Antônio/SP: uma abordagem etnobotânica. Revista Brasileira de Plantas Medicinais 3(1):51–60

CEDEFES (2008) Comunidades Quilombolas de Minas Gerais no século XXI: História e resistência. Autêntica, Belo Horizonte

Chao A, Gotelli NJ et al (2014) Rarefaction and extrapolation with Hill numbers: a framework for sampling and estimation in species diversity studies. Ecol Monogr 84:45–67

Colwell RK (2013) EstimateS: Statistical estimation of species richness anda shared species from samples. Version 9.9. Online at https://www.purLoclc.org/estimates.

Crepaldi MOS, Peixoto AL (2010) Use and knowledge of plants by "Quilombolas" as subsidies for conservation efforts in an area of Atlantic Forest in Espírito Santo State, Brazil. Biodivers Conserv 19(1):37–60

Cunha MMC, Gondim RSD et al (2015) Plantas etnobotânico de plantas medicinais comercializadas em São Luiz, Maranhão, Brasil. Scientia Plena 11(12):1–12

Cupertino MC (2012) Juventude rural quilombola: identificação, reconhecimento e políticas públicas. Dissertação, Universidade Federal de Viçosa

da Silva ER (2013) Comunidade negra rural de Lagoa Santa: história, memória e luta pelo acesso e permanência na terra (1950–2011). Dissertation, Pontifícia Universidade Católica de São Paulo

da Silva NCB, Regis ACD et al (2012) Uso de plantas medicinais na comunidade quilombola da Barra II - Bahia, Brasil. Boletin Latinoamericano y del Caribe de Plantas Medicinales y Aromáticas 11(5):435–453

de Almeida ES, Miranda CAS (2015) História oral, comunidade quilombola e preservação da saúde: narrativas e rememoração. In: X Encontro Regional Nordeste de História Oral. História Oral, Educação e Mídias, 10 August to 13 August. Salvador, Bahia, Brazil

de Castro EG (2005) Entre Ficar e Sair: uma etnografia da construção social da categoria jovem rural. Dissertation, Universidade Federal do Rio de Janeiro

de Nunes RO (2016) Prospecção etnofarmacológica de plantas medicinais utilizadas pela população remanescente de quilombolas de Rolim de Moura do Guaporé, Rondônia, Brasil. Dissertation, Universidade Federal de Viçosa, Minas Gerais

de Oliveira LH (2006) Nas malhas da incerteza: Comportamento e estratégias camponesas na freguesia de Guarapiranga (1750–1820). Dissertation, Universidade Federal de Juiz de Fora, Juiz de Fora

de Siqueira AM (2014) Etnofarmacologia na comunidade quilombola São Sebastião da Boa Vista, Município de Santos Dumont/MG. Dissertation, Universidade Federal de Juiz de Fora, Juiz de Fora

Depommier D (2003) The tree behind the forest: ecological and economic importance of traditional agroforestry systems and multiple uses of trees in India. Trop Ecol 44(1):63–71

Di Stasi LC, Hiruma-lima CA (2002) Plantas Medicinais na Amazônia e na Mata Atlântica, 3rd edn. UNESP, São Paulo

do Oliveira OFV, da Gondim MJC (2013) Plantas medicinais utilizadas pela população de Caldas Novas, GO e o conhecimento popular sobre a faveira (*Dimorphandra mollis Benth Mimosoideae*). Revista Brasileira de Agroecologia 8(1):156–169

do Silva GAN (2014) Os laços da escravidão: população, reprodução natural e família escrava em uma vila mineira. Piranga, 1850–1888. Dissertação, Universidade Federal de São João del-Rei

dos Lisboa MS, Pinto AS et al (2017) Estudo etnobotânico em comunidade quilombola Salamina/Putumujú em Maragogipe, Bahia. Revista Fitos 11(1):48–61

Ferrão BH, Oliveira HB et al (2014) Importância do conhecimento tradicional no uso de plantas medicinais em Buritis, MG, Brasil. Ciência e Natura 36(1):321–334

Ferreira FMC, de Lourenço FJC, Balizada DP (2014) Levantamento etnobotânico de plantas medicinais na comunidade quilombola, Carreiros, Mercês – Minas Gerais. Revista Verde de Agroecologia e Desenvolvimento Sustentável 9(3):205–212

Ferreira LB, Rodrigues MO et al (2016) Etnobotânica das Plantas Medicinais Cultivadas nos Quintais do Bairro de Algodoal em Abaetetuba/PA. Revista Fitos 10(3):220–372

Flor ASSO, Barbosa WLR (2015) Sabedoria popular no uso de plantas medicinais pelos moradores do bairro do sossego no distrito de Marudá – PA. Revista Brasileira de Plantas Medicinais 17(4):757–768

Forzza RC et al (2010) Lista de Espécies da Flora do Brasil. Jardim Botânico do Rio de Janeiro

Freitas AG, Costa V et al (2012) A. Atividade antiestafilocócica do *Plantago major* L. Rev Bras 12(1):64–65

Gomes TB, de Bandeira FPSF (2012) Uso e diversidade de plantas medicinais em uma comunidade quilombola no Raso da Catarina, Bahia. Acta Bot Bras 26(4):794–806

Hsieh TC, Ma KH, Chao A (2016) iNEXT: an R package for interpolation and extrapolation of species diversity (Hill numbers). To appear in Methods in Ecology and Evolution.

IBGE (2018) Cidades. http://www.ibge.gov.br. Accessed 16 May 2021

Ki-Zerbo J (ed) (2010) História Geral da África I. UNESCO, Brasília

Leite IB (2000) Os quilombos no Brasil: questões conceituais e normativas. Etnográfica IV(2):333–354

Linhares JFP (2015) Uso e conservação de plantas medicinais nativas por comunidades quilombolas no município de Alcântara, Maranhão. Dissertação. Universidade Estadual Paulista Júlio de Mesquita Filho

Liporacci SN, Simão DG (2013) Levantamento etnobotânico de plantas medicinais nos quintais do Bairro Novo Horizonte, Ituiutaba, MG. Revista Brasileira de Plantas Medicinais 15(4):529–540

Liporacci HSN, Hanazaki N, Ritter MR, Araújo, E. de L (2017) Where are the Brazilian ethnobotanical studies in the Atlantic Forest and Caatinga? Rodriguésia. 68(4):225–240.

Machado TT, Dória KMAVSD (2017) Principais ervas medicinais utilizadas nos Quilombos do Camburi e da Caçandoca, Ubatuba – SP. Unisanta Bioscience 6(2):145–152

Machado AT, Santilli J (2008) Agrobiodiversidade com enfoque agroecológico: implicações conceituais e jurídicas. Embrapa Informação Tecnológica; Embrapa- Secretaria de Gestão e Estratégia, Brasília

Massarotto NP (2009) Diversidade e uso de plantas medicinais por comunidades quilombolas Kalunga e Urbanas, no Nordeste do Estado de Góias-GO, Brasil. Dissertação, Universidade de Brasília

Merétika AHC, Peroni N et al (2010) Local knowledge of medicinal plants in three artisanal fishing communities (Itapoá, Southern Brazil) according to gender, age, and urbanization. Acta Bot Bras 24(2):386–394

Messias MCTB, Menegatto MF et al (2015) Uso popular de plantas medicinais e perfil socioeconômico dos usuários: um estudo em área urbana em Ouro Preto, MG, Brasil. Revista Brasileira de Plantas Medicinais 17(1):76–104

Mota RS, Dias HM (2012) Quilombolas e recursos florestais medicinais no sul da Bahia. Interações (Campo Grande) 13(2):151–159

Neto FRG, Almeida GSSA et al (2014) Estudo Etnobotânico de plantas medicinais utilizadas pela Comunidade do Sisal no município de Catu, Bahia, Brasil. Revista Brasileira de Plantas Medicinais 16(4):856–865

Oliveira ER, Menini L (2012) Levantamento etnobotânico de plantas medicinais utilizadas pelos moradores do povoado de Manejo, Lima Duarte – MG. Revista Brasileira de Plantas Medicinais 14(2):311–320

Palheta IC (2015) Quintais urbanos e plantas medicinais: um estudo etnobotânico no bairro São Sebastião, Abaetetuba-PA. Dissertation, Universidade do Estado de Pará

Pinto EPP, Amorozo MCM et al (2006) A. Conhecimento popular sobre plantas medicinais em comunidades rurais de mata atlântica - Itacaré, BA, Brasil. Acta Bot Bras 20(4):751–762

Rodrigues AP, Andrade LHC (2014) Levantamento etnobotânico das plantas medicinais utilizadas pela comunidade de Inhamã, Pernambuco, Nordeste do Brasil. Revista Brasileira de Plantas Medicinais 16(3):721–730

Rodrigues AC, Villa PM et al (2019) Fine-scale habitat differentiation shapes the composition, structure and aboveground biomass but not species richness of tropical Atlantic forest. J For Res 4:1–13

Rogério ITS (2014) Levantamento etnofarmacológico de plantas medicinais na comunidade quilombola de São Bento, Santos Dumont, Minas Gerais. Dissertation, Universidade Federal de Juiz de Fora

Rothenburg WC (2008) Direitos dos Descendentes de Escravos (Remanescentes das Comunidades de Quilombos). Revista Internacional de Direito e Cidadania 2:189–206

Santos A (2012) O sabor da história: práticas alimentares e identidade quilombola. Intratextos. Special Number 3:54–71

Schiffler MF, Nascimento JL (2012) Cantares Ticumbis: resistência e memória na literatura oral de comunidades quilombolas. In: I Congresso Nacional Africanidades e Brasilidades, 26 to 29 June. Vitória. Espírito Santo, Brazil

Silva CSP, Proença CEB (2008) Uso e disponibilidade de recursos medicinais no município de Ouro Verde de Goiás, GO, Brasil. Acta Bot Bras 22(2):481–492

Tefera BN, Kim YD (2019) Ethnobotanical study of medicinal plants used as antimalarial and repellent by Sidama people of Hawassa Zuria district, southern Ethiopia. J Complement Med Res 10(1):13–26

Veloso HP, Rangel Filho L et al (1991) Classificação da vegetação brasileira, adaptada a um sistema universal. Departamento de Recursos Naturais e Estudos Ambientais, http://www.ibge.gov.br. Accessed 16 May 2021

Venâncio RP (1997) Os Últimos Carijós: Escravidão Indígena em Minas Gerais: 1711–1725. Revista Brasileira de História 17(34):165–181

Vendruscolo GS, Mentz LA (2006) Levantamento etnobotânico das plantas utilizadas como medicinais por moradores do bairro Ponta Grossa, Porto Alegre, Rio Grande do Sul, Brasil. Iheringia 61:83–103

Viertler RB (2002) Métodos antropológicos como ferramenta para estudos em etnobiologia e etnoecologia. In: Amorozo MCM et al (eds) Métodos na coleta e análise de dados em etnobiologia, etnoecologia e disciplinas correlatas. UNESP, Rio Claro, pp 11–29

Vinagre C, Vinagre S et al (2019) The use of medicinal plants by the population from the Protected Landscape of "Serra de Montejunto", Portugal. J Ethnobiol Ethnomed 15(1):1–30

Viu AFM, Viu MAO et al (2010) Etnobotânica: uma questão de gênero? Revista Brasileira de Agroecologia 5(1):138–147

Zank S, Ávila JVC et al (2016) Compreendendo a relação entre saúde do ambiente e saúde humana em comunidades Quilombolas de Santa Catarina. Revista Brasileira de Plantas Medicinais 18(1):157–161

Chapter 10
From *Mulungu* to *Mamulengo*: The Sharing of Knowledge Among Teachers, Academic Researchers, and *Mamulengueiros* (Traditional Puppeteers) in a Participatory Workshop

Maria Carolina Sotero, Ângelo Giuseppe Chaves Alves, and Maria Franco Trindade Medeiros

Abstract Ethnoscientists are ethically and politically committed to discussing the benefits of research carried out with local communities, including ways to share their results. One of the forms of sharing for such research is the participatory workshop. This chapter describes the authors' experience with such a workshop, from its conception to its evaluation. The workshop was held in Glória do Goitá, Pernambuco (Brazil), with the participation of 49 initial fundamental education teachers, 8 public administrators, 5 artists (traditional puppetry specialists), and 5 researchers. Data referring to the influence of the local artistic-cultural context (*mamulengo* art) on the connection between local and scientific knowledge in the classroom were shared. During the workshop, the artists explained the origins of *mamulengo* and its basic characteristics, while the researchers explained the morphological and physiological characteristics of the plant used to make the dolls used by the artists. Data obtained from interviews were presented in an accessible way. At the end of the activities, those present expressed thanks for the sharing of information from the research and for moments of dialogue about the local culture and its associated natural resource, as well as for the opportunity to strengthen ties between themselves and the university.

M. C. Sotero (✉)
Secretaria de Educação de Pernambuco, Escola Várzea Fria, São Lourenço da Mata, PE, Brazil
e-mail: mcarolsotero@gmail.com; carol.sotero@yahoo.com.br

Â. G. C. Alves
Departamento de Ecologia, Universidade Federal Rural de Pernambuco, Recife, PE, Brazil

M. F. T. Medeiros
Departamento de Botânica, Laboratório Interativo em Etnobotânica (LinE), Museu Nacional, Universidade Federal do Rio de Janeiro (MN/UFRJ), Rio de Janeiro, RJ, Brazil

M. F. T. Medeiros, B. de Sá Haiad (eds.), *Aspects of Brazilian Floristic Diversity*,
https://doi.org/10.1007/978-3-031-07453-0_10

Keywords *Erythrina* · Ethnoscience · Local knowledge · Scientific knowledge · Basic education

10.1 Participatory Workshop as a Sharing Strategy: Ethical and Political Pact

The object of study of most researchers in the field of ethnosciences is local or traditional ecological knowledge. This knowledge comes from experiences lived through beliefs, perceptions, and traditions and from knowledge constructed and legitimized throughout history by human groups in relation to natural resources (Albuquerque and Alves 2014; Medeiros and Albuquerque 2012). Therefore, to access their objects of study, ethnoscientists must, directly or indirectly, be in contact with the holders of this knowledge – people.

This meeting between academic researchers and their local field collaborators or participants may be seen also as a meeting between different cultures and expectations, between the *emic* and the *etic* dimensions of the phenomena being studied. This asymmetry within in the nature of the investigative process, as treated by Kloppenburg Jr. (1991), has allowed some experiences to be guided by the misconduct of researchers while working with local communities over the years.

To avoid possible harm as a result of this interaction, research procedures must comply with legal guidelines contained in national and international guiding documents and be submitted to an analysis of their ethical principles in accordance with what is established in each country and in light of codes of ethics and international agreements.

Reflection on legal aspects in ethnobiological and ethnoecological research covers different issues about the protection of intellectual property rights related to traditional knowledge associated with biodiversity, including the sharing of possible benefits generated from research such as by sharing or the so-called "return" activities with communities (Posey 1990; Albuquerque et al. 2010; Soldati and Albuquerque 2016).

Ethnobiological and ethnoecological research must, therefore, follow these national and international ethical and legal principles aimed at respecting people who have knowledge in relation to biodiversity (Soldati and Albuquerque 2016; Zank et al. 2019).

Among the ethical and legal frameworks generated from discussions about good research practices and mechanisms for protecting communities' intellectual property rights, the International Society of Ethnobiology Code of Ethics (2006) plays a central role in the search of ethical commitment. The code establishes 17 principles that guide the conduct of researchers when carrying out their research with indigenous, traditional, and local communities.

The principles of "active participation," "full disclosure," and "reciprocity, mutual benefit, and equitable sharing" address, among other aspects, the prior analysis of results before dissemination through the communities, the dissemination and application of results in an adapted manner for these populations, and the enjoyment of those involved in any possible benefits arising from the investigation, respectively (International Society of Ethnobiology 2006).

In addition to the need to comply with these legal obligations regarding the ethical aspects of research, ethnobiologists and ethnoecologists also establish a "pact" to share the results with the studied communities, and this can be done by means of sharing or return actions (Albuquerque et al. 2014). It is important for every researcher to understand the sharing of benefits arising from research as not only as a legal obligation, but as an ethical and moral presupposition (Albuquerque et al. 2010).

Sharing activities are understood as ethical and political activities constructed and thought out in a dialectical way between research participants who seek to contribute to local development, considering the problems and difficulties of the community and of collective interest (Albuquerque et al. 2014).

Participatory workshops can serve as sharing activities for the researched communities (Albuquerque et al. 2014), as well as a technique for collecting data from ethnobiological surveys that use participatory methods (Sieber and Albuquerque 2010).

Participatory workshops can be considered as a way of sharing benefits in a non-monetary way. In this way of sharing, it is possible to promote the training of local populations regarding the sustainable use of resources, the exchange of training teams, technology transfer, and increased local scientific capacity with the participation of local experts (Elisabetsky 2003 cited by Albuquerque et al. 2010).

Considering participatory workshops as a means of structuring the experience of the present authors during the performance of a sharing activity, this chapter will describe the experience from its conception and planning to its execution and evaluation of the achieved results.

The experience in question is the result of activities carried out by our research group in a small town in the countryside of Northeast Brazil. There is a unique cultural context in the place, with the main object of its artistic expression being made from a local plant resource.

The activity involved university researchers (the authors included), initial fundamental education (years 1–5) teachers, artists (local puppetry specialists), and public administrators.

First, a summary of the performed ethnobiological research will be presented, as well as a characterization of the place and its biocultural context. Some of the main findings of the research will also be presented, aiming to scale the importance of sharing these results for each social actor involved. Then, the step-by-step stages of elaboration and execution of the workshop will be presented, with impressions and observations of this experience.

This chapter aims to assist researchers in the construction and execution of a participatory workshop-type sharing activity, as well as demonstrate its relevance as a technique for sharing and reviewing collected information – the opportunity for connection between social actors and dissemination and enhancement of traditional or local knowledge.

10.2 From *Mulungu* to *Mamulengo*: Knowing the Artistic-Cultural Context of the Research Site

The ethnobiological research that generated the results shared in the workshop was part of the doctoral thesis of author Maria Carolina Sotero, under the supervision of Dr. Maria Medeiros and Dr. Ângelo Alves.

This thesis aimed to analyze the articulations between local and scientific knowledge in the school environment, in the teaching↔learning process, as well as the influence of a specific artistic-cultural context on these articulations as reflected in the daily practice of local teachers.

For the development of this research, it was necessary to find a place that offered basic education to its population, in a broad way and under the influence of a striking artistic-cultural context. These characteristics were found in the municipality of Glória do Goitá, state of Pernambuco, Brazil.

Glória do Goitá encompasses an area of 231,831 km^2, 58.8 km from the state capital (Agência Estadual de Planejamento e Pesquisa de Pernambuco 2014). Founded in 1878, Glória do Goitá has typical characteristics of a "município" in the Brazilian interior, with a small population, estimated at 30,504 inhabitants in 2018, and an economy mainly focused on agriculture (IBGE 2019).

The rural area of the municipality makes up most of its territory, comprising of several communities, each with its own peculiarities. These areas surround the tiny city center, which is made up of public administration buildings, small shops, an open market, old houses that still preserve traces of colonial architecture, and new, simpler housing on the outskirts.

Right in the heart of the city center, there is a beautiful, ancient, and imposing building by local standards. It is the former flour market, now transformed into a museum dedicated to the main artistic-cultural activity in the city: the Museu do Mamulengo (Mamulengo Museum) (Fig. 10.1a).

But after all, what is *mamulengo*?

Mamulengo is the name given to a type of popular dramatic representation, which also gives its name to the wooden puppets used to represent it (Borba-Filho 1966) (Fig. 10.1b).

An artist who performs this dramatization, and who also makes the puppets, is called a *mamulengueiro*. To present their show, a *mamulengueiro* sets up a tent outdoors or in a room inside and manipulates their puppets to represent stories, mostly improvised, representing local people and institutions in a rather sarcastic tone, using music and dance (Borba-Filho 1966) (Fig. 10.1c).

Fig. 10.1 (**a**) Museu do Mamulengo and headquarters of the Associação Cultural dos Mamulengueiros e Artesãos de Glória do Goitá, Pernambuco, Brazil. (**b**) *Mamulengo* dolls made of *mulungu* wood (*Erythrina velutina* Willd.) with their extravagant ornaments. (**c**) *Mamulengueiro*, inside the museum, exhibiting his puppets and *mulungu* wood in the background for producing new pieces. (**d**) A member of the local mamulengueiros' association carving *mulungu* wood to produce puppets

The Mamulengo Museum of Glória do Goitá is administered by a nongovernmental organization called Associação Cultural dos Mamulengueiros e Artesãos de Glória do Goitá (Cultural Association of the Mamulengueiros and Craftspeople of Glória do Goitá). The Association was founded in 2003. Its main objective is to foster the *mamulengo* culture, through the training of young and adult people in the art of making and handling puppets. It also organizes puppetry performances and encourages local puppeteers to participate in fairs and cultural exhibitions promoted by other organizations (Associação Cultural dos Mamulengueiros e Artesãos de Glória do Goitá 2003).

There, the artists sculpt, exhibit, and sell their pieces, receive visitors, and offer *mamulengo* workshops for the general public.

In addition to the peculiarity of the fact that *mamulengo* portrays local reality, this artistic-cultural expression also interacts with the socioecological system of the place through the use and manipulation of a specific natural resource for making the puppets.

Mamulengo puppets are made from the wood of a typical tree species of the region, *Erythrina velutina* Willd., locally called *mulungu*, and are dressed in calico clothes and extravagant accessories (Santana and Lucena-Filho 2012) (Fig. 10.1d).

Mulungu (*E. velutina*) belongs to the family Fabaceae. The etymology of the Latin name alludes to the color and texture of its flowers, with the term "erythrina" (from the Greek, *erythros*) meaning "red" and "velutina" (from Latin) referring to velvet (Carvalho 2008).

The species has an arboreal habit and deciduous behavior and is endowed with thorns. It can reach up to 15 m in height and 80 cm diameter at breast height (DBH) when mature (Fig. 10.2a). It has porous, soft wood with low natural durability (Figure 10.2b, c). Its flowers are hermaphroditic and visited by European or Africanized bees (*Apis mellifera*) and by carpenter bees (*Xylocopa* spp.), with anemochoric and zoochoric pollination, mainly by birds (Carvalho 2008).

The leaves of *mulungu* are trifoliate, with alternate phyllotaxis. The inflorescences measure 12–20 cm in length, the flowers vary from orange to red, and the fruits are legume-like, containing one to three seeds, which are dark red and red-orange in color (Fig. 10.2d) (Carvalho 2008).

Erythrina velutina occurs naturally in the Northeast (states of Alagoas, Bahia, Ceará, Paraíba, Pernambuco, Piauí, Rio Grande do Norte, Sergipe) and Southeast (state of Minas Gerais) regions of Brazil (Martins 2021). It is a pioneer, occurring preferably in secondary formations, with very irregular and discontinuous distribution (Carvalho 2008).

Mulungu is not listed as a species in danger of extinction (Martinelli and Moraes 2013). Species of the genus *Erythrina* are used in folk medicine for tranquilizing, sedation, insomnia control, and inflammation treatment (Silva et al. 2013) (Fig. 10.2b, c).

10.3 Teachers, Managers, and *Mamulengueiros*: Getting to Know the Social Actors Involved in the Workshop

The first local people with whom the authors had contact in the field were the artists at the Mamulengo Museum. Unstructured interviews with these puppeteers made it possible to understand the scenario and the importance of the local culture of *mamulengo* in that municipality, as well as its influence on the life of the population.

Understanding the biocultural context from the cosmovisions of *mamulengueiros* was essential to assist the authors in structuring the ethnobiological research.

Mamulengueiros were the local experts, as they knew the main natural resource (*mulungu*) and its management well. They had been experiencing the art of

Fig. 10.2 (**a**) A *mulungu* tree (*Erythrina velutina* Willd.); (**b**) *mulungu* bark sold at the municipal market in Glória do Goitá; (**c**) trunk of *mulungu* with evidence of cutting for the use of bark; (**d**) *mulungu* seeds

mamulengo on a daily basis through generations, thus embodying its history and relationship with the people of Glória do Goitá in present time.

After knowing the artistic-cultural context in a macro way, through the eyes of the *mamulengueiros*, it was necessary to understand its influence (if any) on education in the municipality. Thus, contacts were established with municipal managers: coordinators and technicians from the local Secretariat of Education.

During these contacts, it was possible to understand the structure of education in the municipality and present the research objectives and its potential to contribute to municipal education, as well as discuss strategies for fieldwork.

Finally, contact was made with the teachers of the city, people who were in fact the focus of data collection, as the authors of this research sought to understand the influence of the artistic-cultural context on the practice of these professionals.

All 28 public schools administered by the municipality that were offering the initial fundamental education level (years 1–5), aimed at children between 6 and 12 years of age, were visited. These schools were in both urban and rural areas.

Teachers were invited to participate in a semi-structured interview, with audio and photographic recording, when authorized. Almost all teachers at this level of education in the city were interviewed (97.5%, $n = 79$), with only two being left out of the sample.

The interview script contained questions about the socioeconomic profile of the teachers, their knowledge about the artistic-cultural context of *mamulengo*, and the use of the theme in their classes.

To fulfill the research objectives, it was necessary to investigate whether knowledge and use of the theme of *mamulengo* in classes influenced the teachers in the way they welcomed, used, or addressed local knowledge in general (about plants, folk medicine, animals, myths, rituals, etc.).

To understand the relationship of teachers with local knowledge in general, a question involving a hypothetical situation was employed. In this way, teachers were asked what attitude they would have if, during a class, a student expressed local knowledge associated with flora.

10.4 What Knowledge Did We Want to Share?

After systematization, data analysis, and discussion involving the specialized literature, we had relevant results to be shared with the people participating in the research, presented briefly below.

For the teachers, it was essential to have a dialogue with them about the possible influence of the artistic-cultural context on their actions aimed at the connection between knowledges.

We found that the socioeconomic profile of the teachers and the artistic-cultural context of *mamulengo* seemed, in fact, to influence the knowledge and application of this context in the classroom, as well as the actions of the teachers regarding local knowledge in general.

In general, the teachers who most knew and applied knowledge about *mamulengo* were those who most positively reacted to the hypothetical question we asked them, indicating the possibility of carrying out actions to connect local and scientific knowledge. These teachers stated that they would value the student's speech, complement it with more information, and make connections with the scientific content they were addressing.

For the *mamulengueiros*, it was important to inform them that their actions to disseminate and maintain this cultural heritage could reach schools, teachers, and students. And there was an indication that the knowledge of, and interaction with, the artistic-cultural context they disseminated seemed to make teachers more likely to connect local and scientific knowledge, thus contributing to the teaching-learning process.

For the municipal managers, it was interesting to inform them about the influence of socioeconomic factors (age, level of education, length of service) and labor (classroom and school structure) on the possibility of teachers to know and apply information on the context of *mamulengo* in the classroom, consequently also establishing connections between local and scientific knowledge.

The results showed that the teachers who most knew and used the topic of *mamulengo* in their classes, and consequently who made more connections between knowledges, were those who had a higher degree in formal education, with greater length of service, and who worked in more structured schools.

Based on this information, the Secretariat of Education of the municipality could devise strategies aimed at reducing the disparity in characteristics among groups of teachers, such as updating and training programs, teacher development, and improving the physical structure of schools.

Finally, we also wanted to show municipal managers the relevance of the artistic-cultural context to education in the municipality, encouraging them to adopt public policies that allow the continuity of local puppetry artists' activities in the municipality and specifically in actions with education.

10.5 Sharing Workshop: Exchanging Knowledge and Cosmovisions

The sharing workshop had already been planned as a sharing or return-to-community activity since the construction of the research project. In addition to the ethical and political commitment, previous research experiences by our research group showed that this activity should be carried out.

The planning of the workshop took place prior to the interviews with teachers in the municipality, in meetings between the first author Maria Carolina Sotero, her advisors Dr. Maria Medeiros and Dr. Ângelo Alves, and other members of the Ethnoecology Study Group (GEE) of the Federal Rural University of Pernambuco (UFRPE). On these occasions, possible dates, format, event schedule and logistics, and organization were discussed.

The dates and location were eventually determined in agreement with the local Secretariat of Education, and, following the principle of mutual benefit, the workshop became part of the official teacher training calendar, with dates already pre-established by the municipal administration.

To facilitate the mobilization of teachers to participate in the workshop, an invitation was made at the end of each data collection interview. This strategy made it unnecessary to return to schools, which were often quite distant, and gave personalization to the invitation.

Shortly prior to the workshop, the teachers received a virtual invitation via message using the *WhatsApp* application. As the workshop was a partnership with the Secretariat of Education, teachers were also encouraged to participate through an invitation from the municipality itself.

According to Patzlaff and Peixoto (2009), it is difficult to establish a priori what form of sharing or return is suitable for each community. However, since most of the participants in the study were teachers, who are already used to the participative workshop format, we believed that this form of sharing strategy would be the most appropriate to put forward.

A total of 67 participants were present in the participatory workshop, 49 of whom were teachers from the municipality (21 urban and 28 rural), 7 technicians and the secretary of the Secretariat of Education, 5 artists from the Mamulengo Museum, and 5 researchers from GEE-UFRPE.

Of all the teachers previously interviewed ($n = 79$), 62% attended the workshop. Considering the limited time available for face-to-face interaction with the teachers (only during the interview) and the absence that sometimes occur in teacher education programs, we considered the workshop to have had good adherence.

Difficulty in mobilization and participation in a workshop-type sharing activity was reported by Marques et al. (2010). According to this author, the low workshop participation by inhabitants of the studied riverside community was due to a possible lack of identification of informants with the research objectives or with the chosen form of return. We believe that these factors did not have a significant impact on participation in our study.

The fact that the main participants were teachers brought an advantage to the objective of disseminating and applying the research results. Teachers are sharers by definition, which makes it possible for the sharing action to become efficient in reaching a greater number of people, with rapid dissemination of information (Albuquerque et al. 2014). In this way, the knowledge shared with them during the workshop has the potential to be passed on to their peers and their students (not only for the current academic year, but throughout their entire professional life).

The workshop took place in the building of the Secretariat of Education of Glória do Goitá, on September 4, 2019 (with urban teachers), and on November 11, 2019 (with rural teachers).

Upon arriving at the workshop, the participants were welcomed by the GEE-UFRPE team and invited to sign the attendance list, after which they received an event t-shirt. A breakfast was also offered to better welcome and thank the participation of those present.

Author Maria Carolina Sotero opened the workshop by thanking everyone for attending and presenting the workshop schedule. The person who was in charge of the Secretariat of Education, Maria de Fátima Santana, also thanked the municipality's teachers for joining the activity and the partnership with the Mamulengo Museum and UFRPE.

The artists of the Mamulengo Museum were invited to explain the art of the *mamulengo* to the audience. The artists' speech encompassed the origin of *mamulengo* culture in the region, its historical master artists, its basic characteristics, the way in which the puppets were made, the history of the local Association of Mamulengueiros, and the challenges they face in maintaining their activities (Fig. 10.3).

Mamulengueiros also had the opportunity to present participants with a sample of a *mamulengo* show. The artists set up a tent and, cradling musical instruments, improvised a short performance of some typical characters that they normally use to present (Fig. 10.4).

These moments in the workshop were opportunities to bring the *mamulengueiros*, teachers, academic researchers, and public administration closer together.

The artists were able to demystify some prejudices of the teachers in relation to the art of *mamulengo*, arising from the fact that the presentations were originally aimed only at an adult audience and always included scenes of murders, profanity, insults, and texts with sexual, sexist, and racist insinuations. Some local teachers who had experienced *mamulengo* performances in the past recognized that they previously associated that kind of traditional show with slang and scenes inappropriate for children.

Fig. 10.3 Explanations by members of the Association of Mamulengueiros during the workshop "From *mulungu* to *mamulengo*: the teacher's role in connecting knowledges" in Glória do Goitá, Pernambuco

Fig. 10.4 Presentation by *mamulengueiros* during one of the workshop sessions of "From *mulungu* to *mamulengo*: the teacher's role in connecting knowledges" in Glória do Goitá, Pernambuco

The *mamulengueiros* explained that there are different moments in the presentation of the original *mamulengo*, with parts aimed at children and families and others at a more adult audience. The artists also informed that, currently, they are making adaptations, with the aim of bringing *mamulengo* closer to more people by changing scenes, texts, gestures, and language according to the target audience.

During the workshop, the *mamulengueiros* also took the opportunity to publicize the cultural projects developed by the Association of Mamulengueiros at the Mamulengo Museum and invited the teachers to experience these activities there together with their students.

After everyone got involved in the theme of the artistic-cultural context of *mamulengo*, it was time to present participants with information about the natural resource used in making the puppets – *mulungu* (*E. velutina*) wood.

The presentation about *mulungu* was under the responsibility of Sofia Moura, a biological sciences undergraduate student. For the scientific content to be understood by all participants, the student prepared the presentation in a more accessible language, striving for the care that a researcher must take regarding the proper adaptation of the content and methodology used in the research, taking into account the previous formal education experience, way of life, customs, and beliefs of the participants (Albuquerque et al. 2014). Respecting the needs of this adaptation, information was presented regarding the geographic distribution, morphological and physiological characteristics, and forms of use found in the scientific literature about the plant.

On these occasions during the workshop, the participants were very curious about the information presented, and it was exciting for us to experience the involvement and exchange of knowledge about the resource between academic researchers and local teachers, managers, and *mamulengueiros*.

Finally, author Maria Carolina Sotero presented, also with the scientific language adapted for everyone to understand, the preliminary results of her doctoral research in which the participants were involved.

First, information from the literature on the importance of connecting local and scientific knowledge in the teaching-learning process was presented. Then, our hypothesis about the influence of the artistic-cultural context (in this case the *mamulengo*) on the teachers' attitudes toward local knowledge was presented (Fig. 10.5).

The results of the research and the specific contributions for each type of participant (*mamulengueiros*, managers, academic researchers, and local teachers) were then discussed. Participants were able to observe the importance and influence of each one of them in the maintenance of the artistic-cultural context of *mamulengo* and, consequently, in the reception of related local knowledge in the classroom. Valuing the knowledge of field participants is an important element for a community's acceptance and recognition of academic researchers' own knowledge, as well as encouraging self-recognition of local participants as experts on a given topic by their peers and by the community itself (Patzlaff and Peixoto 2009; McAlvey et al. 2021).

At the end of the presentations, representatives of teachers, managers, academic researchers, and *mamulengueiros* expressed thanks for the opportunity of participating in the workshop and for sharing the results generated by the scientific research. We believe that this mutual gratitude among the parties involved in the research is the result of the authors' and other participants' commitment since the beginning of their involvement in the research project. These should be discussed in depth to shed light on the rights and duties of each person or institution involved in the investigation (Lin 2006). Even though this ethical awareness is widespread, Quinteiro et al. (2013) found that a specific strategy devised to the sharing of research results was not present in most works in the field of ethnobotany analyzed by them. According to these authors, the research, in general, brought proposals and suggestions for sharing or return activities, but they did not actually present the respective results, characteristics, applicability, and difficulties.

Fig. 10.5 Presentation of thesis results by the first author during one of the workshop "From *mulungu* to *mamulengo*: the teacher's role in connecting knowledges" in Glória do Goitá, Pernambuco

The teachers specifically gave thanks for the opportunity to get closer to *mamulengo* culture and the artists of the Mamulengo Museum and for the explanations about *mulungu* (*E. velutina*) and the importance of local knowledge in the teaching-learning process.

The *mamulengueiros* gave thanks for the opportunity to publicize their activities and for the authors' interest in studying the cultural context in the municipality, as well as for the information presented that demonstrated the importance of preserving and publicizing their art.

The managers of the Secretariat of Education gave thanks for the partnership with UFRPE and for the sharing of results that had the potential to assist in intervention actions to improve the quality of the education offered.

The satisfaction of those present for knowing and discussing the research results and, even more, our satisfaction in sharing them were notable.

10.6 Final Considerations

In this chapter, we intended to publicize our experience in designing, organizing, and carrying out an activity for sharing research results in the form of a participatory workshop. In our view, sharing or return activities do not necessarily need to be large, resource-intensive events. Academic researchers can, during the execution of their activities, enable several small "returns" to researched communities, such as for the preparation of folders and booklets (Medeiros and Andreata 2003; Giraldi and Hanazaki 2010), distribution of plant seedlings (Sátiro et al. 2019), construction of a garden of medicinal plants, collaborations in courses and lectures, construction of maps and English classes, and so on, as reported in the studies analyzed by Patzlaff and Peixoto (2009).

However, we believe that channeling energy and resources for such activities since the very beginning of the construction of the research project, is important to prioritize their realization.

We must remember that when we are "immersed" in our fields of research, datasheets, or readings, facing deadlines for articles, dissertations, and theses, we are prone to fail to fulfill some planned actions. At this point, the performance of sharing or return activities is threatened. It is always necessary to keep in mind the ethical and political commitment we have to the people who were involved in our research, prioritizing the carrying out of return activities that match the contribution of the participants to the research and to the scientific field as a whole.

Participatory workshops, in addition to performing their role as a sharing strategy, can provide new data for the researchers or for other, smaller research projects in their group. Workshops can also be instruments for training younger scientists, who can make presentations with the adaptation of scientific language or the reception of the participants. They also provide an opportunity to yet adjust some information from scientific research that may be incongruous.

During workshops, academic researchers have the opportunity to strengthen ties with the local community, opening doors for other activities, such as courses, field classes, new research, and partnerships for publications.

Participatory workshops are quick ways to transmit information to researched groups that can change the way participants interact with natural resources. Imagine that a researcher has the result that a certain plant species is more effective than another or that the form of management adopted by a local community is supposedly putting a certain species at risk. In this way, sharing results can lead to a strongly participative debate, and the connections between publishing, disseminating information, and generating local solutions can be strengthened.

We should stress that the academic researcher's role in a participatory workshop, with the function of sharing research results, is not that of a "behavior judge" that stares at a local community. The workshop must be a dialogic moment and not verticalized. In no way can it be directed toward a simplistic debate of the university vs. community or scientific knowledge vs. local knowledge.

Freire (1996) states that teaching requires availability for dialogue, with respect to the differences between educators and students. According to the author, teachers need to be confident about the issues they propose to discuss, avoiding the false assumption that they know everything or that they are superior, but with the conviction that they know something and that they ignore something, that everyone should be prone to continuous learning. We believe that this should also be the researcher's attitude toward workshop participants.

Acknowledgments We would like to thank the Programa de Pós-Graduação em Etnobiologia e Conservação da Natureza (PPGEtno-UFRPE); the Programa de Pós-Graduação em Ciências Biológicas (Botânica), UFRJ; the Departamento de Botânica do Museu Nacional, UFRJ; the Secretariat of Education from Glória do Goitá, especially teacher Fátima Santana and Joelma Farias; the Association of Mamulengueiros, especially Masters Titinha, Jaci, Bel, and Bila, musician Paulo dos Oito Baixos, and cultural producer Pablo Dantas; researchers Ezequiel Silva Júnior, Rayane Santos, and Sofia Moura; and the teachers who participated in the interviews and workshop, as well as their directors and coordinators who welcomed us into the schools. We would also like to register our gratitude and tribute (in memoriam) to Master Zé Lopes, for sharing with us his knowledge and joy of the art of *mamulengo*.

References

Agência Estadual de Planejamento e Pesquisa de Pernambuco (2014) Perfil Municipal Glória do Goitá

Albuquerque UP, Alves AGC (2014) O que é etnobiologia? In: Albuquerque UP, Alves RRN (eds) Introdução à etnobiologia. NUPEEA, Recife, pp 17–22

Albuquerque UP, Hanazaki N, Melo JG (2010) Problemas e perspectivas na publicação de trabalhos em etnobiologia. In: Albuquerque UP, Lucena RFP, Cunha LVFC (eds) Métodos e técnicas na pesquisa etnobiológica e etnoecológica. NUPEEA, Recife, pp 437–462

Albuquerque UP, Araújo TAS, Soldati GT, Fernandes LRRMV (2014) "Returning" ethnobiological research to the communities. In: Albuquerque UP, Lucena RFP, Cunha LVFC, Alves RRN (eds) Methods and techniques in ethnobiology and ethnoecology. Springer, New York, pp 451–463

Associação Cultural dos Mamulengueiros e Artesãos de Glória do Goitá (2003) Estatuto da Associação Cultural dos Mamulengueiros e Artesãos de Glória do Goitá. Glória do Goitá

Borba-Filho H (1966) Fisionomia e espírito do mamulengo: o teatro popular do Nordeste. Companhia Editora Nacional, São Paulo

Carvalho PER (2008) Mulungu (Erythrina velutina). Embrapa Florestas, Colombo, Circular Técnica 160

Elisabetsky E (2003) Direitos de propriedade intelectual e distribuição equitativa de benefícios no contexto de inovação tecnológica. In: Anais do I Simpósio de Etnobiologia e Etnoecologia da região Sul: aspectos humanos da biodiversidade. Sociedade Brasileira de Etnobiologia e Etnoecologia, Florianópolis, pp 170–174

Freire P (1996) Pedagogia da autonomia: saberes necessários à pratica educativa. Paz e Terra, São Paulo

Giraldi M, Hanazaki N (2010) Uso e conhecimento tradicional de plantas medicinais no Sertão do Ribeirão, Florianópolis, SC, Brasil. Acta Bot Bras 24(2):395–340

IBGE (2019) IBGE Cidades Glória do Goitá Panorama. https://cidades.ibge.gov.br/brasil/pe/gloria-do-goita/panorama. Accessed 26 Oct 2021

International Society of Ethnobiology (2006) ISE Code of Ethics (with 2008 additions). http://ethnobiology.net/code-of-ethics/. Accessed 26 Oct 2021

Kloppengurg J Jr (1991) No hunting. Cult Survival 15:14–18

Lin CM (2006) Plantas medicinais na Reserva Extrativista Chico Mendes (Acre): uma visão etnobotânica. Editora UNESP, São Paulo

Marques M, Lacerda V, Giraldi M et al (2010) Valorização do conhecimento etnobotânico dos moradores do Sertão do Ribeirão, Florianópolis/SC – Brasil. Extensio: R Eletr de Extensão Ano 7:47–58

Martinelli G, Moraes M (2013) Livro vermelho da flora do Brasil, 1st edn. Andrea Jakobsson: Instituto de Pesquisas Jardim Botânico do Rio de Janeiro, Rio de Janeiro

Martins M (2021) Erythrina. In: Flora do Brasil 2020. http://floradobrasil.jbrj.gov.br/reflora/floradobrasil/FB29679. Accessed 26 Oct 2021

McAlvay A, Armstrong C, Baker J et al (2021) Ethnobiology phase VI: decolonizing institutions, projects, and scholarship. J Ethnobiol 41:170–191. https://doi.org/10.2993/0278-0771-41.2.170

Medeiros MFT, Albuquerque UP (2012) Dicionário Brasileiro de Etnobiologia e Etnoecologia. NUPPEA, Recife

Medeiros MFT, Andreata RHP (2003) Cartilha sobre as plantas medicinais utilizadas pelos sitiantes da Reserva Particular do Patrimônio Natural Rio das Pedras, Mangaratiba, RJ. In: Siqueira JC (org) Educação ambiental: resgate de valores socio-ambientais do Município de Mangaratiba, RJ, PUC-Rio, Rio de Janeiro, pp 49–64

Patzlaff R, Peixoto A (2009) A pesquisa em etnobotânica e o retorno do conhecimento sistematizado à comunidade: um assunto complexo. Hist Ciênc Saúde Manguinhos 16:237–246. https://doi.org/10.1590/s0104-59702009000100014

Posey DA (1990) Intellectual property rights: what is the position of ethnobiology? J Ethnobiol 10(1):93–98

Quinteiro M, Tamashiro A, Moraes M (2013) Formas de retorno da pesquisa etnobotânica à comunidade no paradigma da complexidade ambiental e Educação Ambiental. Revista Brasileira de Educação Ambiental (RevBEA) 8:91–99. https://doi.org/10.34024/revbea.2013.v8.1803

Santana E, Lucena-Filho S (2012) O processo folkcomunicacional como estratégia para o desenvolvimento local: o caso da Associação de Mamulengueiros de Glória do Goitá, Pernambuco. Revista Internacional de Folkcomunicação 10:100–110

Sátiro LN, Vieira JH, Rocha DF (2019) Uso místico, mágico e medicinal de plantas nos rituais religiosos de Candomblé no agreste alagoano. Revista Ouricuri 9(2):45–61

Sieber S, Albuquerque UP (2010) Métodos participativos na pesquisa etnobiológica. In: Albuquerque UP, Lucena RFP, Cunha LVFC (eds) Métodos e Técnicas na Pesquisa Etnobiológica e Etnoecológica. NUPPEA, Recife, pp 85–106

Silva M, Santana A, Pimentel R et al (2013) Anatomy of leaf and stem of Erythrina velutina. Rev Bras 23:200–206. https://doi.org/10.1590/s0102-695x2013005000013

Soldati GT, Albuquerque UP (2016) Ethnobiology, ethics, and traditional knowledge protection. In: Albuquerque UP, Alves RRN (eds) Introduction to ethnobiology. Springer, Cham, pp 83–89

Zank S, Ludwinsky R, Blanco G, Hanazaki N (2019) Protocols and ethical considerations in ethnobiological research. In: Albuquerque UP, Lucena RFP, Cunha LVFC, Alves RRN (eds) Methods and techniques in ethnobiology and Ethnoecology. Springer, New York, pp 229–253

Chapter 11
Biocultural Heritage Through Museological Narrative as a Way of Return on Research in Historical Ethnobotany

Maria Franco Trindade Medeiros

Abstract This chapter develops a reflexive narrative of conceptual, experiential, and propositional structure. It will consider as a guiding axis, in a first take, the biocultural heritage and the space of the museum. Next, three distinct experiences of return in the field of ethnobotany will be reported, which is believed to be a living and driving force as a creative process of museums as well as within museums themselves. These experiences will lead us to the exhibition spaces created in the Benedictine Monasteries of the cities of Olinda (state of Pernambuco, Brazil) and Rio de Janeiro (state of Rio de Janeiro, Brazil), as well as video-lecture communication as part of the activities that integrated the participation of the National Museum/ Federal University of Rio de Janeiro during the 19th National Museum Week. In this way, it is intended that this placement of experience can be useful as a reference model for the establishment of the museological narrative as a means of return and thus sharing of scientific research with the public, in this case, the users of the museum. The exhibition will converge to a transnaturality that will associate archivist contents with elements of areas transversal to the sciences and arts.

Keywords Collective benefit · Data governance · Ethics · Museums · Responsibility

M. F. T. Medeiros (✉)
Departamento de Botânica, Laboratório Interativo em Etnobotânica (LinE), Museu Nacional, Universidade Federal do Rio de Janeiro (MN/UFRJ), Rio de Janeiro, RJ, Brazil
e-mail: mariaftm@mn.ufrj.br

M. F. T. Medeiros, B. de Sá Haiad (eds.), *Aspects of Brazilian Floristic Diversity*,
https://doi.org/10.1007/978-3-031-07453-0_11

11.1 Initial Reflections on Principles of Ethics in Ethnobotany Related to Museology

In the context of the retrospective view of the interaction that was established between the human person and plants, the historical ethnobotany works from the memory traces that refer to this interaction (Medeiros 2021a) which, like Prance et al. (2007) refer to, is an interaction of survival. From this perspective, ethnobotany makes use of different material sources for the development of a research such as textual, iconographic, and photographic documents (Medeiros 2009, 2010). This historical focus allows, therefore, the inclusion of multiple evidences, which makes research broadly interdisciplinary. Therefore, knowledge from areas such as botany, history, archival science, and museology is combined with each other.

One of the structural aspects of ethnobotany research focuses on the issue of return of the results achieved with the society in which the study was developed. Its existence integrates the thought around the principles that seek to point to what would be a good direction of researchers who praise the development of research with ethical and legal bases aimed at the commitment to the participating societies. This issue integrates a social and politic commitment in which will be discussed among researchers and different social and ethnical groups what would be the possible ways of return.

When we address this matter that involves a legal basis about governance of traditional knowledge and the requirement of formal agreements that associate this knowledge to the genetic resources, we face to its fundament in the legal and global institutional construction of the so-called Convention on Biological Diversity (CBD). This document was elaborated from the discussions that occurred in the occasion of Rio Earth Summit (United Nations Conference on Environment and Development – UNCED) in Rio de Janeiro (Brazil), and it was opened for signatures in 1992. The main objective of CBD was to regularize the access to genetic heritage and benefit-sharing provisions, in addition to expose "sustainable development" as a guiding. This legal milestone has as fundaments assumptions of the nature, development, and circulation of knowledge, both those considered traditional and those scientific. The most prominence consequence of this legal milestone was to redefine the relations among people and knowledge and, for that, apply the terms *holders* of traditional knowledge and *sovereignty* of traditional stages over genetic resources (Cunha 2009).

Integrating the principles of the Convention on Biological Diversity (1992) with those of the Code of Ethics of the International Society of Ethnobiology (2006), the premises of the participation of interlocutors in the process of composing results, access and analysis of the data dissemination material and its sharing, in an equitable manner, ensuring a circulation of ideas through dissemination adjusted to the desire of participants and the real possibilities of the researchers, and the sharing of benefits arising from the research, these appear as basic elements in the question of the return of research.

Going toward these premises, it is also important to mention the Declaration of *Belém*, created in 1988 on the occasion of the 1st International Congress of Ethnobiology, in the city of Belém (state of Pará, Brazil). This document brought in its lines the recognition of the importance of indigenous and non-indigenous traditional peoples, their hall of knowledge, and management practices in favor of conservation of biodiversity and natural resources, and, in addition to this central contribution of the Declaration, there was also the circumscription of the exercise of the ethnobiologist researchers (and ethnobotanists) in the dissemination process, return of the results of their research in the native language.

Dialoguing directly with these documents that highlight respect for different social and ethnic groups are also the *Tkarihwaié:ri* Code of Ethical Conduct (2011) and the Code of Ethics of the Latin American Society of Ethnobiology (2015). These actions are the result of a concern with respect for the cultural and intellectual heritage of these groups, in the same way that they also turn to the conservation and sustainable use of biodiversity. They bring in their scope the need to have an effective, consented, and approved participation by indigenous and local communities in investigations about their knowledge, resources, and territories.

Going forward in history, 30 years after the first congress held in 1988, which has already been commented above, in 2018, *Belém*+30 brought together the XVI Congress of the International Society of Ethnobiology; the XII Brazilian Symposium on Ethnobiology and Ethnoecology; the IX State Fair of Science, Technology and Innovation; and the 1st World Fair of the Sociobiodiversity, also in the city of Belém (state of Pará, Brazil). At this congress in 2018, the *Belém*+30 Declaration was presented, which brought an assessment of the challenges and achievements of the Declaration of *Belém* elaborated in 1988 and considered the perspective of different traditional peoples and communities, defending respect for their territories, freedom of management, and self-determination of their ways of life.

In addition to these documents mentioned, since the twentieth century other covenants, codes, declarations, laws, and agendas have also recognized the rights of social and ethnic groups, the importance of traditional knowledge for the conservation of natural resources, and biocultural heritage and were in defense of the equitable sharing of benefits. Among these actions we can also mention as an example the Universal Declaration on Human Rights (United Nations 1948), the Convention No.169 on Indigenous and Tribal Peoples (ILO 1989), the Convention for the Safeguarding of the Intangible Cultural Heritage (UNESCO 2003), the 2005 Convention on the Protection and Promotion of the Diversity of Cultural Expressions (UNESCO 2015), the United Nations Second International Decade of the World's Indigenous Peoples (2005–2014), the Universal Declaration on Cultural Diversity (UNESCO 2001), the Universal Declaration on Bioethics and Human Rights (UNESCO 2005a, b), the *Bonn* Guidelines on Access to Genetic Resources and Fair and Equitable Sharing of the Benefits Arising out of their Utilization (CBD 2002), the *Akwé: Kon* Guidelines (2004), the United Nations Declaration on the Rights of Indigenous Peoples (2007), and the *Nagoya* Protocol on Access to Genetic Resources and the Fair and Equitable Sharing of Benefits Arising from their Utilization (2010).

What is called into question when we think about the issue of commitment between researchers and their interlocutors, whether originating peoples or other different social and ethnic representations, what we propose to think about is the definition of the viable possibilities of return together with these people. Social and ethnic groups must integrate this process in an active and informed way, either by the information shared or by the possible profits generated from them, as Moreira (2005) adds.

In the case of historical ethnobotany, this commitment to think about the results and their dissemination among the communities participating in the research can be developed through actions of return that prioritize social actors, their epistemologies found in memory traces, and respecting their particular cultural contexts.

Given the multifaceted characteristic of ethnobotany, essentially historical, combined with the imperative need to establish the return activity in these surveys, the following question was posed: How then to integrate museology into the process of returning research into historical ethnobotany?

From this inquiry, this reflection will consider in the foreground the biocultural heritage and the museum space. Next, we will report return experiences in the field of the historical ethnobotany, aiming that they can serve as a reference model for the establishment of the museological narrative as a means of return, as well as for the sharing of scientific research with the visiting public that effectively constitutes the user of the museum.

11.2 Some Considerations About Cultural Heritage and Museums

Understanding people-nature interactions in the progressive discovery of Brazilian biodiversity is not easy. It walks between writings, Nankin ink illustrations, traditions, oils on canvas, watercolors, landscapes, sculptures, stories, objects of all sorts of varied ethnicities, memories of plants, and objects.

In the ambit of the symbolic and the imaginary, the memory can be accessed in the materiality of the registers and the visibility of the images. Thus, several images and registers become documents of social memory when recalled in the collective plan.

In the set of information that reveals the meanings of memory, biocultural heritage, material, or virtual objects constitute the elements that will mediate social relations and actions.

In this search for past know-how that transcends the times, we are in the face of the situation that trying to identify memory through biocultural heritage is not guarantee that we will access it completely. What we find are the vestiges, the traces of memory printed on the biocultural heritage that may be sheltered, for example, in the places of physical or virtual memory.

In these places we find objects, vestiges, and testimonies, which can be, for example, books, documents, or a variety of objects that can represent an individual, places, phenomena, a class, a family of objects, or objects that are not perceptible to the human person or that do not allow their displacement. These objects are supports that will potentially lead us to some remembrance, some feeling, a fact, a particular person, an ethnic group, and so on.

Among the places of memory that safeguard these traces, the museum, which could be understood as a space of "concentrated reality" (Wagensberg 2005), can shelter in its collections a multiplicity of objects of varied natures and typologies. These objects can evoke reflections and questions through narrative strategies and also have the potential to confer materiality and visibility to realities that sometimes are configurated as inaccessible to human perception.

Considering that the objects in a museum are parted from their primary context, and therefore consequently from their original reality, when transferred to the museum space, these objects are inserted into another reality. In this new ambience, the intention is that from the objects it is possible to document their original reality from which they were displaced (Mensch 1992).

Thus, it is emphasized in this way that the master line of the musealization of objects is the process that transforms it into a document. This musealization would introduce elements of the original context of objects, such as references related to spatiality, temporality, and significance, in a contemporary reality to the user public, the exhibition, and the museum (Meneses, 1992).

Through museological narratives an order and classifications of the objects are proposed. These objects represent the fragments of the universe of material culture, and this organization sends us to a broader reality full of complexifications. What is being placed is that museums, when turning to objects, lead us to the proposition of a system of ordering and narrative that is endowed with an internal coherence that gathers the fragments of reality and recontextualizes them. From this process, it will be in museums that we will have the possibility to visualize the realities of these objects and immerse ourselves in a greater context of their integrity from their recontextualization in this new space, that of exhibitions, collections, or technical reserves.

Within the scope of ethnobotany, these vestiges of biocultural heritage, when put under analysis, can lead us to reflection and bring us information about the meaning and historical and contextualized application of plant species in different societies, as we will discuss below.

11.3 Taking as a Model of Return the Practices of Exhibition in the Museum: Physical and Virtual Environments

Objects that comprised the daily life of past societies can also serve as starting point for the establishment of studies and reveal important questions about the meaning and use of Brazilian vegetation in past centuries through the analysis of these objects, that is, through the detailed analysis of our biocultural heritage.

We can find this type of register dealing with the plants used as raw material during the nineteenth century in the hospital of Saint Benedict located in the Monastery of Saint Benedict of Olinda (state of Pernambuco, Brazil) and in the *botica* (pharmacy) and infirmary of the Monastery of Saint Benedict of Rio de Janeiro (state of Rio de Janeiro, Brazil).

The actuation of this hospital, *botica*, and infirmary of Saint Benedict, which operated on the premises of the Benedictine Monasteries of Olinda and Rio de Janeiro, along the nineteenth century, was related to health and disease issues, taking care of the religious community itself, slaves, and the lay people who resorted to this place, through the medicinal use of plant species and animals (Alencar et al. 2010; Medeiros 2007; Medeiros et al. 2011; Medeiros and Albuquerque 2012; Medeiros and Alves 2020).

Saint Benedict, Italian, founder of the Benedictine Order in the fifth century, in his rule of life left to the religious who entered the monasteries as well as to those who also currently enter the monasteries the so-called Rule of Saint Benedict, where he says in Chapter 36 on the care that monks must have for the sick (The Rule of Saint Benedict 2003). Carrying out this activity, the Benedictine monks of Olinda and Rio de Janeiro put into practice in the *botica* and infirmary the guidance of their founding father, Saint Benedict.

Considering the informational set formed by the collection of recipes of these *boticas* and infirmaries, medicinal plants, as well as the Benedictine actuation in Brazil and therapeutic practices of the nineteenth century, our motivation for the organization of exhibitions pointed to some questions: Is it possible an ethnobotany that situates people in time and space? If we agree that yes, that is possible, does memory articulate itself with this perspective? Being that, would not the museological narrative be the great vehicle that approximates memory, biocultural heritage, and ethnobotany?

All these reflections lead us to a discussion about the meaning of memory. In the exercise of thinking about memory, what guide our thoughts is to perceive how the relations between memory, biocultural heritage, and ethnobotany can be carried out by the museological narrative.

We think about museological narratives in which the "axis of the musealization" is the process of transformation of the objects, such as prescriptions, information, plants, and therapeutic practices, in short, how these objects become documents that introduce us to references of other times and meanings in a contemporaneity that is that of the exhibition itself and its user.

We gathered, reorganized, and recontextualized the objects, attributing visibility to these past realities, thinking about today's visiting public. The organization of exhibitions of a social and educational nature aimed at the general public contemplated the information generated from research conducted within the scope of ethnobotany. Thus, in Codex 81 (according to the organization of the Benedictine Archive of Olinda), which was transcribed and printed in 1966, under the tittle *Receituário of Joaquim Jerônimo Serpa* (Prescription Book of Joaquim Jerônimo Serpa) (Schmalz 1966) and Codices 138 and 139 (according to the organization of the Archive of the Monastery of Saint Benedict of Rio de Janeiro) were taken as central documents.

These exhibitions, through the study of manuscripts gathered and preserved in the Archives of the Monasteries of Saint Benedict of Olinda and Rio de Janeiro, referred us to the history of the Benedictine actuation action, as well as expressed one of the characteristics of Brazilian culture – the importation of ideas, values, and practices from other origins related to the use of plants from the cultural network that forged the formation of healing practices in Brazil, associated with native plants of the country itself.

The exhibition projects were then elaborated aiming at the plant resource and sought to redesign a post-biographical itinerary of the Benedictine hospital, *botica*, and infirmary during the nineteenth century, from the sources consulted. The proposed itineraries are related to an integrative narrative of their own temporality, that is, the period between 1823 and 1829, having as reference Olinda, and the years 1837–1880, in this case, Rio de Janeiro.

These exhibitions converged to a transnaturality that integrates archivist content with elements in areas transversal to the sciences and arts. A selected cast of medicinal plants native to Brazil constituted the essential element of these scientific exhibitions, which revealed to the general public the research with historical sources safeguarded by the monasteries of Olinda and Rio de Janeiro. Visitors to the exhibitions could see that the plants on display reflected a usual medical practice during the nineteenth century in the monasteries, which were now hosting the exhibitions. As such, the plants presented had become an object of research that went through the processes of collection (selection), classification, conservation, and documentation so that they could become an element of exposure.

In a museum of natural history, dried plants can be taken as a representative object of their species, being therefore part of the set of objects that can be exhibited in the space of a museum, as a dead and representative object, as systematized by Alberti (2005). Thus, the strategy adopted in the two exhibitions for presentation of the plants was in the form of exsiccates (Fig. 11.1). The exsiccates consist of parts of plants taken from a plant (species) containing plant vegetative structures (root, stem, and/or leaf) and/or reproductive structures (flowers and/or fruits), which have undergone a herborization processing, which includes pressing, drying, and assembling the botanical material (Fidalgo and Bononi 1984).

Together with the parts of plants were provided data such as popular name in the nineteenth century, botanical identification (species and family), geographical distribution of the plant, part of the plant used, and therapeutic indication. This way, visitors were able to contemplate the biological structure of medicinal plants and know their application in the past.

Working the exhibitors at height that allowed a good visualization of these for all visitors, also considering children from 5 years, this information was written in a simplified calligraphy, but that was close to that of the original document. The cubes that served as exhibitors of the information and presentation of the parts of the plants were coated with botanical illustrations printed on paper and had a transparent antireflective glass protection placed on top of each of the cubes as a means of

Fig. 11.1 Visualization of exsiccate as an exhibiting form of medicinal species used in the Monastery of Saint Benedict of Olinda during the nineteenth century

protection of the botanical material exposed (Fig. 11.1). Also integrated the museological narrative, a set of black and white photographs as well as colorful photos organized and displayed in several large frames, which evoked a large picture frame. In these compositions presented, arrows were also drawn along the photos, which indicated and made available to the visitor the images of the façade of the building where the hospital operated, images of the floor plan of the building and reproduction of historical illustrations, as well as current photographs that clearly situated the location of the old *botica* and infirmary. In addition to this hall of imagery information, there was also the display of current photographs of medicinal plants, as well as the reproduction of details of the historical sources consulted (Fig. 11.2).

As elements that were exclusive of the exhibition held on the premises of the Benedictine Monastery of Rio de Janeiro, we mentioned the organization of a bilingual panel to open the exhibition, the creation of a space for the visitor to sit down and watch the display of a video, whose narrative included other information that provided an immersion in the history about the use of plants in medicine exercised by this monastery in times past (Fig. 11.3). There was also the incorporation of an exhibiting table on which parts of medicinal plants used in the *botica* and infirmary were exposed (Fig. 11.4). These parts were fruits, seeds, barks, leaves, and flowers of medicinal plants that were arranged in faience on the table, referring to the usual packaging in the *boticas* of the nineteenth century. The botanical identification and

Fig. 11.2 Visualization of the imagery arrangement of medicinal species used in the Monastery of Saint Benedict of Olinda in the nineteenth century

medicinal application at the time of operation of the *botica* were informed by means of written identification labels, positioned near each part of the exposed plant.

The first exhibition was installed in the Benedictine Monastery of Olinda itself, during the period from July 10 to 24, 2011, on the occasion of a wider exhibition – *São Bento no Tempo: Fatos e Fotos do Mosteiro de Olinda* (Saint Benedict in Time: Facts and Photos of the Monastery of Olinda – which brought together the entire monastery to discuss its memory in commemoration of its 425 years of foundation. This exhibition – *Plantas Medicinais no Hospital Beneditino Oitocentista, Olinda – Pernambuco – Brasil* (Medicinal Plants at the nineteenth-century Benedictine Hospital, Olinda – Pernambuco – Brazil) – was visited by 3000 people. The second

Fig. 11.3 Visualization of the space dedicated to the video display about the history of the use of plants in medicine exercised by the Monastery of Saint Benedict of Rio de Janeiro in the nineteenth century

scientific exhibition was installed in the space of the Benedictine Monastery of Rio de Janeiro, during the period from July 20 to 26, 2013, integrating the cultural corridor that took place in Rio de Janeiro during the World Youth Day with Pope Francis. *Farmácia Botânica dos Monges, Mosteiro de São Bento do Rio de Janeiro – Brasil* (Botanical Pharmacy of Monks, Monastery of Saint Benedict of Rio de Janeiro – Brazil) had the registered visitation of 3284 people.

After these two exhibition experiences, we organized a return action by digital means, in 2021. This action was implemented through a video-lecture that integrated the participation activities of the National Museum of the Federal University of Rio de Janeiro in the *19ª Semana Nacional de Museus* (19th National Week of Museums), whose central theme was *O Futuro dos Museus: Recuperar e Reimaginar* (The Future of Museums: Recover and Reimagine) in the same year.

Due to the pandemic by SARS-CoV-2, in 2021, the *Semana* (Week) was entirely in digital format. Thus, our video-lecture was made available on the National Museum's YouTube channel on May 17, 2021, and has been available for viewing even since (Medeiros 2021b). It is worth mentioning that this event has been coordinated by *Instituto Brasileiro de Museus – Ibram* (Brazilian Institute of Museums), and it takes place every year, since 2003, as a cultural season in celebration of International Museum Day (May 18) (Ibram 2021).

Fig. 11.4 Visualization of the exhibiting table on which parts of medicinal plants used in the Monastery of Saint Benedict of Rio de Janeiro in the nineteenth century were presented

The thematic perspective of this chapter was based on the dynamics we have developed for this reflective experience with the general public and brought up this discussion of the possibility of working ethnobotany, memory, and museology, taking as a guiding thread the curatorial experiences in the Benedictine Monasteries of Olinda and Rio de Janeiro.

11.4 Final Appreciations

Strengthening and deepening the interactions between ethnobotany, memory, and museological narrative with the creation of a field of convergence between museums and society, what was promoted through the actions of return of shows and video-lecture were environments for the exercise of human experience, whether social, environmental, or historical.

The circulation of ideas among individuals from different instances of the society, from different countries of origin, created a field of possibility so that they could access a portion of Brazilian memory and be instigated to recognize the historical importance of plants in their life in society.

The shows, therefore, pointing out some of the plants observed in the medical practice of Saint Benedict Hospital and Saint Benedict *Botica* and Infirmary sought to stimulate the imagination and the senses and, above all, to rescue and preserve an important portion of the memory associated with the biocultural heritage of Brazil.

Thus, we intend that this placement of the experience can serve as a reference model for the establishment of a museological narrative as a means of return and, therefore, of sharing scientific research with the general public, in this case, the users of the museum. The exhibition converged to a transnaturality, bringing memory as a point of inflection, a dialogical arc between the constitutive elements of the narrative around the biocultural heritage, establishing that objects – plants and information about their use – serve as a means for the construction of a narrative, an act of thought, and for the reconstitution of aspects that involved the modus vivendi of a given society, its know-how, its rituals, its therapeutic practices, and other aspects that involve human life.

In this way, this continuous historical process, within the perspective of our time, with the help and facility of virtual means, can be increasingly rescued, recovered, and presented to society as a whole, promoting the interaction of this same society with its own history and, above all, the belief in its freedom so that, when analyzing this history, it can benefit from this history in the projection of its future.

The return of research in historical ethnobotany through scientific exhibitions evokes the repositories of experiences consistently, awakening the public to get in touch with the memory of a place or with the network of information, which constitute the knowledge, wisdom, and practices that have been bequeathed to us and that we live today.

Acknowledgments We thank the Monastery of St. Benedict of Rio de Janeiro and the Monastery of St. Benedict of Olinda, in the figure of Abbot Filipe da Silva for the support in carrying out documentary research and scientific-cultural exhibitions; the *Núcleo de Comunicação e Eventos do Museu Nacional/Universidade Federal do Rio de Janeiro* (Communication and Events Center of National Museum of the Federal University of Rio de Janeiro) for mediation for the realization of the video-lecture during the *19ª Semana Nacional de Museus* (19th Week of Museums); and Maria José Franco Trindade Medeiros for the English version of the text.

References

A Regra de São Bento (2003) 3rd edn. Lumen Christi, Rio de Janeiro

Alberti SJMM (2005) Objects and the museum. ISIS 96:559–571

Alencar NL, Medeiros PM, Medeiros MFT (2010) Medicinal plants prescribed in the hospital of the São Bento Monastery between 1823 and 1824 in Olinda Northeastern Brazil. Open Complement Med J 2:74–79

Arquivo do Mosteiro de São Bento de Olinda. Codex 81. Receituário, 1823–1829

Arquivo do Mosteiro de São Bento do Rio de Janeiro. Codex 138. Livro de receita de medicamentos para os monges e escravos do Mosteiro de São Bento do Rio de Janeiro e demais pessoas necessitadas de auxílio terapêutico, 1837 a1840

Arquivo do Mosteiro de São Bento do Rio de Janeiro. Codex 139. Livro de receita de medicamentos para os monges e escravos do Mosteiro de São Bento do Rio de Janeiro e demais pessoas necessitadas de auxílio terapêutico. Abade Frei Marcellino do Coração de Jesus

Convention on Biological Diversity (1992) United Nations. https://www.cbd.int/doc/legal/cbd-en.pdf. Accessed 23 Sept 2021
da Cunha MC (2009) "Culture" and Culture. Prickly Paradigm Press, Chicago
Declaração de Belém (1988) https://8c072c10-5be5-477b-abb5-d1b9bf88126f.filesusr.com/ugd/99885a_f95c2f4f07d24c54baa7b9a3e39939f9.pdf. Accessed 23 Sept 2021
Declaração Belém+30 (2018) https://www.museu-goeldi.br/noticias/declaracao-belem-30-pela-valorizacao-dos-povos-e-comunidades-tradicionais-do-mundo/declaracao-belem-30.pdf. Accessed 23 Sept 2021
Fidalgo O, Bononi VLR (1984) Técnicas de coleta, preservação e herborização de material botânico. Instituto de Botânic, São Paulo
Instituto Brasileiro de Museus (2021) Semana Nacional de Museus. https://www.gov.br/museus/pt-br/acesso-a-informacao/acoes-e-programas/semana-nacional-de-museus/semana-nacional-de-museus. Accessed 23 Sept 2021
International Labour Organization (ILO) (1989) Indigenous and Tribal Peoples Convention, C169. https://www.ilo.org/dyn/normlex/en/f?p=NORMLEXPUB:12100:0::NO::P12100_ILO_CODE:C169. Accessed 23 Sept 2021
International Society of Ethnobiology (2006) Código de Ética da ISE (com adições em 2008). http://www.ethnobiology.net/ethics.php. Accessed 23 Sept 2021
Medeiros MFT (2007) Botica monachorum: Farmácia botânica dos monges, Mosteiro de São Bento do Rio de Janeiro/Brasil. PhD thesis, Universidade Federal do Rio de Janeiro
Medeiros MFT (2009) Etnobotânica histórica: princípios e procedimentos. Nupeea/Sociedade Brasileira de Etnobiologia e Etnoecologia, Recife
Medeiros MFT (2010) A interface entre a história, a etnobiologia e a etnoecologia. In: Medeiros MFT (ed) Aspectos históricos na pesquisa etnobiológica. Nupeea, Recife, pp 10–15
Medeiros MFT (2021a) Historical ethnobiology. Elsevier, Cambridge
Medeiros MFT (2021b) Narrativa Museológica em torno do Patrimônio Biocultural. Vídeo-palestra presented at the 19a. Semana Nacional de Museus, Instituto Brasileiro de Museus, 17–21 May 202. https://youtu.be/HNLNip_RfJg. Accessed 23 Sept 2021
Medeiros MFT, Albuquerque UP (2012) The pharmacy of the Benedictine monks: The use of medicinal plants in Northeast Brazil during the nineteenth century (1823 1829). Journal of Ethnopharmacology 139:280–286
Medeiros MFT, Alves RRN (2020) Nineteenth century zootherapy in Benedictine monasteries of Brazil. Anais da Academia Brasileira de Ciências 92:2–11
Medeiros MFT, Andreata RHP, Senna-Valle L (2011) As oficinas de enfermaria e botica do Mosteiro de São Bento do Rio de Janeiro durante o século XIX. In: Peixoto AL, Silva IM (eds) Saberes e usos de plantas: legados de atividades humanas no Rio de Janeiro. Ed. PUC-Rio, Rio de Janeiro, pp 15–37
Meneses, UB de (1992) A exposição museológica: reflexões sobre pontos críticos na prática contemporânea. Ciências em Museus 4:103–120
Moreira TC (2005) O acesso ao patrimônio genético e aos conhecimentos tradicionais associados e os direitos de comunidades indígenas e locais no Brasil. In: Ming LC, Carvalho I, Vasconcellos MC, Radomski MI, Costa MAG (orgs) Direitos de recursos tradicionais: formas de proteção e repartição de benefícios, SBEE –UNESP, Botucatu, p 55–86
Nagoya Protocol on Access to Genetic Resources and the Fair and Equitable Sharing of Benefits Arising from their Utilization to the Convention on Biological Diversity: text and annex (2010) Secretariat of the Convention on Biological Diversity. https://www.cbd.int/abs/doc/protocol/nagoya-protocol-en.pdf. Accessed 23 Sept 2021
Prance GT et al (2007) Ethnobotany, the science of survival: a declaration from Kaua'i. Econ Bot 61:1–2. https://doi.org/10.1007/BF02862367
Schmalz AC (1966) Receituário (1823–1829) de Joaquim Jerônimo Serpa (Hospital de São Bento de Olinda). Arquivo Público Estadual, Imprensa Oficial, Recife
Secretariat of the Convention on Biological Diversity (CBD) (2002) Bonn Guidelines on Access to Genetic Resources and Fair and Equitable Sharing of the Benefits Arising out of their Utilization. https://www.cbd.int/doc/publications/cbd-bonn-gdls-en.pdf. Accessed 23 Sept 2021

Secretariat of the Convention on Biological Diversity (CBD) (2004) Akwé: Kon Voluntary Guidelines for the Conduct of Cultural, Environmental and Social Impact Assessment regarding Developments Proposed to Take Place on, or which are Likely to Impact on, Sacred Sites and on Lands and Waters Traditionally Occupied or Used by Indigenous and Local Communities Montreal. Accessed 23 Sept 2021

Secretariat of the Convention on Biological Diversity (CBD) (2011) Tkarihwaié:ri – Code of Ethical Conduct to Ensure Respect for the Cultural and Intellectual Heritage of Indigenous and Local Communities Relevant to the Conservation and Sustainable Use of Biological Diversity. https://8c072c10-5be5-477b-abb5-d1b9bf88126f.filesusr.com/ugd/99885a_4239cfdd92c34a1ea1b49870794378b4.pdf. Accessed 23 Sept 2021

Sociedad Latinoamericana de Etnobiología (2015) Código de Ética para la Investigación, la Investigación-Acción y la Colaboración Etnocientífica en América Latina. Versión dos. Etnobiología 13(1):1–34. https://8c072c10-5be5-477b-abb5-d1b9bf88126f.filesusr.com/ugd/99885a_040a3cd4c7f64ed7ac902cefa396dcbb.pdf. Accessed 23 Sept 2021

UN Educational, Scientific and Cultural Organisation (UNESCO) (2001) Universal Declaration on Cultural Diversity. http://www.unesco.org/new/fileadmin/MULTIMEDIA/HQ/CLT/pdf/5_Cultural_Diversity_EN.pdf. Accessed 23 Sept 2021

UN Educational, Scientific and Cultural Organisation (UNESCO) (2003) Convention for the Safeguarding of the Intangible Cultural Heritage. https://ich.unesco.org/doc/src/2003_Convention_Basic_Texts-_2020_version-EN.pdf. Accessed 23 Sept 2021

UN Educational, Scientific and Cultural Organisation (UNESCO) (2005a) Preliminary draft convention on the protection and promotion of the diversity of cultural expressions. https://www.refworld.org/docid/435cbdac4.html. Accessed 23 Sept 2021

UN Educational, Scientific and Cultural Organisation (UNESCO) (2005b) Universal Declaration on Bioethics and Human Rights. https://en.unesco.org/themes/ethics-science-and-technology/bioethics-and-human-rights. Accessed 23 Sept 2021

UN Educational, Scientific and Cultural Organization (UNESCO) (2015) The 2005 Convention on the Protection and Promotion of the Diversity of Cultural Expressions. UNESCO, Paris

United Nations (1948) Universal declaration of human rights. https://www.un.org/en/about-us/universal-declaration-of-human-rights. Accessed 23 Sept 2021

United Nations (2005–2014) Second International decade of the World's Indigenous Peoples. https://www.un.org/development/desa/indigenouspeoples/second-international-decade-of-the-worlds-indigenous-people.html. Accessed 23 Sept 2021

United Nations (General Assembly) (2007) Declaration on the Rights of Indigenous People. https://www.un.org/development/desa/indigenouspeoples/declaration-on-the-rights-of-indigenous-peoples.html. Accessed 23 Sept 2021

Van-Mensch P (1992) Towards a methodology of museology. PhD thesis, University of Zagreb

Wagensberg J (2005) The "total" museum, a tool for social change. História, Ciências, Saúde–Manguinhos 12:309–321

Part IV
Final Part

Chapter 12
Conclusion of the Reflections on *Aspects of Brazilian Floristic Diversity: From Botany to Traditional Communities*

Maria Franco Trindade Medeiros and Bárbara de Sá-Haiad

Abstract Considering the issues addressed in this book, such as aspects of plant reproduction, taxonomic and palynological approaches, ecological issues, and ethnobotanical reflections, the proposition of this chapter is to bring an overview of the discussions developed in each one of the previous chapters presented. Covering macro and micro aspects, our conclusion will address the main reflection made through these different approaches, that is, the great concern with the conservation of flora. Thus, our intention is to emphasize the importance of the issues brought by each botanical research line in order to produce science knowledge. In addition, this closing chapter aims to point out some directions that could inspire and guide further studies.

Keywords Flora · Cultural heritage · Conservation · Brazil

12.1 A Brief Book Picture to Be Considered

Aspects of Brazilian Floristic Diversity: From Botany to Traditional Communities presents an eclectic approach since each of its chapters deals with a different botanical aspect. Going through important themes surrounding botany and conservation, a variety of treatments gave us a broader immersion when micro and macro issues lead us to the reflections emerged by scientific knowledge produced.

The reflections discussed throughout the book were concerns about flowering phenology, sexual system and flower sexuality, phytoplankton richness variation

M. F. T. Medeiros (✉)
Departamento de Botânica, Laboratório Interativo em Etnobotânica (LinE), Museu Nacional, Universidade Federal do Rio de Janeiro (MN/UFRJ), Rio de Janeiro, RJ, Brazil
e-mail: mariaftm@mn.ufrj.br

B. de Sá-Haiad
Departamento de Botânica, Laboratório de Anatomia Vegetal, Museu Nacional, Universidade Federal do Rio de Janeiro (MN/UFRJ), Rio de Janeiro, RJ, Brazil

M. F. T. Medeiros, B. de Sá Haiad (eds.), *Aspects of Brazilian Floristic Diversity*,
https://doi.org/10.1007/978-3-031-07453-0_12

over a large spatial scale, palynology, palynotaxonomy, ecological palynology, flora conservation, historical ethnobotany, traditional communities ethnobotany, and biocultural heritage.

Desiring to bring an overview of the aspects discussed, we can say that all these themes reflected a thematic guide. That means that this structural thinking had a proposal to lead us on a travel through micro and macro scales that would be addressed in each of these themes. So, what we had in our hands was the beautiful possibility of entering into a deeper understanding of Brazilian floristic diversity.

This proposed journey did not signal a near end to all possibilities of approach within botany, but rather made a great picture about some aspects of Brazilian floristic diversity, always subject to further investigations. Considering this, the *leitmotiv* was to create a path through by which we could achieve a cognitive understanding of a variety of aspects, including broader current issues such as impacts of climate change on plants, animal and human interaction with flora, and both flora and cultural heritage conservation. Certainly, with the multiple results and discussions, the core of the issue in relation to the infinite character of biodiversity research has been reaffirmed.

Therefore, the ultimate intention to present such a varied range of knowledge within botany was to inspire future studies. When we placed together researchers and scientific production on different aspects of Brazilian floristic diversity, we offered to the reader the opportunity to meet the concepts of each botanical research line considered in this book. These research lines represent the work of the different research groups, working in a Brazilian institution with the unifying proposal to know, protect, and conserve plant diversity.

The design of ideas contemplated not only the literature data but also the experiences achieved by short- or long-term investigations. The aspects developed some understandings about the significance of structure, functionality, expression, strategies, and cellular processes in the reproductive development of plants, in a better taxonomic distinction of plant families, in a better understanding of how the environment affects phytoplankton species richness in freshwater ecosystems, and in conservation tactics to guide decision-makers on the selection of plant species for biological conservation control. Also, the circulation of ideas within this book showed us the importance of historical sources for the dynamics of the use of plants over time indicating how the plants were represented in past times and how people have given us and still give us precious information about the traditional use of medicinal plants.

Closing the discussions, we made a reflection on the ethical conduction of investigations, observing the return of the results to local or traditional communities and at the same time seeking preservation and biocultural conservation.

Over the time of reading all the ideas presented in this book, we move from animal-plant interaction to what concerns the ethical aspects of research with human-plant interaction for survival. We conclude the book emphasizing museums, where plants can be considered key objects for exhibition in this environment, opening the mind of visitors to the importance of botany. It is a space where the interaction between the knowledge produced by the biodiversity researchers and visitors takes place.

Thus, we hope that from a space, for example, a museum, or other multiple forms of language, not only researchers but also all people in general will be aware of Brazilian floristic diversity and awaken for its conservation.

Index

M. F. T. Medeiros, B. de Sá Haiad (eds.), *Aspects of Brazilian Floristic Diversity*, https://doi.org/10.1007/978-3-031-07453-0

The manufacturer's authorised representative in the EU is Springer Nature Customer Service Centre GmbH, Europaplatz 3, 69115 Heidelberg, Germany. If you have any concerns regarding our products, please contact ProductSafety@springernature.com

Printed and bound by CPI Group (UK) Ltd, Croydon, CR0 4YY
30/09/2026
02225242-0007